Physical Chemistry

Physical Chemistry

Edited by **Bernard Wilde**

NY RESEARCH
P R E S S

New York

Published by NY Research Press,
23 West, 55th Street, Suite 816,
New York, NY 10019, USA
www.nyresearchpress.com

Physical Chemistry
Edited by Bernard Wilde

International Standard Book Number: 978-1-63238-488-1 (Hardback)

The publisher's policy is to use permanent paper from mills that operate a sustainable forestry policy. Furthermore, the publisher ensures that the text paper and cover boards used have met acceptable environmental accreditation standards.

Trademark Notice: Registered trademark of products or corporate names are used only for explanation and identification without intent to infringe.

Printed in the United States of America.

Contents

             Can Be Used as a Substitute to the Langmuir-Octet Rule in Understanding
             Interconnections between Atoms in Ions and Molecules   **248**
             Geoffroy Auvert

Chapter 24   **Thermal and Photochemical Effects on the Fluorescence Properties of Type**
             I Calf Skin Collagen Solutions at Physiological pH   **259**
             Julian M. Menter, Latoya Freeman, Otega Edukuye

             Permissions

             List of Contributors

Preface

Physical chemistry utilizes the principles of physics to study atomic, subatomic, macroscopic and particulate phenomena. Intermolecular forces, reaction kinetics, surface chemistry, electrochemistry, thermodynamics, colligative properties, electrochemical cells are some of the topics covered under this discipline. This book is a complete source of knowledge on the present status of this important field. The ever growing need of advanced technology is the reason that has fueled the research in this field in recent times. It is a vital tool for all researching or studying in the field of physical chemistry as it gives incredible insights into emerging trends and concepts.

This book is a result of research of several months to collate the most relevant data in the field.

When I was approached with the idea of this book and the proposal to edit it, I was overwhelmed. It gave me an opportunity to reach out to all those who share a common interest with me in this field. I had 3 main parameters for editing this text:

1. Accuracy – The data and information provided in this book should be up-to-date and valuable to the readers.

2. Structure – The data must be presented in a structured format for easy understanding and better grasping of the readers.

3. Universal Approach – This book not only targets students but also experts and innovators in the field, thus my aim was to present topics which are of use to all.

Thus, it took me a couple of months to finish the editing of this book.

I would like to make a special mention of my publisher who considered me worthy of this opportunity and also supported me throughout the editing process. I would also like to thank the editing team at the back-end who extended their help whenever required.

Editor

Charge Transfer Mechanism and Spatial Density Correlation of Electronic States of Excited Zinc (3d^9) Films

Li Chen[*], Mitsugi Hamasaki, Hirotaka Manaka, Kozo Obara

Country Graduate School of Science and Engineering, Kagoshima University, Kagoshima, Japan
Email: [*]chenli2009@live.jp

Abstract

In material science, half filled 3d orbital of transition metals is essentially an important factor controlling characteristics of alloys and compounds. This paper presents a result of the challenge of excitation of inner-core electron system with long lifetime of zinc films. The advanced zinc films with excited inner-core electron, $3d^n$ (n = 9, 8). We report experimental results of XPS measurements of 9 points in the sample along vertical direction, respectively. The most pronounced futures are existence of satellites, which are about 4 eV higher than the main lines. According to the charge transfer mechanism proposed by A. Kotani and K. Okada, it was clarified that the origins of these peaks are $\underline{c}3d^9\underline{L}$ for the main peak and $\underline{c}3d^9$ for the satellite, respectively. From the energy difference, δE_B, and peak intensity ratio, I_+/I_-, between $\underline{c}3d^9$ and $\underline{c}3d^{10}\underline{L}$, the energy for charge transfer, Δ, and mixing energy, T, were estimated. In the region where the intensity of $\underline{c}3d^{10}\underline{L}$ becomes large, Δ becomes small, $1.2 < \Delta < 2.7$, and T becomes small, too, $0.1 < T < 0.9$, respectively. In this calculation, we supposed $U_{dc} = 5.5$ eV and $U_{dd} = 5.5$ eV. In the analysis along vertical direction, intensity profile of $Zn3d^9$ showed odd functional symmetry and that of $Zn3d^{10}\underline{L}$ showed even functional symmetry. Only the intensity profile of C1s (288 eV) showed the same spatial correlation with $Zn3d^9$. In our experiment, the sample also showed high mobility of the constituting elements. These suggest that charge conservation in excited zinc atom suggests combination between $Zn3d^9$ and C^{2-}.

Keywords

XPS, Zn3d^9, Charge Transfer, Spatial Symmetry of Excited States

---

[*]Corresponding author.

1. Introduction

In developing new materials, there are nanotechnology, researching and development of composite materials, the surface material modification and so on. However, these methods are controlling the outer shell electrons, but in material creation, just controlling the outer shell electrons is insufficient. Our research is exciting inner shell electrons of materials (Zinc). This is the way that excitation can affect the nature and performance of materials. There are some important elements in the formation of excimer such as Cu, Zn and Ga. In particular they have been investigated by many researchers, which just proofed the existence of inner-shell holes and the lifetime of atomic excitation state with very short lifetime [1] [2]. Although, our research is creating excited states with the permanent lifetime and learning the mechanism formation processes. The quantum dynamic approach to the condensation process of excited zinc films by ion-recombination has been concluded by M. Hamasaki and M. Obara [3] [4]. In this paper, the permanent lifetime of excited zinc thin films also has been substantiated by XPS.

2. Sample Preparation

The experiment apparatus which was developed just by our research group is an integral evaporation system with transmission electron spectroscopy evaluation (Electron-assisted PVD). Detailed sample preparation procedure has been published in [3] [4]. Main futures are followings: 1) substrate is 6.5^{φ} mm sapphire enclosed by gold electrode; 2) the surface of sapphire was irradiated with electrons of constant energy; 4) Background pressure is 10^{-5} Pa; 5) substrate temperature is from room temperature 30°C to 100°C; 6) The incident angle of electrons was 45° from the substrate surface. Then the 0.1 g zinc atoms were deposited on the insulate area from the effusion cell at 600°C. The energy dependence of condensed materials was measured. Discrete energy dependence was observed at 10 eV, 90 eV, 100 eV, 140 eV, and 230 eV. These energies are related to the binding energies of zinc atom, 3d (10 eV), 3p (90 eV), and 3s (140 eV). Observed discreet energy dependence was classified into single excitation and double excitation. Since initial condition of the ion-recombination process is Zn^+ and Zn^-, transition processes should depend on the selection rule, $\Delta l = \pm 1$ [5]. Observed strong diffuse scattering intensity and enhancement of the intensity of Bragg diffraction are related to the transition at Zn^+ and Zn^-, respectively. The sample shown in **Figure 1** was deposited at 100°C and 230 eV.

We discussed experimental results of XPS measurements of 9 points in the sample with 6.5^{φ} mm along vertical and horizontal direction, respectively, and it's different between vertical and horizontal direction. In this paper, we report the experimental results at vertical. The direction of the incident electron for the surface bias and zinc ionization is parallel to the horizontal direction. Zinc films could conclude Zn what is the reaching object element, Al (substrate of the special excited zinc film is constituted by sapphire, Al_2O_3), Si (the effusion cell is constituted by SiO_2), C (adsorption from environment) and O (Al_2O_3, SiO_2 and adsorption from environment). The standard point of these data is $Au4f_{7/2}$ (84.0 eV), $Au_{5/2}$ (87.7 eV) and C1s (285.0 eV).

3. Results of XPS

Figure 1 shows measurement points on the exited zinc film surface and spatial distribution of XPS spectra of C1s, O1s, Zn2p3/2, Al2p, Zn3p and Si2p. In C1s spectra, two peaks are recognized. The separation energy between the two peaks depends on the location. O1s peak profiles are relatively wide, which were fitted by three peaks. Zn2p3/2 profiles include two peaks. Intensity of peaks depended on the location, and the two peaks had different dependence on the location. The peaks of B at $Zn2p_{3/2}$, mostly exist at the center region of the substrate. Peaks of Zn3p include Zn3p3/2, Zn3p1/2 and shift peaks of Zn3p3/2 and Zn3p1/2. It is difficult to fit out four peaks of those, clearly. Analysis of element, Zinc, is focused on Zn2p3/2. The peaks of Si2p spectra almost can't exhibit clearly. Al2p spectra showed strong broad single peak. From the viewpoint of correlation, spectra of C1s at peak A and peak A of $Zn2p_{3/2}$ suggest high correlation. And O1s and Al2p also suggest high correlation. However these correlations include spatial different intensity profiles. In next section we show the results of peak analysis.

4. Analysis of XPS Spectra

Elements of XPS spectrum from N1 to N9 were exhibited in **Figure 1**, respectively. Silicon was scarcely, and Si hadn't been fitted. In **Figure 2**, XPS spectra of C1s, O1s, Zn2p3/2 and Al2p at N5 were fitted out, but Si2p. The

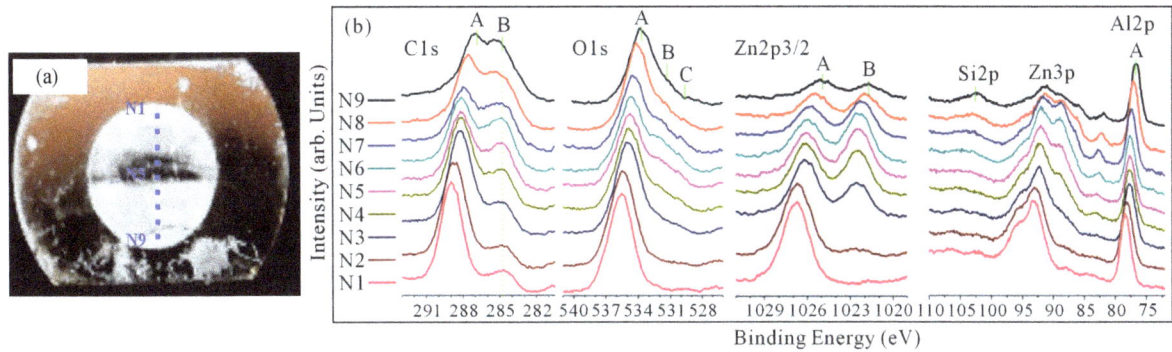

Figure 1. XPS spectrum of C1s, O1s, Zn2p3/2 and Al2p depended after 44 weeks. (a) showed N1-N9 9 points along vertical, and N5 at the center of sample. (b) showed the results of XPS measurements of the 9 points.

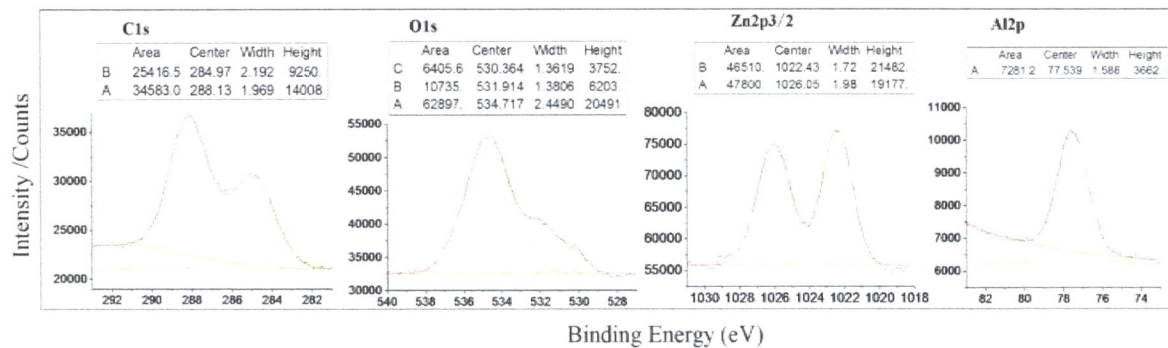

Figure 2. XPS spectrum of C1s, O1s, Zn2p3/2 and Al2p at N5 was fitted out.

spitting peaks and characteristic factors were indicated out:

1) C1s was spitted to two (A and B) peaks: Binding energy of peak center of B was 285.0 eV, what showed as the factor of "Center" in table of **Figure 2**-C1s. The standard point of these data is C1s (285.0 eV). The width at half maximum height of peak B was 2.2 eV, what exhibited as factor of "Width". The integral value of intensity of peak B was 25416, what exhibited as factor, "Area". The intensity of peak center of B was 9250, what exhibited as factor, "Height". The factors of fitted peak A also showed in the table at **Figure 2**-C1s.

2) O1s was a boarding peak. At high binding energy side, there was a strong satellite marked to A. From 530eV to 534eV, there should be two different combinative states oxygen, at least. The factors of fitted peaks of oxygen were exhibited in the table of **Figure 2**-O1s.

3) Spectrum of Zn2p3/2 was fitted to two peaks, clearly. The factors of fitted peaks of zinc were exhibited in the table of **Figure 2**-Zn2p3/2.

4) Spectrum of Al2p was single. The factors of fitted peaks of alumina were exhibited in the table of **Figure 2**-Al2p.

The fitting results of binding energy of peaks from N1 to N9 were showed in **Figure 3**; and the fitting results of intensity of fitting out peaks from N1 to N9 were showed in **Figure 4**.

Figure 3 showed the spatial change of the binding energy of C1s, O1s, Zn2p3/2 and Al2p from N1 to N9. The peaks at 285.0 eV of C1s are due to adsorbed carbon oxide, which was fixed as the standard of the spectra. C1s at 288 eV region showed stepwise structure. The stepwise structures were observed in all spectra with higher binding energies. The other peaks with lower energies of O1s and $Zn2p_{3/2}$ showed monotonous decrease as changing the position from N1 to N9.

Figure 4 showed the spatial profiles of integrated peak intensities of analyzed peaks. Intensity profiles of two peaks at A and B of C1s showed complemented structures with odd symmetry. The summation of these peaks was almost constant. Intensity profiles of O1s include two groups. The symmetry of these profiles is all even symmetry. However, the profiles with lower energy, peak-B and Peak-C of O1s, showed even symmetrical, on the other hand, the profiles with higher energy, peak-A of O1s, showed the sum of even symmetrical and odd symmetrical. Intensity profiles of Zn2p3/2 include even symmetry of the profile with lower energy, peak-B of

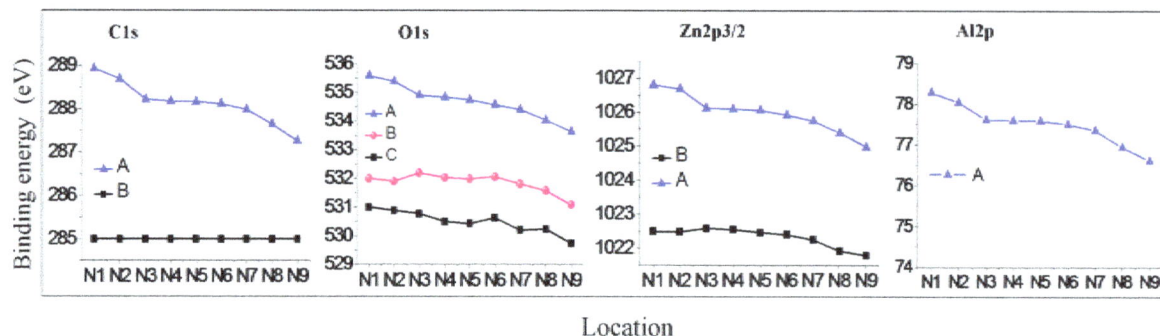

Figure 3. Spatial changes of the binding energy of C1s, O1s, Zn2p3/2 and Al2p.

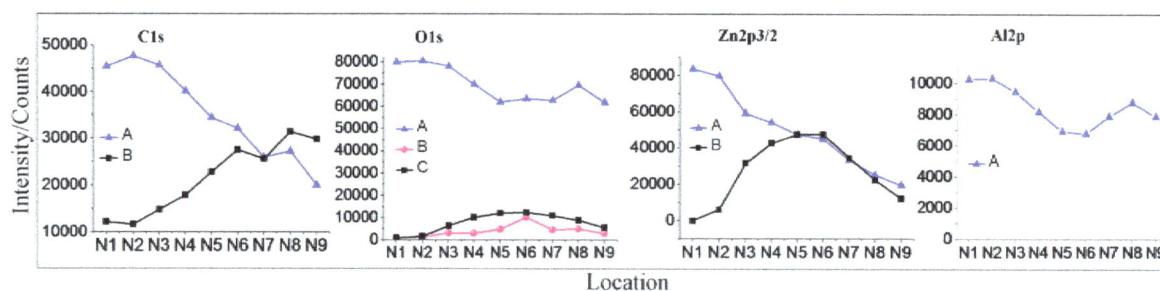

Figure 4. XPS spectrum intensities of C1s, O1s, Zn2p3/2 and Al2p spatial symmetry.

$Zn2p_{3/2}$, and odd symmetry of that with higher energy, peak-A of $Zn2p_{3/2}$. The summation of these is not constant. The intensity in the half from N6 to N9 decreased linearly. The intensity profile of Al2p showed the sum of even symmetrical and odd symmetry at high energy side of normal Al_2O_3 (74.4 eV).

5. Discussion

We discussed the states of special excited zinc film and the spatial symmetry of zinc, and we calculated the important independence parameters, U_{dc}, T, and Δ of charge transfer mechanism.

5.1. The States of Zinc (This Part Has Done a Lot of Modifications)

From **Figure 5(b)**, the XPS spectrum of surface slightly oxidized mental zinc (Zn, ZnO) and N5 in **Figure 1** was showed. There were two satellites at the high binding energy side of Zn2p at N5. To elucidated the states of zinc in our special film. Here we consider the 2p XPS of transition mental compounds, firstly. Experimental data observed by Rosencwaig *et al.* (1971) [6]. There is no satellite peak in the Zn2p of ZnF_2 where the 3d shell is filled. However, the satellite of 2p XPS from CuF_2 to MnF_2 occurs on the higher binding energy side of the main peak. It is now well established that it originates from the charge transfer between the ligand 2p and metal 3d orbital. From other researches, the properties of strong correlation and high temperature superconductivity directly related to half filled 3d orbital of the transition metals. However, how can distinguish the 3d full filled states and 3d half filled state? **Figure 5(a)** showed X-ray photo spectroscopy of Cu2p and O1s in Cu, Cu_2O and CuO [7]-[9]. Metal Cu and Cu_2O with $3d^{10}$ showed almost the same spectrum of Cu2p, but CuO with $3d^9$ (orbital of 3d is not full filled.) showed different spectrum of Cu2p to Cu and Cu_2O. The peak at high binding energy side was identified as the peak, what due to $2p^5 3d^9$ final state that the movement of the holes did not occur. The peak of low binding energy side could be fitted to two peaks. The fitting peak at high binding energy side is final state of $2p^5 3d^{10}\underline{L}$, what the movement of hole was occur from Cu3d to O2p with the photoelectron emission. Components what form the sharp rise at the low binding energy side, is $2p^5 3d^{10}$ final state, that is derived from a hole in the valence band is moved to O2p band of adjacent site.

However, 3d orbit of ZnO and Zn is full filled; and new peak at high energy side isn't exhibited. The electron transition from O2p to Zn3d couldn't happen, and there is no satellite peaks to identify the $3d^{10}\underline{L}$ state of high binding energy side in the X-ray photoelectron spectroscopy. If 3d is half filled, new peaks at the high binding

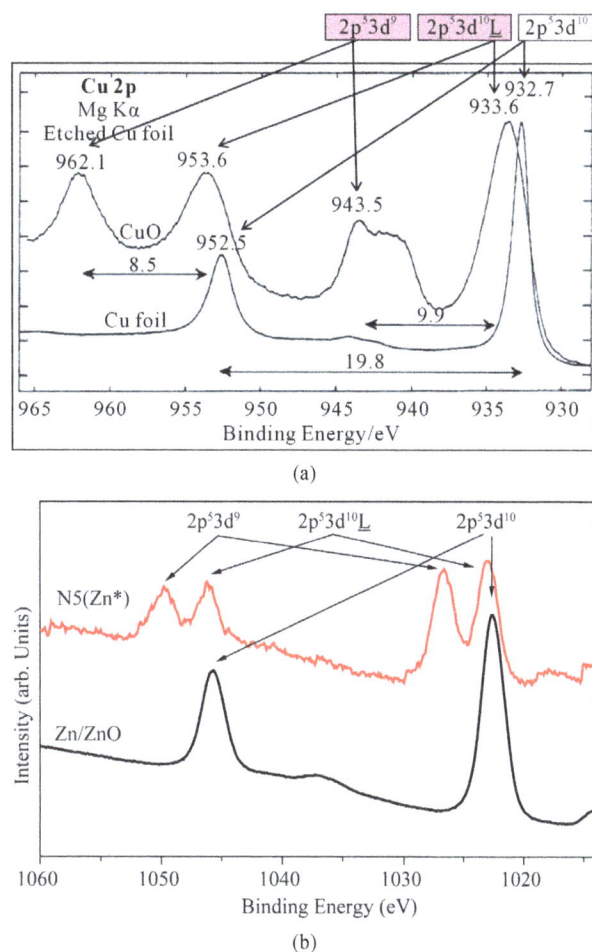

Figure 5. (a) 2p XPS spectrum of Cu foil and CuO [8] [9]; (b) 2p XPS spectrum of Zn/ZnO and $N_5(Zn^*)$, N_5 in **Figure 1** at center of the sample (Zn^*: Excited state of zinc).

energy side could be exhibited, what is caused by electron transition between ligand to transition metal. From this we can infer that, in the **Figure 1**, **Figure 2** and **Figure 5(b)**, the new peaks at the high binding energy side of Zn^* could be exhibited, what caused by half filled of Zn^* $3d^n$ ($n < 10$) orbits. We considered the peak A of Zn2p3/2 at high energy side was originated to final state of $Zn2p^53d^9$, and the peak B of Zn2p3/2 must originated to the final states of $Zn2p^53d^{10}$ or $Zn2p^53d^{10}\underline{L}$. In other experiments, the relative intensity of peak A and peak B was changed by irradiation of strong X-ray, and it was reported between $3d^9$ and $3d^{10}\underline{L}$ [6] [7] [10]-[13].

We considered peak A of $Zn2p_{3/2}$ originated by final state of $Zn2p^53d^9$, peak B of $Zn2p_{3/2}$ originated by the final state of $Zn2p^53d^{10}\underline{L}$.

5.2. The Spatial Symmetry of Excited Zinc Film

The correlation between analyzed profiles is useful to decide the interaction between each element. From **Figure 4**, the electron state of peak-A in C1s correlates with the electron state of peak-A in Zn2p3/2. The electron state of Al2p correlates with the electron state of peak-A of O1s. The state of peak-C in O1s correlates with peak-B of Zn2p3/2. These correlations suggest the combination of the origins of these states. The satellites at high binding energy side of zinc were odd symmetry, and the peaks at low binding energy side of zinc were even symmetry. We can realize that the initial state of $Zn3d^9$ is odd symmetry, but the initial state of $Zn3d^{10}\underline{L}$ shows even symmetry. The different spatial symmetry of $Zn3d^9$ and $Zn3d^{10}\underline{L}$ originated from different force [14].

The external fields on earth just gravitational field, magnetic field and electric field can be considered. From sample preparation, there is an even symmetry electric field be made out by electron incident. The even symmetry of $Zn3d^{10}\underline{L}$ correlated to electric field. There was micro electric current occurred on the sapphire substrate, because of electron incident, and the direction of the micro electric currents were same. The same direction micro electric current made the same direction of magnetic field. So, the odd symmetry of $Zn3d9$ could correlate to magnetic field.

5.3. Theoretical Calculation of CT (This Part Has Done a Lot of Modifications)

The correspondence between initial state, final state and XPS spectrum of charge transfer process was showed in **Figure 6**. Factors of charge transfer energy (Δ), mixing energy (T) and coulomb interaction energy between 3d and inner holes (Q) are the most important factors in charge transfer process. In **Figure 6**, we show an ionic state energy diagram of XPS. From charge transfer theory by A. Kotani and K. Okada [7] [15] [16], there are initial state of d^n and $d^{n+1}\underline{L}$ ($n < 10$), and energy of $d^{n+1}\underline{L}$ state is ΔeV bigger than d^n state. Mixed valence state of d^n is marked as "G.S.", and energy is smaller than d^n of initial state. The final state originated to emission of inner core electroscopes. If energy of $\underline{c}d^n = E_0$, energy of $\underline{c}d^{n+1}\underline{L}$ equal $E_0 + \Delta - Q$.

About T, Δ, U_{dd} and Q, the following series of equations is established:

1) Ground state and final state showed as following:

$$G.S. |g\rangle = \cos\theta_g |d^9\rangle - \sin\theta_g |d^{10}\underline{L}\rangle \quad \tan\theta_g = \frac{E_g}{T} = \frac{\sqrt{\Delta^2 + 4T^2} - \Delta}{2T} \tag{1}$$

$$F.S. \begin{cases} |f_1\rangle = \cos\theta_f |d^9\rangle - \sin\theta_f |d^{10}\underline{L}\rangle \\ |f_2\rangle = \cos\theta_f |d^9\rangle + \cos\theta_f |d^{10}\underline{L}\rangle \end{cases} \begin{cases} \tan\theta_f = \frac{\sqrt{\Delta_f^2 + 4T^2} - \Delta_f}{2T} \\ \Delta_f = \Delta - Q \end{cases} \tag{2}$$

2) The peak state energy of f_1 and f_2, peak interval of δE_B, the intensity ratio of peaks showed as:

$$\text{Final state energy} : E_{f1,2} = \frac{\Delta_f \pm \sqrt{\Delta_f^2 + 4T^2}}{2} \tag{3}$$

$$\text{Peak interval} : \delta E_B \equiv E_{f2} - E_{f1} = \sqrt{\Delta_f^2 + 4T^2} \tag{4}$$

$$\text{Intensity ratio} : R = \frac{I_{f2}}{I_{f1}} = \tan^2\left(\theta_f - \theta_g\right) \tag{5}$$

From **Figures 1-3**, we got the peak intervals and intensity ratio from source data of XPS. In our zinc film, the ligand should be C. The value of Q is not very dependent on the metal ions, and Q is reduced with electrical negative degrees of ligand to transition metal compound [7] [15]. In this calculation, we supposed $Q = 5.5$ eV, equal coulomb interaction energy of d-d, U_{dd}.

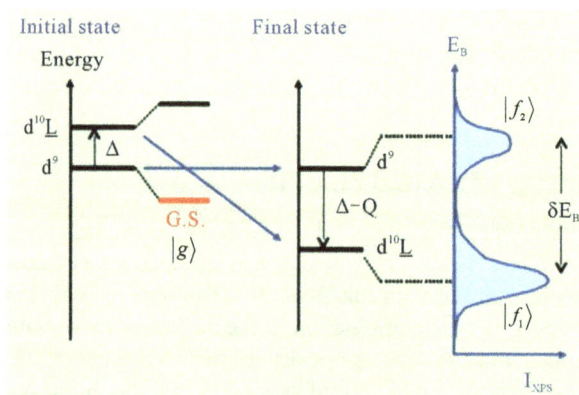

Figure 6. The correspondence between initial state, final state and XPS spectrum of charge transfer process [15] $\underline{L}$: An electron transited from ligand to metal, where must be hole exist.

Table 1. The independent parameters of U_{dc}, Δ and T were calculated from N1 to N9.

| | The independent parameters (Δ, T, U_{dc}) of Zn^* films | | | | | | | | | Ground state | | Final state | | |
	δE_B	I_+/I_-	U_{dc}	Udd	Δ	Δf	(θ_f)	(θg)	T	Eg	d^9	$d^{10}\underline{L}$	$2p^5d^9$	$2p^5d^{10}\underline{L}$	$2p^5d^{10}\underline{L}+Eg$
N1	**4.31**	60	**5.5**	5.5	**1.20**	−4.30	87.35	4.72	**0.10**	−0.01	0.00	1.20	0.00	−4.30	**−4.31**
N2	**4.20**	7.31	**5.5**	5.5	**1.34**	−4.16	81.72	12.28	**0.31**	−0.07	0.00	1.34	0.00	−4.16	**−4.23**
N3	**3.55**	1.78	**5.5**	5.5	**2.23**	−3.27	68.82	15.82	**0.69**	−0.19	0.00	2.23	0.00	−3.27	**−3.46**
N4	**3.55**	1.19	**5.5**	5.5	**2.36**	−3.14	64.98	17.54	**0.83**	−0.26	0.00	2.36	0.00	−3.14	**−3.40**
N5	**3.60**	0.91	**5.5**	5.5	**2.43**	−3.08	62.48	18.87	**0.94**	−0.32	0.00	2.43	0.00	−3.08	**−3.40**
N6	**3.52**	0.92	**5.5**	5.5	**2.50**	−3.00	62.39	18.18	**0.92**	−0.30	0.00	2.50	0.00	−3.00	**−3.30**
N7	**3.50**	0.91	**5.5**	5.5	**2.54**	−2.96	61.85	18.22	**0.94**	−0.31	0.00	2.54	0.00	−2.96	**−3.27**
N8	**3.47**	1.05	**5.5**	5.5	**2.50**	−3.00	63.22	17.52	**0.88**	−0.28	0.00	2.50	0.00	−3.00	**−3.28**
N9	**3.19**	1.55	**5.5**	5.5	**2.67**	−2.83	65.28	14.40	**0.73**	−0.19	0.00	2.67	0.00	−2.83	**−3.02**

From the energy difference, δE_B, and peak intensity ratio, I_+/I_-, between $2p^53d^9$ and $2p^53d^{10}\underline{L}$, the Δ and T were estimated. The results were exhibited in **Table 1**. In the region where the intensity of $2p^53d^{10}\underline{L}$ becomes large, Δ becomes small, 1.2 eV < Δ < 2.7 eV, and T becomes small, too, 0.1 eV < T < 0.9 eV, respectively.

5.4. The Structural Instability of Excited Zn Film

Charge conservation in excited zinc atom suggests combination between two $Zn3d^9$ and C^{2-} ions. The size of C^{2-} is almost comparable to that of oxygen atom. If the electrons transfer from C^{2-} to $3d^9$, the size of the carbon shrinks to 0.154 nm from 0.25 nm. Therefore the charge transfer induces a vacancy due to the rapid diffusion of carbon. In our experiment, the sample showed high mobility of the constituting elements.

6. Conclusions

This paper presents a result of the challenge of the excitation of inner-core electron system with long lifetime of excited zinc films. The advanced zinc films with excited inner-core electron, $3d^9$, were formed by surface ion recombination process controlled by the collision energy. We report experimental results of XPS measurements of 9 points along vertical direction from N1 to N9. The most pronounced futures are existence of the satellites, which are about 4 eV higher than the main lines. This is the proof of the existence of 3d-hole, $Zn3d^9$, which is the same mechanism of CuO. The measured XPS spectra apparently depend on the location in the sample. These satellites are due to the charge transfer mechanism of electron from the ligand to $Zn3d^9$. According to the charge transfer mechanism proposed by A. Kotani and K. Okada, it was clarified that the origins of these peaks are $\underline{c}3d^{10}\underline{L}$ for the main peak and $\underline{c}3d^9$ for the satellite, respectively. The samples contain Zn, Al, O, C and Si. The surface densities of these elements depend on the location of the sample. In the analysis along the vertical direction, intensity profile of $Zn3d^9$ showed odd functional symmetry and that of $Zn3d^{10}\underline{L}$ showed even functional symmetry. Only the intensity profile of C1s (288 eV) showed the same spatial correlation with $Zn3d^9$. This is the reason that the ligand for $Zn3d^9$ is *Carbon, not Oxygen*. The other elements roughly show the even functional symmetry. We analyzed experimental data by using Kotani and Okada's theoretical model. From the energy difference, δE_B, and peak intensity ratio, I_+/I_-, between $2p^53d^9$ and $2p^53d^{10}\underline{L}$, the energy for charge transfer, Δ, and mixing energy, T, were estimated. In the region where the intensity of $2p^53d^{10}\underline{L}$ becomes large, Δ becomes small, 1.2 < Δ < 2.7, and T becomes small, too, 0.1 < T < 0.9, respectively. In this calculation, we supposed U_{dc} = 5.5 eV. The excited zinc films also showed high mobility of the constituting elements which correlated to C^{2-}.

We conclude that the excited zinc films have 3d-hole, $3d^9$, with almost permanent lifetime and $3d^9$ combine with carbon. Although the detailed mechanism of the long lifetime of the inner-core excited zinc atoms is not clear at the present time, this material suggests the high potential for the advanced application in wide area [17].

References

[1]　Hay, P.J., Dunning Jr., T.H. and Raffenetti, R.C. (1976) *Journal of Chemical Physics*, **65**, 2679.

[2]   Wei, M., Boutwell, R.C., Garrett, G.A., Goodman, K., Rotella, P. and Schoenfeld, W.V. (2013) *Material Letters*, **97**, 11-14. http://dx.doi.org/10.1016/j.matlet.2013.01.090

[3]   Hamasaki, M., Yamaguchi, M., Kuwayama, M. and Obara, K. (2011) A New Approach for Sustainable Energy Systems Due to the Excitation of Inner-Core Electrons on Zinc Atoms Induced by Surface-Ion-Recombination. *AIP Conference Proceeding*, **1415**, 43-50. http://dx.doi.org/10.1063/1.3667216

[4]   Obara, M., Hamasaki, M., Manaka, H. and Obara, K. (2011) Slowly Relaxing Structural Defects of Zinc Films with Excited States Induced by Ion Recombination Processes. *Advanced Materials Research*, **277**, 11-20. http://dx.doi.org/10.4028/www.scientific.net/AMR.277.11

[5]   Callis, P.R. and Scott, T.W. (1983) Perturbation Selection Rules for Multiphoton Electronic Spectroscopy of Neutral Alternant Hydrocarbonsa). *Journal of Chemical Physics*, **78**, 16. http://dx.doi.org/10.1063/1.444537

[6]   Rosencwaig, A., Wertheim, G.K. and Guggenheim, H.J. (1971) Origins of Satellites on Inner-Shell Photoelectron Spectra. *Physical Review Letters*, **27**, 479. http://dx.doi.org/10.1103/PhysRevLett.27.479

[7]   Rosencwaig, A. and Wertheim, G.K. (1971) Origins of Satellites on Inner-Shell Photoelectron Spectra. *Physical Review Letters*, **27**, 479-481. http://dx.doi.org/10.1103/PhysRevLett.27.479

[8]   Tahir, D. and Tougaard, S. (2012) Electronic and Optical Properties of Cu, CuO and Cu_2O Studied by Electron Spectroscopy. *Journal of Physics*: *Condensed Matter*, **24**, 175002. http://dx.doi.org/10.1088/0953-8984/24/17/175002

[9]   Tanaka, K. (1998) X-Ray Photoelectron Spectroscopy.

[10]   Schon, G. (1973) ESCA studies of Cu, Cu_2O and CuO. *Surface Science*, **35**, 96-108. http://dx.doi.org/10.1016/0039-6028(73)90206-9

[11]   Tsuda, N., Nasu, K. and Fujimori, A. (2010) Electronic Conduction in Oxides. Heidelberg. 223.

[12]   Schon, G. (1973) Auger and Direct Electron Spectra in X-Ray Photoelectron Studies of Zinc, Zinc Oxide, Gallium and Gallium Oxide. *Journal of Electron Spectroscopy and Related Phenomena*, **2**, 75-86. http://dx.doi.org/10.1016/0368-2048(73)80049-0

[13]   de Groot, F. and Kotani, A. (2008) Core Level Spectroscopy of Solids, 1-40, 71-75, 145-160, 182-185.

[14]   Fazelzadeh, S.A. and Ghavanloo, E. (2012) Nonlocal Anisotropic Elastic Shell Model for Vibrations of Single-Walled Carbon Nanotubes with Arbitrary Chirality. *Composite Structures*, **94**, 1016-1022. http://dx.doi.org/10.1016/j.compstruct.2011.10.014

[15]   Okada, K. (2004) Changes in the Electronic State of the Copper Oxide and Cu 2p XPS by Doping.

[16]   Veal and Paulikas (1985) Final-State Screening and Chemical Shifts in Photoelectron Spectroscopy. *Physical Review B*, **31**, 5399. http://dx.doi.org/10.1103/PhysRevB.31.5399

[17]   Yan, S.C., Yu, H., Wang, N.Y., Li, Z.S. and Zou, Z.G. (2011) ESI for Chemical Community.

2

Characterization and Adsorption Study of Thymol on Pillared Bentonite

Mohamed El Miz[1*], Samira Salhi[1], Ikrame Chraibi[2], Ali El Bachiri[1],
Marie-Laure Fauconnier[3], Abdesselam Tahani[1]

[1]LACPRENE, Faculté des Sciences, Université Mohamed 1er, Oujda, Morocco
[2]Département de Géologie, Faculté des Sciences, Université Mohamed 1er, Oujda, Morocco
[3]Unité de Chimie Générale et Organique, Gembloux Agro-Bio Tech, Université de Liège, Gembloux, Belgique
Email: [*]elmiz.mohamed@gmail.com, a.tahani@ump.ma

Abstract

Pillared clay (PILC) was prepared from Moroccan clay and characterized, and its aqueous thymol adsorption capacities were studied using a batch equilibrium technique. So, we tested the encapsulation of thymol by aluminum pillared clay (PILC). The PILCs displayed a total surface area of 270 m^2/g, a total pore volume of 0.246 cm^3/g and an average pore diameter of 8.9 Å, which corresponds to the size of Al_{13} forming the pillars between the clay layers. The adsorption capacity shown by the PILCs for thymol from water is close to 319 $mg \cdot g^{-1}$ for low solid/liquid ratio (0.2%). This result suggests that the PILCs have both hydrophobic and hydrophilic characteristics, as a result of the presence of silanol and siloxane groups formed during the pillaring and calcination of the PILCs. The experimental data were analyzed by the Freundlich and the Langmuir isotherm types for low values of equilibrium concentration. The rise of the isotherm in this range of concentrations was related to the affinity of thymol for clay sites, and the equilibrium data fitted well with the Freundlich model with maximum adsorption capacity of 319.51 mg/g for a ratio $R_{S/L}$ = 0.2%. Pseudo-first and pseudo-second-order kinetic models were tested with the experimental data and pseudo-first order kinetics was the best for the adsorption of thymol with coefficients of correlation R^2 ≥0.986, and the adsorption was rapid with 90% of the thymol adsorbed within the first 20 min.

Keywords

Clays, Bentonite, Thymol, Adsorption, Desorption, Kinetics, Pillared Clay

---

[*]Corresponding author.

1. Introduction

In recent years, material sciences have involved the studies related to the production of materials having a controlled pore structure and improvement of porous materials found in the nature. The production principle is to hold the inorganic layers a part from each other, introducing a bulky guest agent between them. When the used materials are clay minerals, the resulting materials after pillaring are called Pillared Layered Clays (PILCs). The host solid and the production conditions have rather important effects on the quality of the product. Due to its high cation exchange capacity, swelling properties, and large sheets, montmorillonite type clay have an important place in the production of pillared clays.

Any material which could enter between the layers and has a thermal stability and suitable dimensions can be used as pillaring agent [1]-[5].

In the case of aluminum, there is a lot of information about the oligocations formed in solution that facilitates studies on its properties. Pillaring of clays with Al is thought to be an ion exchange process of the major pillaring agent $[Al_{13}O_4(OH)_{24}(H_2O)_{12}]^{7+}$, the so called keggin or Al_{13} ions [6] [7] were able, through the XRD analysis, to confirm the intercalation of aluminum polycations Al_{13} with basal spacings of about 18 Å at room temperature and which are transformed into Al_2O_3 oxide after calcination at 500°C.

The intercalated species act as props (pillars) that keep the clay layers apart and prevent them from collapsing under vacuum, at higher temperatures, or under specific conditions [8].

So, the pillaring process generates micro and or meso-porosity in the inter lamellar spaces of clays and are also called nano materials. Pillared clays have remarkable adsorption properties that are related both to the geometrical features of the porous spaces and to specific interactions of the pillars and the clay layer [8].

Adsorption and pH-dependent ion exchange capacities of PILCs have been extensively studied [9]-[16]. These modified clay minerals (PILCs) have been used in attempts to improve controlled release formulations of pesticides [17].

Elsewhere, the adsorption properties of PILCs on active molecules such as thymol have not yet been established not even its application to reversible encapsulation.

Thymol (2-isopropyl-5-methylphenol, IPMP) is known for its bactericidal effect (has a lethal activity of micro-organisms) [18]-[21]. It has been shown to be an efficient acaricide molecule against the Varroa destructor, an external parasitic mite that attacks honey bees [22].

Thymol is present in thyme essential oil and volatile essential oils of diverse plants. The advantage of using thymol is based on its low cost and easy to be handled because it is not toxic. Its use is limited by the temperature of the medium; thymol is only effective at temperatures varying between 15°C and 30°C and has a low solubility in water 1 g/L equivalent to 6.6×10^{-3} mol/L.

In our previous work [23], we have studied the adsorption of thymol on sodium bentonite (montmorillonite) and the maximum amount adsorbed was 177 $m^2 \cdot g^{-1}$ showing a certain affinity of thymol for anionic clay sites.

The aim of this work is the study of characterization, adsorption isotherm and adsorption kinetics of thymol on pillared bentonite (modified montmorillonite clay), for multiple applications related to the technology of encapsulate volatile material in clay mineral nano-capsules.

2. Material and Method

2.1. Purification and Intercalation Method

The bentonite used in this study is a bentonite rich with montmorillonite clay type that was provided from the region of Nador (northest of Morocco) and used in a purified forme.

Bentonite clays were homoionized with a NaCl solution 2 M, with a solid: liquid ratio of 1/50 (10 g clay/500 mL NaCl). The ion exchange was realized at 25°C and was repeated three times. After each process, the clay was washed with distilled water until no chloride ions were found. To avoid the rapid evaporation of the water, the ion exchanged samples were slowly dried at 35°C for 24 h [24]. A portion of the sample was used to prepare pillared-bentonite.

The simplified pillaring method proposed in this paper was carried by intercaling of the pillaring agent in a aqueous sodic clay suspension 5% (%mass), ageing, washing, drying and calcination of the intercalated clay at 350°C.

The pillaring solution containing $[Al_{13}O_4(OH)_{24}(H_2O)_{12}]^{7+}$ cations was obtained by adding 250 ml of $AlCl_3$

(0.4 M) stepwise to 550 ml of NaOH (0.4 M). The final neutralization ratio which is defined as $[OH]_{Tot}/[Al(III)]_{Tot}$ was 2.4 and the solution was agitated for 12 h at room temperature. The resultant solution of pH = 4.5 was added to the clay suspension 2% (1 g clay/100 ml H_2O) and stirred for 6 h at room temperature. The pillared clay form was then centrifuged, filtered, and dried at 60°C in air. Calcination was performed at 350°C for 6 h. The degree of intercalation of the pillaring cations was determined by XRD, by analyzing variations of d (001) in oriented clay-aggregate specimens.

2.2. Characterization

The natural samples purified and modified clay are subjected to analysis and identification by X-ray diffraction (XRD), infrared spectroscopy (IR) and Thermal analysis.

X-ray diffractograms were recorded in a Shimadzu XRD diffractometer D6000 stations working on the monochromatic copper Kα1 radiation (1.54 Å) (**Figure 1**).

The textural characteristics of clays before and after pillaring were determined from N_2 adsorption/desorption isotherms at 77°K using micrometrics ASAP 2000 volumetric adsorption-desorption apparatus and surface Area and Pore size Analyzer (**Figure 2** and **Figure 3**). The surface areas were calculated using the multi-point BET method, and a relative pressure (P/P_0) between 0.00095 and 0.9917 was applied.

Infra Red (I.R) spectra were acquired using a Shimadzu Fourier Transform spectrometer over a range varying from 400 to 4000 cm^{-1} with a resolution of 2 cm^{-1}, and the samples were prepared in the form of a dispersion in a vial KBr (1/200 by weight) (**Figure 4** and **Figure 5**).

Thermal analysis was carried out in a SHIMATZU D6000 coupled to a DC ampler and temperature controller. Data from DTA-TG were obtained in all cases at a heating rate of 5°C/min between 30°C and 1000°C and under N_2 atmosphere (**Figure 5** and **Figure 6**).

2.3. Equilibrium Studies

Adsorption of thymol by the pillared bentonite fractions was carried out in batch at different initial concentrations (5 to 760 mg/L). Increasing amounts of clay fractions of 0.05, 0.1 and 0.15 g, were dispersed in 20 ml of each initial solution (C_0) of thymol, and equilibrated in an overhead shaker at room temperature (19°C) for 3 h. The particles were allowed to settle and separated by centrifugation.

After a centrifugation at 10000 rpm for 20 min speed, the amount adsorbed by clay was calculated from the initial concentration and the final concentration of products determined by absorbance measurement using double beam *UV/v* is spectrophotometer (Shimadzu, Model *UV* 6500, Japan) at 273.5 nm.

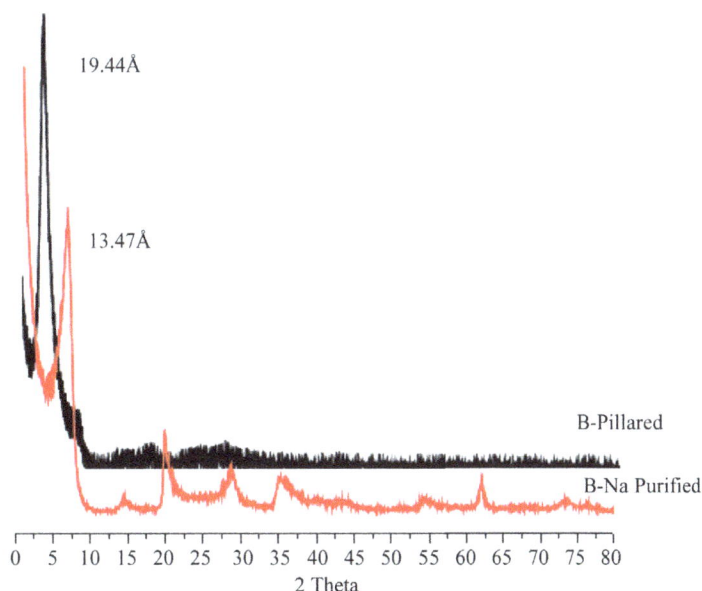

Figure 1. X-ray diffractograms for the indicated samples.

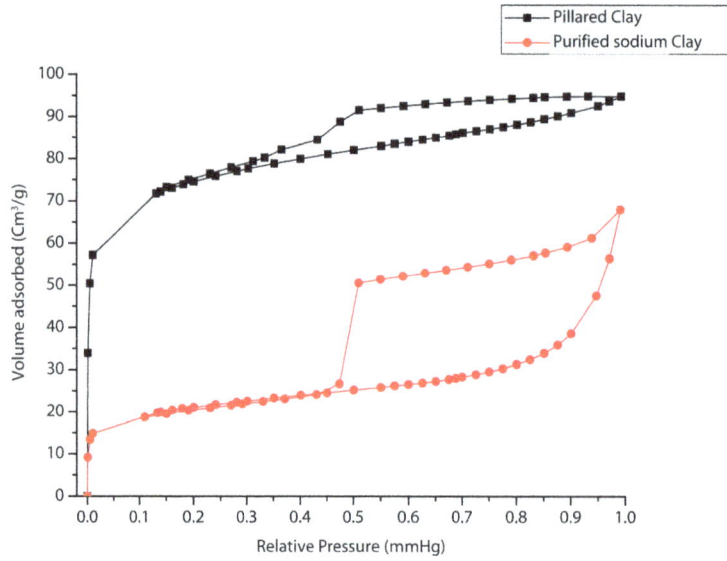

Figure 2. Nitrogen adsorption-desorption isotherm for pillared & purified sodium bentonite.

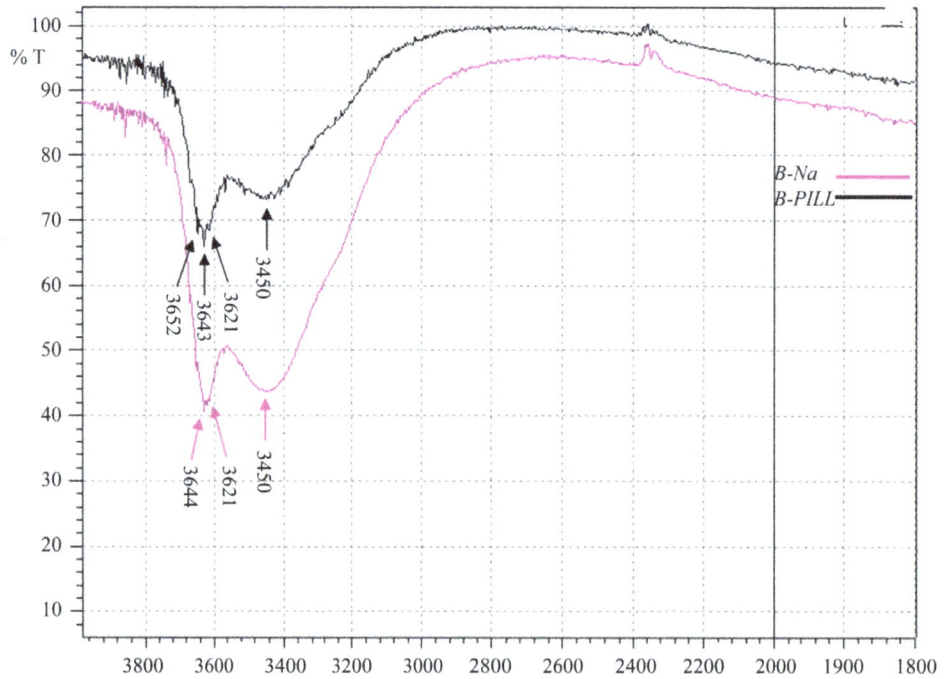

Figure 3. FTIR spectra of B-Na purified and pillared in the range of 4000 - 1800 cm^{-1}.

$$q_e = \frac{V \times (C_0 - C_e)}{m} \tag{1}$$

with: C_0 is the initial concentration (mg/L), C_e is the equilibrium concentration, V (ml) is the total volume of the sample, m (mg) is the mass of pillared clay used and q_e is amount adsorbed (mg per grams) of pillared clay (mg/g). The product under consideration adsorption isotherm is obtained by drawing the curve:

$$q_e = f(C_e) \tag{2}$$

For the adsorption experience, the mass of thymol that was lost during the balancing of the solution was sup-

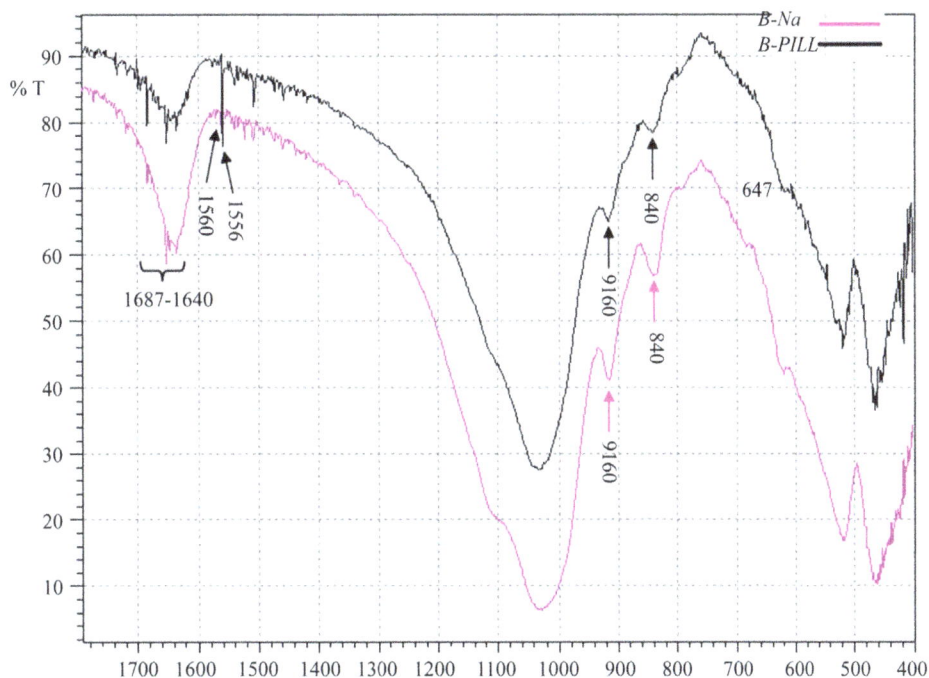

Figure 4. FTIR spectra of B-Na purified and B-Al-PILCs in the range of 1800 - 400 cm^{-1}.

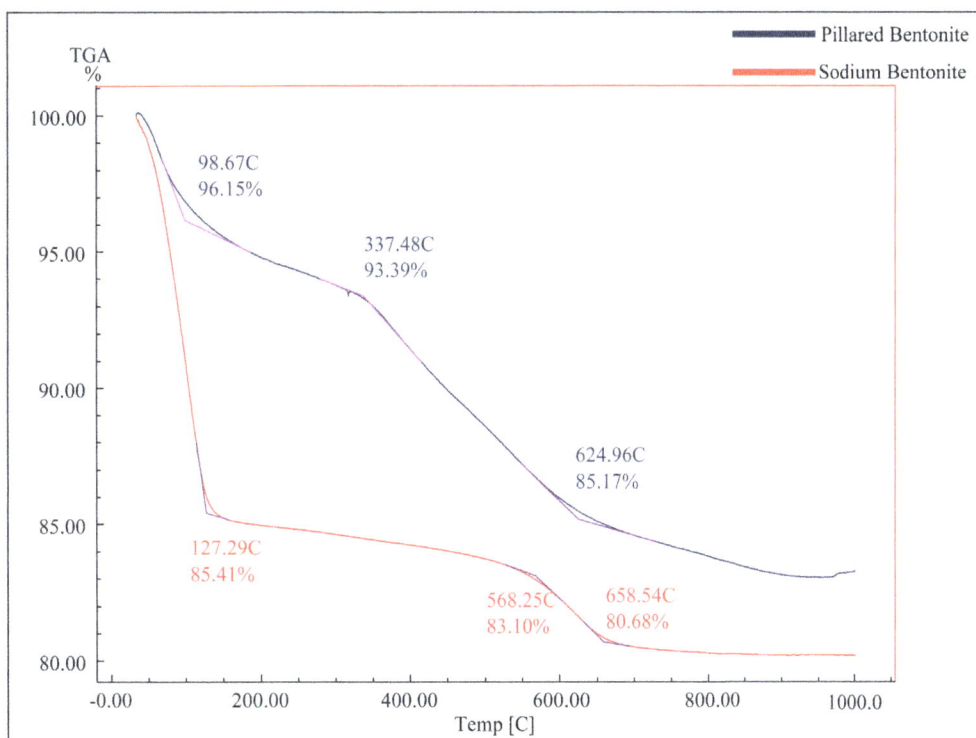

Figure 5. TG analysis of the pillared and purified sodium bentonite.

posed to be adsorbed by pillared clay.

The desorption experiments were carried out after an adsorption step in the same conditions as described above. After adsorption and phase separation by centrifugation, the supernatant solutions were discarded. Volume of solution remaining in the pellets was calculated by weight and the amount of non adsorbed thymol was calculated.

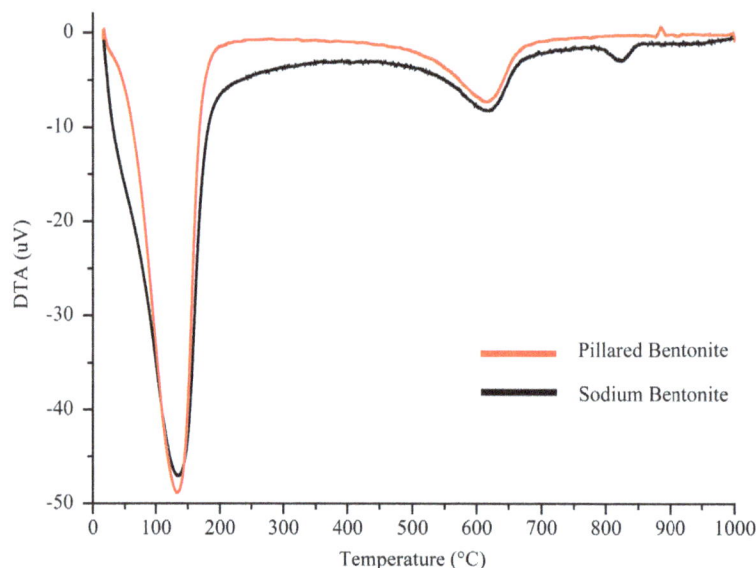

Figure 6. DTG curves of the pillared & purified sodium bentonite.

Account was taken of the thymol remaining in the solution of the moist clay pellet in calculations of the fraction desorbed. The sequential (adsorption/desorption) run were conducted to determine the mobilization factor [25] following the equation:

$$K_{SD} = \frac{M_{ads}}{M_{des}}$$ (3)

where: K_{SD}: the adsorption:desorption ratio; M_{ads}: the amount of solute adsorbed (mg/g); M_{des}: the amount of solute desorbed (mg/g).

2.4. Adsorption Kinetic

For kinetic studies, 0.150 g of pillared Bentonite (B-PILC) was contacted with 200 mL of two different concentrations of thymol $C_1 = 78.3$ mg/L and $C_2 = 93.84$ mg/L. Using water-bath shaker at 20°C. The agitation speed was kept constant at 500 rpm. At predetermined intervals of time, solutions were analyzed for the final concentration of thymol. The amount of adsorption q_t (mg/g), at time t (h), was calculated by:

$$q_t = \frac{V \times (C_0 - C_t)}{m}$$ (4)

where C_t (mg/L) is concentrations of thymol at time t (h).

3. Results and Discussion

3.1. Characterization of Samples

3.1.1. X-Ray Diffraction
The (**Figure 1**) shows the XRD patterns of the sodic and the pillared bentonite obtained.

The XRD analyses show in all cases, a clear shift of the signal at 13.22 Å (corresponding to the smectitic signal for the sodic bentonite), to a value close to 19.44 Å, which represents an increase in the basal space of the pillared clay. This indicates that the chemical modification of the clays led to a successful pillaring process. The basal spacing that expands to about 19.44 Å is equal to the thickness of one clay layer (9.4 Å) plus the height of one Al_{13} cation (9.7 Å).

3.1.2. Textural Characteristics
These analyses were done in order to determine the influence of chemical modification of clays (purified sodium

and pillared bentonite) on there structural characteristics. Before analyses the samples were automatically degassed under vacuum for 4 h at 475 K; the samples mass used are 0.1318 g and 0.1438 g of purified sodium and pillared bentonite.

The adsorption-desorption isotherms are important due to the parameters that can be established: specific surface area, porosity, pore volume, pore size distribution and average pore diameter. More, we can also obtain qualitative information regarding the structure (pores shape, interconexions etc.). To determine the textural characteristics several models are proposed [24].

In order to determine the textural properties, the Langmuir model in a range of relative pressure smaller than 0.07 [26], and the correction to the BET model proposed by [27] were used. The microporous areas were determined by curves using the De Boer's method, and the micropore volumes were determined by curves using the Harkin-Jura equation.

The adsorption-desorption isotherms with N_2 of purified sodium bentonite (B-Na purified) and the pillared bentonite (B-Al-PILCs), are shown in **Figure 2**. The comparison between the tow isotherms, lead to a type IV isotherm with H1 hysteresis for B-Na purified [28] and H3 hysteresis for B-Al-PILCs [29]. This means that in the case of sodium purified bentonite (B-Na), the distribution of pore size is regular, and the increase in the adsorption of N_2 for B-Al-PILCs in comparison to the B-Na purified, as a result of the porosity generated because of the pillaring process.

The pillared material present developed mesoporosity and the hysteresis has a narrow loop with two branches which are almost vertical and parallel.This form of hysteresis is often associated with adsorbents made up of agglomerates and aggregated plane particles forming slit shape and narrow size pore distribution. Relevant data for the sodic purified and pillared bentonite obtained are presented in **Table 1**.

The increase in the superficial area and the porous volumes of the B-AL-PILCs compared to B-Na purified is remarkable [30]. This increase could be related to the high introduction of Al in the clays, and the high homogeneity and intensity of the pillaring signals reached by the methodology used in this paper. The superficial area of the B-AL-PILCs is determined mainly by the microporous area (**Table 1**), which is evident for of a significant textural modification of the B-Na purified.

3.1.3. FT IR Spectroscopy

FTIR spectra of B-Na purified and B-Al-PILCs clays are given in **Figure 3** and **Figure 4**. At high frequency regions of the infrared spectra (**Figure 3**), bands corresponding to the water molecules present in the interlayers and the structural hydroxyl groups in the clay layers were observed in the region between 3350 - 4000 cm^{-1}. The absorption band at 3450 cm^{-1}, corresponds to the symmetric O-H stretching vibration of H-bonded water, (**Figure 3**) [31].

Also, it can be seen that its intensity is dependent on the type and the concentration of the interlayer cations. For B-Al-PILCs, the band locates at 3643 cm^{-1} and the intensity is higher than that of B-Na purified 3644 cm^{-1}. The former is ascribed to the O-H stretching vibration in hydroxy-Al cations while the latter corresponds to the hydroxyl groups involved in water-water hydrogen bands [32]. The samples show a band at 3621 cm^{-1} related to the OH stretching of Si-OH, corresponding to silanol groups. Such hydroxyls are either located at corners and fractures of sheets or are formed by the processes of tetrahedral inversion. For B-Al-PILCs sample, a new band is found at 3652 cm^{-1} (**Figure 3**), which is probably produced by the change of position of the SiOH group in the structure of the smectite. The original position was altered by the entrance of the Al ion in the smectite structure, which is likely to be due to Si-OH species perturbed by pillars, and designated as SiOH*.

The low frequency regions of the infrared spectra of the B-Na purified and B-Al-PILCs are very similar, but the latter shows a very small low intensity band at 550 - 450 cm^{-1} originated from Si-O bending and Al-O stretching vibration and there were no changes in Si-O bending, but a small increase in Al-O stretching in

Table 1. Selected textural properties of the investigated samples.

Samples	S_{BET} (m^2/g)	S_{ext} (m^2/g)	$V_{0.991}$ (cm^3/g)	V_{mic} (cm^3/g)	D_{max} (Å)	D_{med} (Å)
B-Na purified	107.5	81.024	0.123	0.053	14.547	5.32
B-AL-PILL	270.430	83.68	0.246	0.078	17.963	8.9

S_{BET}: specific surface area; $V_{0.98}$: total pore volume; V_{mic}: micropore volume; D_{max}: the pore diameter where the maximum of derivative cumulative volume curves is reached; D_{med}: median value of pore diameter.

intensity by pillaring. This situation was supported by the increase in Al content of pillared samples around 657 cm^{-1} (**Figure 4**). The lattice vibration at 657 cm^{-1} can be ascribed to the Al-O bond of tetrahedrally coordinated Al cations in the center of the Al$_{13}$ pillars [33]. The band center of B-Al-PILCs locates at 1556 cm^{-1} with a higher intensity compared with that of B-Na purified at 1560 cm^{-1}. This should be attributed to an increase of water content in B-Al-PILCs, resulted from the intercalation of hydroxy-Al cations into the clay interlayer space. The infrared spectra of B-Na purified presented bands in the region of 1687 - 1640 cm^{-1} attributed to hydrating water. The aluminum pillaring agent had caused a decrease in the free silica peak intensity which occurred at 840 and 916 cm^{-1} [34].

3.1.4. Thermal Analysis (DTA-TG)

The thermal curves are depicted in **Figure 5** and **Figure 6** in the 30°C - 1000°C range. The curves correspond to the starting purified sodium and pillared bentonite, after equilibrating in a desiccator, at room temperature. The general feature of thermal curves clearly reveals two steps: 1) in the 30°C - 340°C and 2) in the 300°C - 650°C temperature ranges.

The first step with purified sodium and pillared clay has been ascribed to physisorption of the hydrating water, whereas the second step is due to dehydroxylation of silicate structure.

Sometimes, this step occurs dissociated into two which is not well visible here but is clearly shown in the corresponding DTG and, moreover, in the TGA curve (**Figure 5**), denoting dehydroxylation of the silicate structure in two different environments. For Al pillared materials, the net isomorphic substitution in the clay with different bonding strengths between the oligocations and the surrounding oxygen (or hydroxyl) ions can be observed [35]. Dehydroxylation continues between 337°C and 650°C and is also detected to approximately 620°C in an important step. This step is related with the stability of the pillars, since an important decrease in the basal spacing values occurs at this temperature, indicating the collapse of the clay structure. Therefore, the thermogravimetric analyses are in agreement with the aforementioned thermal stability of the synthesized pillared clay up to 650°C.

3.2. Adsorption, Desorption Studies of Thymol on B-Al-PILCs

3.2.1. Adsorption Isotherms, Solid/Liquid Ratio and Concentrations Effects

Under ideal saturated conditions, the solid liquid ratio should not influence the amount of organic or inorganic molecules adsorbed per unit of adsorbent. However, some interested studies have shown that both organic and inorganic contaminant adsorption is dependent on solid-liquid ratio to some degree [36].

The adsorption of thymol was studied at different initial concentrations (5 to 760 mg/L). **Figure 7** shows the result for effect of initial concentration and for various solid/liquid ratios on adsorption of thymol onto B-AL-PILCs. The amount of thymol adsorbed q_e, plotted against the equilibrium concentration C_e on B-AL-PILCs, is given in **Figure 7**.

The equilibrium adsorption of thymol increases with the increase of initial thymol concentration and the decrease of the solid/liquid ratio showing the adsorption process to be dependent on the initial concentration and the content of the solid adsorbent. Initially, the adsorption isotherms of thymol molecules show a rising part whose slope increases when the amount of solid decreases, suggesting a strong affinity of the molecules for the surface sites. The amount of adsorption reaches a limiting value of around 319 mg·g^{-1} for low solid/liquid ratio (0.2%). The pH of the thymol solution will effect on the ability of adsorption. The pH of clay suspension tends to be neutral, *i.e.*: pH 7.5. The thymol (pKa 10.6) in this condition is mainly in a neutral form, therefore, it will stick to the surface of the negatively clay.

The high adsorption capacity for thymol uptake presented by pillared clay may be caused by adsorption by Van Der Waals interactions and by hydratation forces with polar groups on the pillared bentonite (Al-OH and Si-OH) with (OH groups and benzene group) of thymol on the basal plane and on the edges of the layers silicates, knowing that in pH < pKa, the thymol is in neutral form. It is evident that the intercalation by pillaring process, makes greatly influences the adsorption of thymol by the surface orientation of pillared clay, and leads to large areas of the pillared clay surface being exposed allowing the adsorption of thymol.

Tow models of isotherms were tested for their ability to describe the experimental results, namely the Langmuir and Freundlich isotherms.

The Langmuir adsorption model [37], (**Figure 9**) is based on the assumption that maximum adsorption cor-

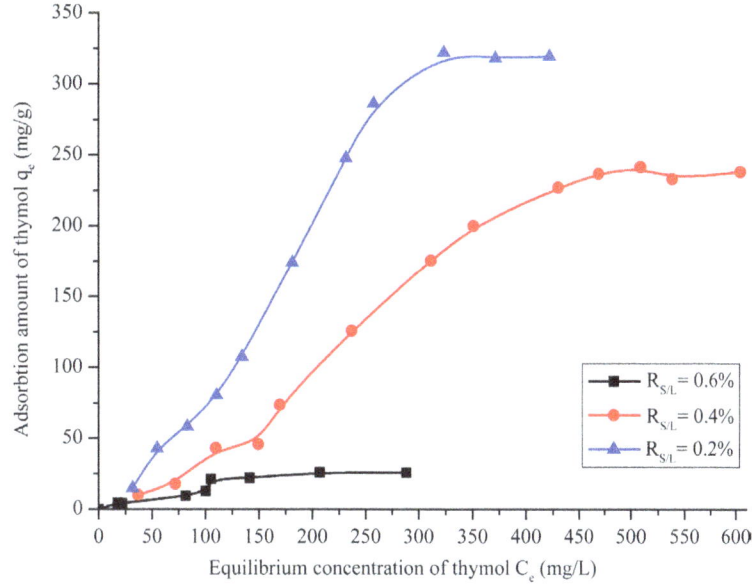

Figure 7. Equilibrium adsorption of thymol on pillared bentonite at different solid/liquid ratio.

responds to a saturated mono-layer of solute molecules on the adsorbent surface, with no lateral interaction between the sorbed molecules. The linear expression of the Langmuir model (**Figure 8**) is given by Equation (5)

$$\frac{C_e}{q_e} = \frac{1}{K_L \times q_m} + \left(\frac{1}{q_e}\right) \times C_e \tag{5}$$

where q_e (mg/g) and C_e (mg/L) are respectively the amount of adsorbed thymol per unit mass of adsorbent and thymol concentration at equilibrium, q_m is the maximum amount of the thymol per unit mass of adsorbent to form a complete monolayer on the surface bound at high C_e, and K_L is a constant related to the affinity of the binding sites (L/mg).

The Langmuir constants q_m and K_L were determined from the slope and intercept of the plot and are presented in **Table 2**.

The essential characteristics of the Langmuir isotherm can be expressed in terms of a dimensionless constant separation factor R_L that is given by Equation (6) [38]:

$$R_L = \frac{1}{1 + K_L \times C_0} \tag{6}$$

where C_0 is the highest initial concentration of adsorbate (mg/L), and K_L (L/mg) is Langmuir constant.

The value of R_L indicates the shape of the isotherm to be either unfavorable $(R_L > 1)$, linear $(R_L = 1)$, favorable $(0 < R_L < 1)$, or irreversible $(R_L = 0)$. The R_L value is <1 indicates that the adsorption model is not conformed.

The Freundlich isotherm (H. Freundlich, 1906) is an empirical equation employed to describe heterogeneous systems. The linear form of Freundlich equation is expressed:

$$\ln q_e = \ln K_F + \left(\frac{1}{n}\right) \times \ln C_e \tag{7}$$

where K_F and n are Freundlich constants with K_F (mg/g) (L/mg), (1/n) is the adsorption capacity of the sorbent and n giving an indication of how favorable the adsorption process. The magnitude of the exponent, $1/n$, gives an indication of the favorability of adsorption.

The Value of K_F and n are calculated from the intercept and slope of the plot (**Figure 9**) and listed in **Table 2**.

The plot of amount adsorbed $(\ln q_e)$ against the equilibrium concentration $(\ln C_e)$. (**Figure 9**) shows that

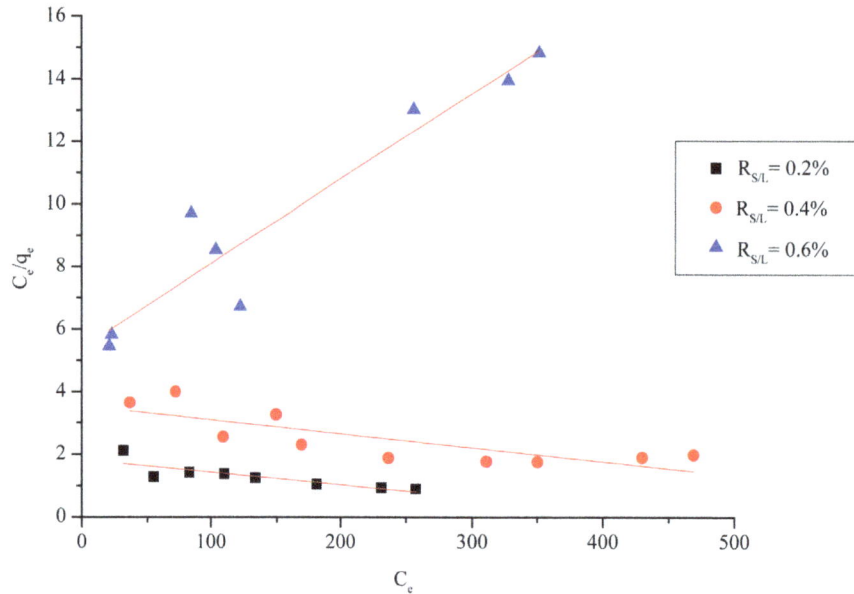

Figure 8. Langmuir adsorption isotherm of thymol on pillared bentonite.

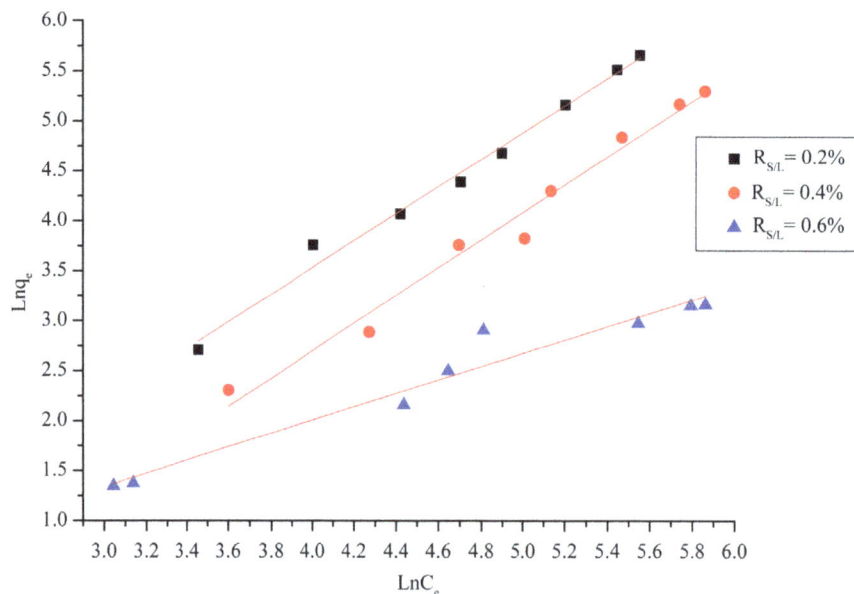

Figure 9. Freundlich adsorption isotherm of thymol on B-AL-PILL.

the adsorption obeys the Freundlich [39] model [40] [41].

The Temkin isotherm [42] has generally been applied in the linear form (**Figure 10**):

$$q_e = B \times \ln A + B \times \ln C_e \tag{8}$$

where $B = R*T/K_T$, K_T is the Temkin constant related to heat of sorption (J/mol); A is the Temkin isotherm constant (L/g), R the gas constant (8.314 J/mol°K) and T the absolute temperature (°K). Therefore, by plotting q_e vs $\ln C_e$ (**Figure 9**) the constants A and B can be determined. The constants A and B are listed in **Table 2**.

As seen in **Table 2**, the Freundlich isotherm fits quite well with the experimental data.

Examination of the plot suggests that the linear Freundlich isotherm is a good model for the sorption of the thymol onto B-AL-PILL.

Table 2. Isotherm parameters for adsorption of thymol by B-AL-PILL.

		Langumiur adsorption isotherm:				
	R^2		Value	Standard error	K_L	q_m
$R_{S/L} = 0.2\%$	0.89246	Intercept	1.82802	0.160	−0.007	--
		Slope	−0.00398	0.001		
$R_{S/L} = 0.4\%$	0.66635	Intercept	3.5448	0.319	−0.015	--
		Slope	−0.00446	0.001		
$R_{S/L} = 0.6\%$	0.60175	Intercept	5.39073	0.716	0.146	36.88
		Slope	0.02711	0.003		
		Freundlich adsorption isotherm:				
	R^2		Value	Standard error	K_F	n
$R_{S/L} = 0.2\%$	0.98763	Intercept	−1.874	0.271	0.15	0.74
		Slope	1.351	0.057		
$R_{S/L} = 0.4\%$	0.97685	Intercept	−2.846	0.404	0.058	0.72
		Slope	1.386	0.08		
$R_{S/L} = 0.6\%$	0.9508	Intercept	−0.657	0.272	0.518	1.5
		Slope	0.666	0.057		
		Temkin adsorption isotherm:				
	R^2		Value	Standard error	A	B
$R_{S/L} = 0.2\%$	0.87889	Intercept	−533.243	86.792	0.023	141.486
		Slope	141.486	17.374		
$R_{S/L} = 0.4\%$	0.85304	Intercept	−403.723	71.983	0.017	99.884
		Slope	99.884	13.689		
$R_{S/L} = 0.6\%$	0.91856	Intercept	−19.119	3.818	0.069	7.154
		Slope	7.154	0.8001		

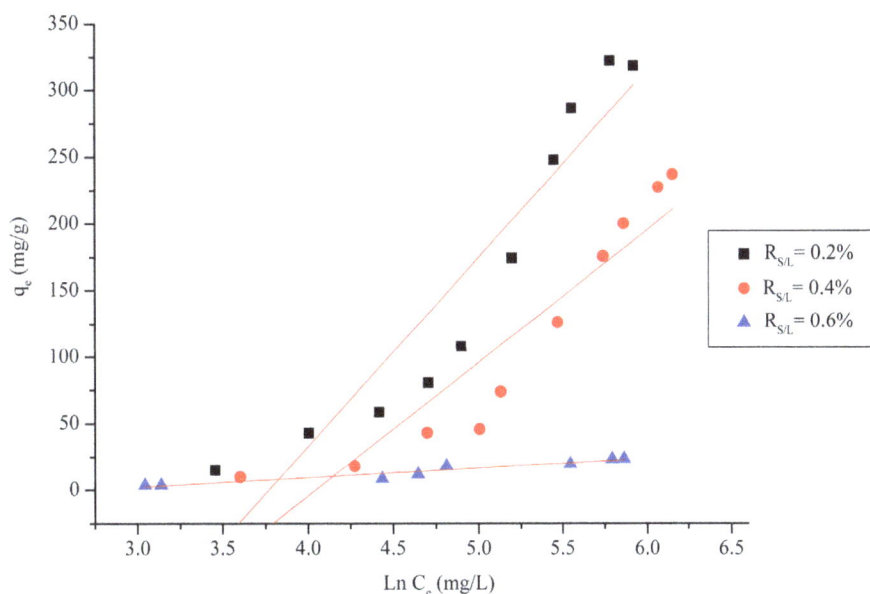

Figure 10. Temkin adsorption isotherm of thymol on pillared bentonite.

Table 2 shows the linear Freundlich sorption isotherm constants, coefficients of determination (R^2) and error values. Based on the R^2 values, (correlation coefficient $R^2 > 0.90$). The linear form of the Freundlich isotherm appears to produce a reasonable model for sorptions in all three ratios, implying the presence of the highly energetic sites were the molecules of thymol were adsorbed. After the point of inflection of the experimental data, the Freundlich isotherm predicted that the equilibrium adsorption capacity should keep increasing exponentially with increasing equilibrium concentration in the liquid phase. However, the experimental adsorption isotherm for thymol presented a plateau at higher equilibrium concentration, implying the saturation of adsorption sites and the maximum filling of the pores. Thus, Freundlich model should not be used for extrapolation of this data to higher concentration (problem of high-saturation), [43].

3.2.2. Desorbability

The adsorption and desorption isotherm of thymol on pillared clays was presented in **Figure 11**.

The thymol can be adsorbed on pillared clays through ligand exchange, hydrogen bonding, and electrostatic, hdrophobic, and Van Der Waals interactions.

In this study, the pillared clay and thymol interaction was partly non-reversible, in water and in the same codtions of equilibrium adsorption:

The adsorption and desorption isotherm show a low desorption amount with $K_{SD} > 0.95$ that's mean that the percentage of desorption amount is about 5%, it is probably caused by the hydrophobicity of thymol makes a strong interactions with the clay, and low with water.

3.3. Adsorption Kinetics

3.3.1. Pseudo First and Second Order

Kinetic study of adsorption is important because it gives the times necessary to reach the equilibrium for adsorption and desorption that are required data for obtaining valuable isotherms and parameter characteristics of molecular displacements and reaction near or on the adsorbent surface (which bring the solute in the adsorbed state).

Kinetic values of adsorption were determined by analyzing adsorptive uptake of the thymol from aqueous solutions by pillared bentonite (B-AL-PILL), in two different concentrations $C_1 = 78.3$ mg/L and $C_2 = 93.84$ mg/L.

All the kinetically experiments were carried out at initial pH (pH: 7.5), and the plots were drawn by calculating the average of closed each other values of the experimental data. Then, equilibrium contact times for thymol used were determined from graph drawing residual concentrations no adsorbed of adsorbate versus times.

The type curves of the kinetics of thymol by pillared bentonite (B-AL-PILL) retention are reproduced in

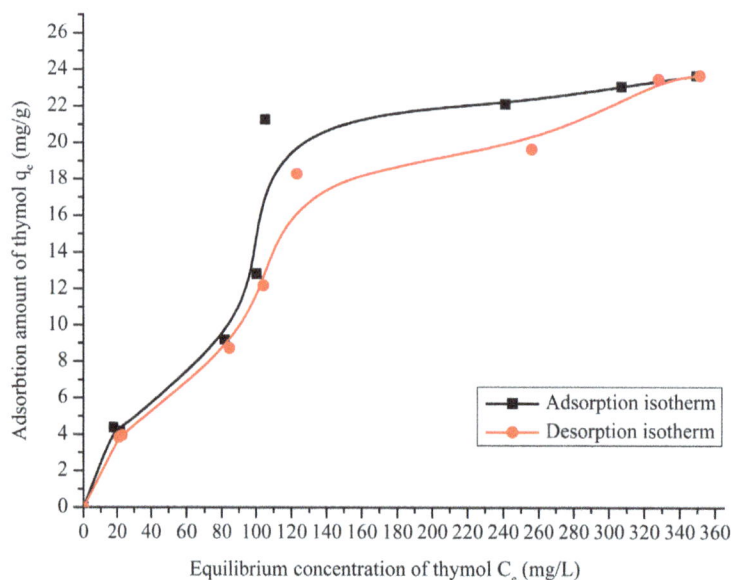

Figure 11. Equilibrium adsorption-desorption of thymol on pillared bentonite.

(**Figure 12**).

The modeling of adsorption kinetics of thymol on pillared bentonite (B-AL-PILCs) was investigated by two common models, namely, the Lagergren pseudo-first order model [44] and Pseudo second order model [45] [46]. Lagergren proposed a method for adsorption analysis which is the pseudo first order kinetic.

For a batch contact time process where the rate of sorption of thymol on to the pillared clay (B-AL-PILCs) surface is proportional to the amount of thymol sorbed from the solution phase, the first-order kinetic equation may be expressed as:

$$\frac{q_t}{dt} = k_1 \times \left(q_e - q_t \right) \qquad (9)$$

After integration and applying boundary conditions, viz that the initial conditions are $\left(q_e - q_t \right) = 0$ at $t = 0$, equation becomes:

$$\ln \left(q_e - q_t \right) = \ln q_e - k_1 \cdot t \qquad (10)$$

A linear plot of $\ln \left(q_e - q_t \right)$ against time with good correlation coefficient allows one to obtain the rate constant, indicating that Lagergren's equation is appropriate to thymol sorption on pillared clay B-AL-PILCs (**Figure 13**). The Lagergren's first-order rate constant $\left(K_1 \right)$ and q_e determined from the model are presented in **Table 3** along with the corresponding correlation coefficient.

The pseudo-second-order kinetics may be expressed as [47] [48]:

$$\frac{t}{q_t} = \frac{1}{k_2 \cdot q_e^2} + \frac{1}{q_t} \qquad (11)$$

where: q_e and q_t are the adsorption capacity at equilibrium and at time t, respectively (mg·g^{-1}), K_1 is the rate constant of pseudo first-order adsorption (L·min^{-1}), K_2 is the rate constant of pseudo second-order adsorption (g·mg^{-1}·min^{-1}). The plot of $\frac{t}{q_t}$ vs. t should give a linear relationship from which q_e and K_2 can be determined from the slope and intercept of the plot, respectively.

The equilibrium adsorption capacity $\left(q_e \right)$, and the second-order constants K_2 (g/mgh) can be determined experimentally from the slope and intercept of plot $\frac{t}{q_t}$ versus t (**Figure 13**).

The conformity between experimental data and the model predicted values was expressed by the correlation coefficients (R^2, values close or equal to 1). A relatively high R^2 value indicates that the model successfully

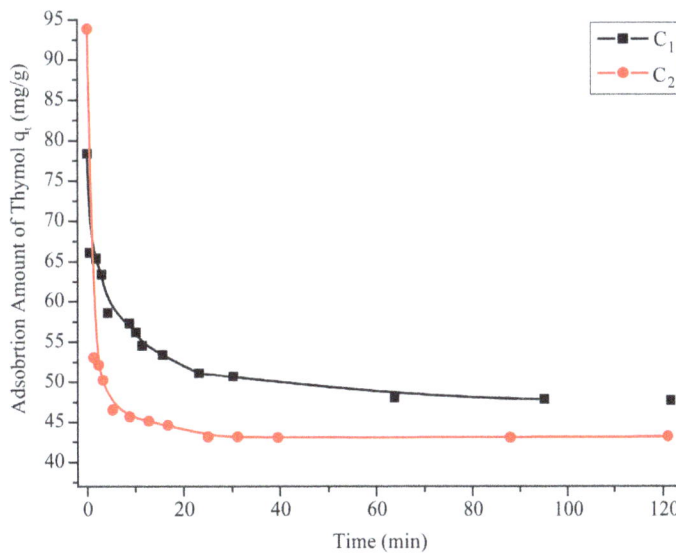

Figure 12. Adsorption kinetics of thymol on B-AL-PILL.

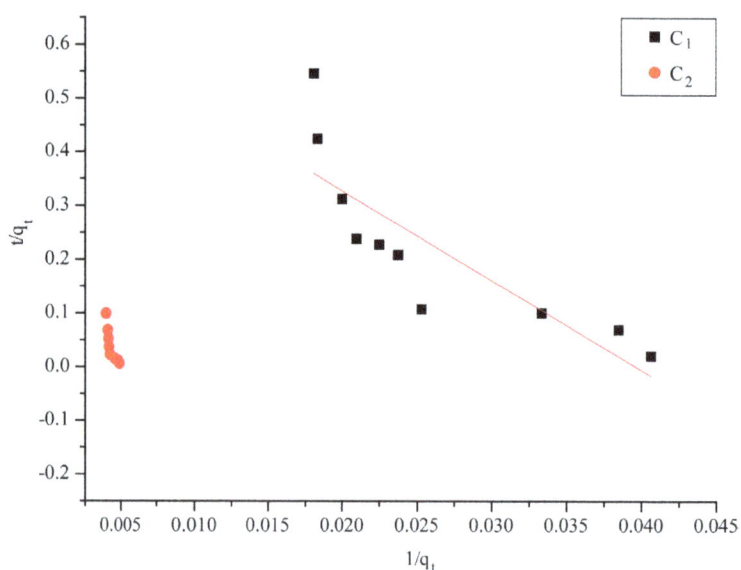

Figure 13. Pseudo-second-order kinetic for adsorption of thymol on pillared bentonite.

Table 3. Comparison of the pseudo-first-order, pseudo second order adsorption rate constants.

C (mg/L)	R^2		Value	Standard error	K_1	q_e
		Pseud first order kinetic				
C_1	0.96591	Intercept	3.59744	0.08349	0.06541	61.076
		Slope	−3.92449	0.22194		
C_2	0.90889	Intercept	4.45222	0.28623	0.18541	253.461
		Slope	−11.12522	1.16763		
C (mg/L)	R^2	Pseudo second order kinetic	Value	Standard error	K_2	q_e
C_1	0.6682	Intercept	0.00932	0.00129	0.0287	61.076
		Slope	−0.22767	0.04583		
C_2	0.67126	Intercept	0.0075	0.00172	0.0207	253.461
		Slope	−1.55553	0.39872		

describes the kinetics of thymol adsorption.

The correlation coefficient R^2 for the pseudo first-order adsorption model (**Figure 14**) has a high value $\left(R^2 = 0.96591\right)$, so the adsorption rate $\dfrac{q_t}{dt}$ is proportional to the first order of $\left(q_e - q_t\right)$.

3.3.2. Adsorption Mechanism

In order to gain insight into the mechanisms and rate controlling steps affecting the kinetics of adsorption, the kinetic experimental results were fitted to the Weber's intra-particle diffusion [39] [49]. The kinetic results were analyzed by the intra-particle diffusion model to elucidate the diffusion mechanism. In this model, the evolution of the amount adsorbed versus the time is expressed as:

$$q_t = K_{id} \times t^{1/2} + C \tag{12}$$

where, C is the boundary layer diffusion effects, K_{id} is the rate constant for intra-particle diffusion. A plot of q_t vs $t^{1/2}$ giving straight line confirms intra-particle diffusion sorption which can be evaluated from the slope of the linear plot of q_t versus $t^{1/2}$ [50] as shown in (**Figure 15**). The intercept of the plot reflects the boundary

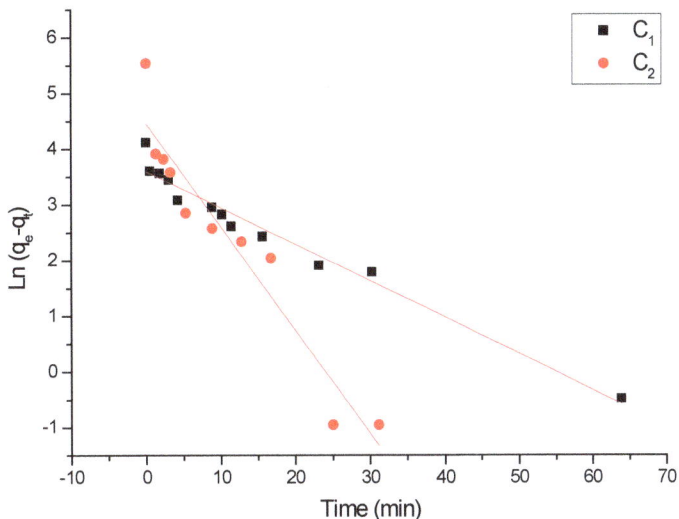

Figure 14. Pseudo-fist-order kinetic for adsorption of thymol on pillared bentonite.

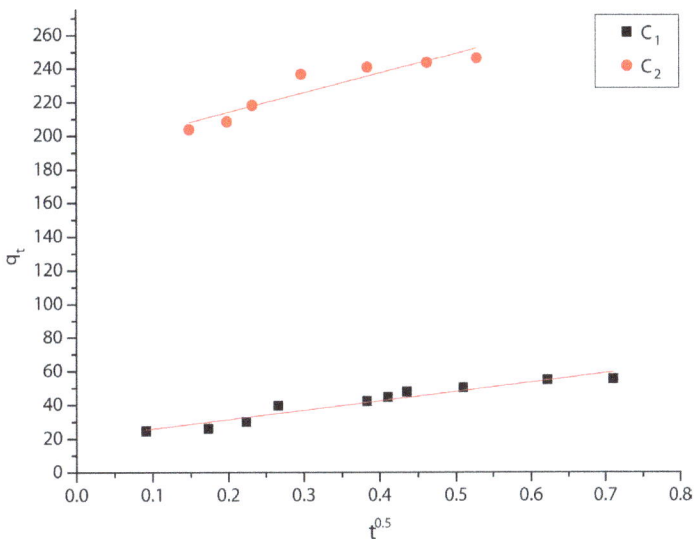

Figure 15. Intraparticle diffusion plots for adsorption of thymol on pillared bentonite.

layer effect. The larger the intercept, the greater and the contribution of the surface sorption in the rate controlling step.

Another type of intra-particle diffusion model is expressed as [49] [50]:

$$R = K_{id} \times t^{a} \tag{13}$$

A linearised form of the equation is given as [51]:

$$\ln R = \ln K_{id} + a \times \ln t \tag{14}$$

where, R is the fraction of the amount adsorbed, t is the contact time (min), "a" is the gradient of linear plots; K_{id} is the intra-particle diffusion rate constant (min^{-1}).

This model was used to analyze the experimental data. The plot of log R vs. log t is shown on (**Figure 16**). The regression equations and R^2 values for intra-particle diffusion model are shown on (**Table 4**).

The values of the constants "K_{id}" and "a", are shown on **Table 4**. The values of "a" and K_{id} were calcu-

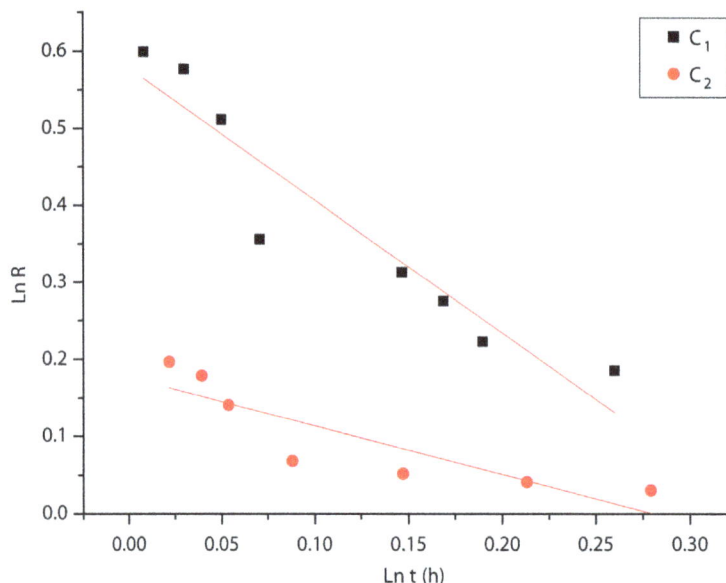

Figure 16. Intraparticle diffusion plots II for adsorption of thymol on pillared bentonite.

Table 4. Comparison of the Intra-particle diffusion model I & II of adsorption.

		Intraparticle diffusion model I for adsorption of thymol				
	R^2		Value	Standard error	K_{id}	C
C_1	0.92511	Intercept	20.34803	2.21853	55.16088	20.34803
		Slope	55.16088	5.20818		
C_2	0.84661	Intercept	190.80805	6.90982	116.34883	190.80805
		Slope	116.34883	19.92007		
		Intraparticle diffusion model II for adsorption of thymol				
	R^2		Value	Standard error	K_{id}	a
C_1	0.88288	Intercept	0.57896	0.03342	1.7841	−1.72199
		Slope	−1.72199	0.23484		
C_2	0.7263	Intercept	0.17696	0.02299	1.0232	−0.63027
		Slope	−0.63027	0.15321		

lated from the slopes and intercepts of the plot respectively. The values of "a" depicts the adsorption mechanism and K_{id} may be taken as a rate factor [52].

A relatively high R^2 value of ***Intraparticle diffusion model I*** indicates that the model successfully describes the mechanism of adsorption kinetics.

If the regression of q_t versus $t^{1/2}$ is linear and passes through the origin, then intra-particle diffusion is the slow rate-limiting step. However, the linear plots (**Figure 14**) at each concentration did not pass through the origin. This indicates that the intra-particle diffusion was not only a rate controlling step. So the multiple nature observed in the intra-particle diffusion plot suggests that intra-particle diffusion is not solely (25 min for equilibrium) rate controlling. External mass transfer of thymol molecules on to pillared clay is also significant in the sorption process, especially at the initial reaction period (80% of amount adsorbed in the first 5 min).

4. Conclusions

The resultant pillared bentonite complexes were characterized using FTIR, with a combination of XRD, and textural analysis. Results obtained from FTIR spectra showed that there were changes in the clay structure with pillaring. The XRD patterns show that the basal spacings of the complexes increase to 19 Å and it was observed that basal spacing, surface area values and the properties against temperature effects of bentonites were improved by pillaring.

A Moroccan bentonite pillared by Al_{13} [OH/Al = 2.4] possessed an important capacity for adsorbing thymol. The quantity of thymol adsorbed was 319.5 mg/g at pH 7.54 and showed high irreversibility (desorption) in water.

The Freundlich isotherms were found to be applicable for the adsorption equilibrium data of thymol on pillared bentonite. The pseudo first order equation gave a better fit to the sorption process than the pseudo second order equation.

The intraparticle diffusion plot confirmed that the sorption process was a controlled particle diffusion. The amount of thymol released (desorbed) from the pillared clay was negligible in water. Mechanisms involved in the adsorption, which explain the high thymol uptake and irreversibility, were the molecule fixed by the adsorbed clay on polar groups (Al-OH and Si-OH) (remaining) sites onto the basal plane and on the layer silicate edges. It is believed that due to these properties, pillared clay, shows much potential as an adsorbent for thymol.

Acknowledgements

The authors are sincerely thankful to CNRST-Morocco (PROTARS p23/66) for its financial support and Professor A. EL MEDIOUNI (FLSH-Oujda) for correcting the English of the article.

References

[1]   Brindley, G.W. and Sempels, R.E. (1977) Preparation and Properties of Some Hydroxy-Aluminum Beidellites. *Clay Minerals*, **12**, 229-237. http://dx.doi.org/10.1180/claymin.1977.012.3.05

[2]   Barrer, R.M. (1989) Shape-Selective Sorbents Based on Clay Minerals: A Review. *Clays Clay Minerals*, **37**, 381-395. http://dx.doi.org/10.1346/CCMN.1989.0370501

[3]   Bergaya, F., Hassoun, N., Barrault, J. and Gatineau, L. (1993) Pillaring of Synthetic Hectorite by Mixed [$Al_{13-x}Fe_x$] Pillars. *Clay Minerals*, **28**, 109-122. http://dx.doi.org/10.1180/claymin.1993.028.1.10

[4]   Bergaya, F. (1994) Focus on Pillaring and Mixed $Al_{13-x}Fe_x$ PILCs. *CEA-PLS Newsletter*, **7**, 11-12.

[5]   Beneke, K., Thiesen, P. and Lagaly, G. (1995) Synthesis and Properties of the Sodium-Lithium Silicate Silinaite. *Inorganic Chemistry*, **34**, 900-907. http://dx.doi.org/10.1021/ic00108a022

[6]   Pinnavaia, T.J., Tzou, M.S., Landau, S.D. and Raythatha, R.H. (1984) On the Pillaring and Delamination of Smectite Clay Catalyst by Polyoxo Cations of Aluminium. *Journal of Molecular Catalysis*, **27**, 195-212. http://dx.doi.org/10.1016/0304-5102(84)85080-4

[7]   Lahav, N., Sham, U. and Shabtai, J. (1978) Cross-Linked Smectites. I. Synthesis and Properties of Hydrox-Aluminum-Montmorillonite. *Clays Clay Minerals*, **26**, 107-115. http://dx.doi.org/10.1346/CCMN.1978.0260205

[8]   Bergaya, F. (1995) The Meaning of Surface Area and Porosity Measurements of Clays and Pillared Clays. *Journal of Porous Materials*, **2**, 91-96. http://dx.doi.org/10.1007/BF00486575

[9]   Tahani, A., Karroua, M., El Farissi, M., Levitz, P., Van Damme, H., Bergaya, F. and Margulies, L. (1999) Adsorption of Phenol and Its Chlorine Derivatives on PILCS and Organo-PILCS. *Journal of Chemical Physics*, **96**, 464-469. http://dx.doi.org/10.1051/jcp:1999153

[10] Cooper, C. and Burch, R. (1999) Mesoporous Materials for Water Treatment Processes. *Water Research*, **33**, 3689-3694. http://dx.doi.org/10.1016/S0043-1354(99)00095-0

[11] Danis, T.G., Albanis, T.A., Petrakis, D.E. and Pomonis, P.J. (1998) Removal of Chlorinated Phenols from Aqueous Solutions by Adsorption on Alumina Pillared Clays and Mesoporous Alumina Aluminum Phosphates. *Water Research*, **32**, 295-302. http://dx.doi.org/10.1016/S0043-1354(97)00206-6

[12] Dyer, A. and Gallardo, V.T. (1990) Cation and Anion Exchange Properties of Pillared Clays. In: Williams, P.A. and Hudson, M.J., Eds., *Recent Developments in Ion Exchange*, Springer Netherlands, Berlin, 75-84. http://dx.doi.org/10.1007/978-94-009-0777-5_8

[13] Dyer, A., Gallardo, V.T. and Roberts, C.W. (1989) Preparation and Properties of Clays Pillared with Zirconium and Their Use in HPLC Separations. Zeolites, Facts, Figures, Future.

[14] Konstantinuo, I.K., Albanis, T.A., Petrakis, D.E. and Pomonis, P.J. (2000) Removal of Herbicides from Aqueous Solutions by Adsorption on Al-Pillared Clays, Fe-Al Pillared Clays and Mesoporous Alumina Aluminum Phosphates. *Water Research*, **34**, 3123-3136. http://dx.doi.org/10.1016/S0043-1354(00)00071-3

[15] Osorio-Revilla, G., Gallardo-Velázquez, T., López-Cortés, S. and Arellano-Cárdenas, S. (2006) Immersion Drying of Wheat Using Al-PILC, Zeolite, Clay and Sand as Particulate Media. *Drying Technology*, **24**, 1033-1038. http://dx.doi.org/10.1080/07373930600776225

[16] Theopharis, G.D., Triantafyllos, A.A., Dimitrios, E.P. and Philip, J.P. (1998) Removal of Chlorinated Phenols from Aqueous Solutions by Adsorption on Alumina Pillared Clays and Mesoporous Alumina Aluminum Phosphates. *Water Research*, **32**, 295-302. http://dx.doi.org/10.1016/S0043-1354(97)00206-6

[17] Nennemann, A., Mishael, Y., Nir, S., Rubin, B., Polubesova, T., Bergaya, F., Damme, H.V. and Lagaly, G. (2001) Clay-Based Formulations of Metolachlor with Reduced Leaching. *Applied Clay Science*, **18**, 265-275. http://dx.doi.org/10.1016/S0169-1317(01)00032-1

[18] Helander, I.K., Alakomi, H.L., Latva-Kala, K., Sandholm, T.M., Pol, I., Smid, E.J. and von Wright, A. (1998) Characterization of the Action of Selected Essential Oil Components on Gram-Negative Bacteria. *Journal of Agricultural and Food Chemistry*, **46**, 3590-3595. http://dx.doi.org/10.1021/jf980154m

[19] Cox, S.D., Mann, C.M., Markhan, J.L., Bell, H.C., Gustafson, J.E., Warmington, J.R. and Wyllie, S.G. (2000) The Mode of Antimicrobial Action of the Essential Oil of *Melaleuca alternifolia* (Tea Tree Oil). *Journal of Applied Microbiology*, **88**, 170-175. http://dx.doi.org/10.1046/j.1365-2672.2000.00943.x

[20] Lambert, R.J.W., Skandamis, P.N., Coote, P.J. and Nychas, G.J.E. (2001) A Study of the Minimum Inhibitory Concentration and Mode of Action of Oregano Essential Oil, Thymol and Carvacrol. *Journal of Applied Microbiology*, **91**, 453-462. http://dx.doi.org/10.1046/j.1365-2672.2001.01428.x

[21] Walsh, S.E., Maillard, J.Y., Russel, A.D., Catrenich, C.E., Charbonneau, D.L. and Bartolo, R.G. (2003) Activity and Mechanism of Action of Selected Biocidal Agents on Gram-Positive and -Negative Bacteria. *Journal of Applied Microbiology*, **94**, 240-247. http://dx.doi.org/10.1046/j.1365-2672.2003.01825.x

[22] Lis-Balchin, M. and Deans, S.G. (1997) Bioactivity of Selected Plant Essential Oils against Listeria Monocytogenes. *Journal of Applied Microbiology*, **82**, 759-762. http://dx.doi.org/10.1046/j.1365-2672.1997.00153.x

[23] El-Miz, M., Salhi, S., El bachiri, A., Wathelet, J.P. and Tahani, A. (2013) Adsorption Study of Thymol on Na-Bentonite. *Journal of Environmental Solution*, **2**, 31-37.

[24] Platon, N., Sminiceanu, I., Miron, N.D., Muntianu, G., Zavada, R.M., Isopencu, G. and Nistor, D. (2011) Preparation and Characterization of New Products Obtained by Pillaring Process. *Revista de Chimie (Bucharest)*, **62**, 799-805.

[25] Rybicka, E.H., Calmano, W. and Breeger, A. (1995) Heavy Metals Sorption/Desorption on Competing Clay Minerals: An Experimental Study. *Applied Clay Science*, **9**, 369-381. http://dx.doi.org/10.1016/0169-1317(94)00030-T

[26] Gregg, S.J. and Sing, S.W. (1982) Adsorption, Surface Area and Porosity. Academic Press, London.

[27] Remy, M.J., Coelho, A.C.V. and Poncelet, G. (1996) Surface Area and Microporosity of 1.8 nm Pillared Clays from the Nitrogen Adsorption Isotherm. *Microporous Materials*, **7**, 287-297. http://dx.doi.org/10.1016/S0927-6513(96)00021-1

[28] Bankoviæ, P., Milutinoviæ, N., Rosiæ, A. and Jovièiæ, N.J. (2009) Structural and Textural Properties of Al, Fe Pillared Clay Catalysts, Russian. *Journal of Physical Chemistry*, **83**, 1485.

[29] Bankovic, P., Nikolic, A.M., Jovic, N., Dostanic, J., Cupic, Z., Loncarevic, D. and Jovanovic, D. (2009) Synthesis, Characterization and Application of Al, Fe-Pillared Clays. *Acta Physica Polonica A*, **4**, 811-814.

[30] Olaya, A., Moreno, S. and Molina, R. (2009) Synthesis of Pillared Clays with Al_{13}-Fe and Al_{13}-Fe-Ce Polymers in Solid State Assisted by Microwave and Ultrasound: Characterization and Catalytic Activity. *Applied Catalysis A: General*, **370**, 7-15. http://dx.doi.org/10.1016/j.apcata.2009.08.018

[31] Madejova, J., Janek, M., Komadel, P., Herbert, H.J. and Moog, H.C. (2002) FTIR Analyses of Water in MX-80 Bentonite Compacted from High Salinary Salt Solutions Systems. *Applied Clay Science*, **20**, 255-271. http://dx.doi.org/10.1016/S0169-1317(01)00067-9

[32] Salerno, P., Asenjo, M.B. and Mendioroz, S. (2001) Influence of Preparation Method on Thermal Stability and Acidity of Al-PILCs. *Thermochimica Acta*, **379**, 101-109. http://dx.doi.org/10.1016/S0040-6031(01)00608-6

[33] Xue, W., He, H., Zhu, J. and Yuan, P. (2007) FTIR Investigation of CTAB-Al-Montmorillonite Complexes. *Spectrochimica Acta Part A: Molecular and Biomolecular Spectroscopy*, **67**, 1030-1036. http://dx.doi.org/10.1016/j.saa.2006.09.024

[34] Tomul, F. and Balci, S. (2007) Synthesis and Characterization of Al-Pillared Interlayered Bentonites. *G. U. Journal of Science*, **21**, 21-31.

[35] Guerra, D.L., Lemos, V.P., Airoldi, C. and Angélica, R.S. (2006) Influence of the Acid Activation of Pillared Smectites from Amazon (Brazil) in Adsorption Process with Butylamine. *Polyhedron*, **25**, 2880-2890. http://dx.doi.org/10.1016/j.poly.2006.04.015

[36] Puls, R.W., Powell, R.M., Clark, D. and Eldred, C.J. (1991) Effects of pH, Solid/Solution Ratio, Ionic Strength, and

Organic Acids on Pb and Cd Sorption on Kaolinite. *Water, Air, and Soil Pollution*, **57-58**, 423-430. http://dx.doi.org/10.1007/BF00282905

[37] Langmuir, I. (1918) The Adsorption of Gases on Plane Surfaces of Glass, Mica and Platinum. *Journal of the American Chemistry Soc*iety, **40**, 1361-1403. http://dx.doi.org/10.1021/ja02242a004

[38] Hall, K.R., Eagleton, L.C., Acrivos, A. and Vermeulen, T. (1966) Pore and Solid-Diffusion Kinetics in Fixed-Bed Adsorption under Constant-Pattern Conditions. *Industrial Engineering Chemistry Fundamentals*, **5**, 212-223. http://dx.doi.org/10.1021/i160018a011

[39] Freundlich, H. (1906) Über dieadsorption in lösungen (Adsorption Insolution). *Zeitschrift für Physikalische Chemie*, **57**, 384-470.

[40] Treybal, R.E. (1968) Mass Transfer Operations. 2nd Edition, McGraw Hill, New York.

[41] Ho, Y.S. and McKay, G. (1998) Sorption of Dye from Aqueous Solution by Peat. *Chemical Engineering*, **70**, 115-124.

[42] Temkin, M.J. and Pyzhev, V. (1960) Recent Modifications to Langmuir Isotherms. *Acta Physicochimica USSR*, **12**, 217-222.

[43] Cardenas, S.A., Velazquiz, T.G., Revilla, G.O., Cortez, M.D. and Perea, B.G. (2005) Adsorption of Phenol and Dichloro-Phenol from Aqueous Solutions by Porous Clay Hetero-Structure (PCH). *Mexican Journal of Chemical Society*, **49**, 287-291.

[44] Lagergren, S. (1898) Zur theorie der sogenannten adsorption geloester stoffe. *Kungliga Svenska Vetenskapsakad, Handlingar*, **24**, 1-39.

[45] Ho, Y.S. and McKay, G. (1998) Sorption of Dye from Aqueous Solution by Peat. *Journal of Chemical Engineering*, **70**, 115-124.

[46] Ho, Y.S. and McKay, G. (1999) The Sorption of Lead(II) Ions on Peat. *Water Research*, **33**, 578-584. http://dx.doi.org/10.1016/S0043-1354(98)00207-3

[47] Hameed, B.H., Salman, J.M. and Ahmad, A.L. (2009) Adsorption Isotherm and Kinetic Modeling of 2,4-D Pesticide on Activated Carbon Derived from Date Stones. *Journal of Hazardous Materials*, **163**, 121-126. http://dx.doi.org/10.1016/j.jhazmat.2008.06.069

[48] Ho, Y.S. and McKay, G. (2000) The Kinetics of Sorption of Divalent Metal Ions onto Sphagnum Moss Peat. *Water Research*, **34**, 735-742. http://dx.doi.org/10.1016/S0043-1354(99)00232-8

[49] Weber Jr., W.J. and Morris, J.C. (1963) Kinetics of Adsorption on Carbon from Solution. *Journal of the Sanitary Engineering Division*, **89**, 31-59.

[50] Srivastava, S.K., Tyagi, R. and Pant, N. (1989) Adsorption of Heavy Metal Ions on Carbonaceous Material Developed from the Waste Slurry Generated in Local Fertilizer Plants. *Water Research*, **23**, 1161-1165. http://dx.doi.org/10.1016/0043-1354(89)90160-7

[51] Igwe, J.C. and Abia, A.A. (2007) Adsorption Kinetics and Intra-Particulate Diffusivities for Bioremediation of Co(II), Fe(II) and Cu(II) Ions from Waste Water Using Modified and Unmodified Maize Cob. *International Journal of Physical Sciences*, **2**, 119-127.

[52] Dermirbas, E., Kobya, M., Senturk, E. and Ozkan, T. (2004) Adsorption Kinetics for the Removal of Chromium(VI) from Aqueous Solutions on the Activated Carbons Prepared from Agricultural Wastes. *Water SA*, **30**, 533-539.

3

The Relation between the Heat of Melting Point, Boiling Point, and the Activation Energy of Self-Diffusion in Accordance with the Concept of Randomized Particles

Vitalyi P. Malyshev, Astra M. Makasheva

Chemical and Metallurgical Institute, Karaganda, Kazakhstan
Email: eia_hmi@mail.ru

Abstract

On the example of typical metals, it's found that the activation energy of self-diffusion is above of the melting heat and below of vaporization heat. This corresponds to the existence of liquid-mobile particle classification based on the concept of randomized particles. A formula for estimating the activation energy of self-diffusion by which it is approximately half of the heat of evaporation of the substance is recommended. We derive the temperature dependence for a fraction self-diffusion's particles.

Keywords

Heat of Fusion, Heat of Boiling, Self-Diffusion, Randomized Particles, Metals

1. Introduction

Thermal state of matter is described by the Boltzmann's distribution for the kinetic energy of the random motion of particles with energy levels from zero to infinite. In this connection it is of interest to determine the location of the self-diffusion particles in the modern graduation randomized particles energy responsible for the stabilization of solid, liquid and gaseous states of matter.

Data on self-diffusion activation energy are used for the analysis of plastic deformation in order to support optimum temperature for heating billet rolling mills. However, until now it was not possible to determine the proportion of particles capable to overcome the energy barrier of the self-diffusion, *i.e.* virtual (reversible) output from the crystal lattice. This feature provides a concept of randomized particles through a clear delineation of these particles on the energy barriers melting and boiling points. In connection with this research objective, it was to establish affiliation self-diffusion particles to one of three classes of randomized particles-crystal-mobile, liquid-mobile and vapor-mobile, determine the absolute share self-diffusion particles in accordance with the general classification of randomized particles. To this end, enough data are extensively used on the thermal characteristics of typical metals, for which they have the most important theoretical and practical significance.

2. The Calculation Part

The aforementioned characteristics of simple substances, as well as their melting point and boiling point are presented in Reference [1], which are taken from data for the typical metal in the presence of all these characteristics (**Table 1**).

Table 1. Reference data on the heat of fusion (ΔH_m), evaporation (ΔH_{ev}) and the activation energy of self-diffusion (E_{sd}), kJ/mol, melting point (T_m, K) and boiling point (T_b, K) of typical metal.

Me	T_m	T_b	ΔH_m	E_{sd}	ΔH_{ev}	$E_{sd}/\Delta H_{ev}$
Ag	1233	2437	11.28	192.17	254.30	0.758
Au	1337	3081	12.68	164.96	335.06	0.492
Be	1560	2723	14.64	147.5	314.76	0.469
Cd	594	1039	6.41	76.2	99.6	0.765
Ce	1071	3530	5.18	85.7	435.1	0.197
Co	1767	3230	15.51	259.58	383.2	0.677
Cr	2150	2945	13.81	247.86	343.98	0.721
Cu	1356	2846	13.05	205	304.4	0.674
Fe	1811	3145	15.19	239.48	340.4	0.704
Hf	2503	4876	21.78	183	575.14	0.318
In	429	2323	3.27	88.6	232.67	0.381
La	1193	3727	6.21	171	430.90	0.397
Li	453	1615	2.887	53.8	147.753	0.364
Mo	2895	4883	27.6	386.02	593.96	0.650
Nb	2742	5115	27.56	397.75	683.73	0.582
Pa	1848	4503	14.64	266.3	460.54	0.578
Pb	600	2018	4.848	116.81	183.5	0.637
Pd	1827	3150	16.69	266.3	361.65	0.736
Pr	1204	3485	6.92	169.5	338.2	0.501
Pt	2045	4100	19.68	278.4	502.80	0.554
Pu	912	3508	2.81	65.7	382.02	0.172
Ra	973	1593	8.38	92.11	162.02	0.569
Ta	3270	5560	31.43	413.2	770.18	0.537
Th	1999	4315	15.64	349.6	585.44	0.597
Ti	1941	3442	17.163	122.67	468.94	0.262
Tl	576	1730	4.312	83.74	162.50	0.515
U	1406	4407	8.518	111.4	446.54	0.205
V	2190	3665	17.59	255.2	457.18	0.558
W	3653	5640	35.30	398	736.69	0.540
Y	1793	3610	17.17	256	392.98	0.651
Zn	692	1180	7.19	92.11	115.38	0.798
Zr	2125	3923	20.07	100.48	579.88	0.173
Average	-	-	13.92	198.01	383.17	0.524

As can be seen from the data, for a variety of metals with melting points from 429 to 3653 K and the boiling point of 1180 to 5640 K, with a heat of melting of from 2.81 to 35.30 kJ/mol, and the boiling heat of 99.6 to 770.18 kJ/mol activation energy for self-diffusion is above heat of melting and below the heat of evaporation of metals. This can be attributed to particles overcoming the barrier activation of self-diffusion, to a class of liquid-mobile particles within the concept of randomized particles.

This concept is based on the Boltzmann's distribution [2], which is relative to the kinetic energy of random thermal motion of the particles is not only true for the gas, but the condensed state of matter [3]. According to this concept [4]-[7], the particles having an energy not higher than the melting heat are responsible for the long-range order of bonds, although a virtual character of their stay in the crystal lattice in the solid state or in the fragments of this lattice (clusters) in a liquid or gaseous states. These particles are called *crystal-mobile* and their share at any temperature according to the Boltzmann's distribution expressed by the relation

$$P_{crm} = 1 - \exp\left(-\frac{\Delta H_m}{RT}\right). \tag{1}$$

Particles with energies above the boiling heat are most randomized, fragmented, and they are named *vapor-mobile*. They are characterized by a zero bond order, and their share is subject to dependence

$$P_{vm} = \exp\left(-\frac{\Delta H_{ev}}{RT}\right). \tag{2}$$

Particles with energies the random motion higher than of the heat of melting, but lower than the heat of evaporation, are responsible for the short-range order of unstructured communication, called *liquid-mobile*. Their share is the difference between one and the shares crystal-mobile and vapor-mobile

$$P_{lqm} = 1 - \left(P_{crm} - P_{vm}\right) = \exp\left(-\frac{\Delta H_m}{RT}\right) - \exp\left(-\frac{\Delta H_{ev}}{RT}\right). \tag{3}$$

Since the sum of share crystal-, liquid- and vapor-mobile particles according to the Boltzmann's distribution is strictly equal to unity, the particles always exist together at all temperatures (except 0 and ∞), and, therefore, in any physical states-solid, liquid and gas, changing them only their ratio and constantly exchanging energy.

The fact that the particles are able to self-diffusion in the solid state and the energy corresponds to the range of the existence of liquid-mobile particles, indicates a further important physical sense of this class of randomized particles. In addition to providing plasticity and solubility of metals, which are discussed in [8], liquid-mobile particles being also diffusing on this basis more specifically reveal its mobile nature and mechanism of the plasticity and solubility of the transition from the state of thermodynamic equilibrium to a nonequilibrium kinetic.

The regular relationship the activation energy of self-diffusion with heat melting and boiling point can be illustrated by statistical analysis of the data in **Table 1**. Thus, the relationship $E_{sd} = f\left(\Delta H_{ev}\right)$ may be found as straightforward nature of the graphic placement of relevant data (**Figure 1**).

To express the linear nature of this relationship is first necessary to ensure the homogeneity of the set $E_{sd}/\Delta H_{ev}$ (see **Table 1**). It can be installed by Nalimov's criterion [9] [10]

$$r_{\substack{max \\ min}} = \frac{\left|\bar{x} - x_{\substack{max \\ min}}\right|}{S(x)\sqrt{\dfrac{n-1}{n}}} \leq r_{cr} \tag{4}$$

$$S(x) = \sqrt{\frac{\sum\left(x_i - \bar{x}\right)^2}{n-1}}, \tag{5}$$

where x_{max}—minimax value of the set, $\bar{x}$—average value, $S(x)$—mean-square error, n—volume of the set.

Normative values are tabulated Nalimov's criterion r_{cr} for 5% level of significance are given in [10], which are approximated in [8], up to 5% to the equation

$$r_{cr} = 1.483\left(n-2\right)^{0.187}. \tag{6}$$

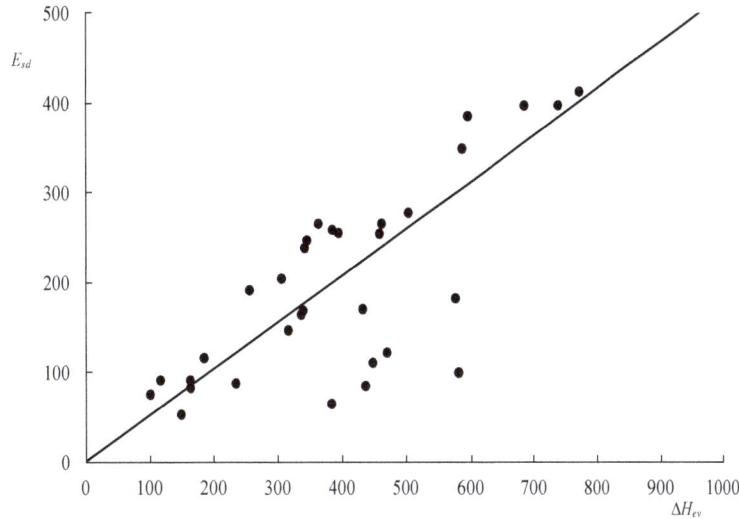

Figure 1. The dependence of the activation energy of self-diffusion of the heat of evaporation for typical metals (kJ/mol). Point—The reference data [1]; The line—According to Equation (8).

Here $f = n - 2$—the number of degrees of freedom Nalimov's criterion.

For a given set of $E_{sd}/\Delta H_{ev}$ most deviant value 0.173 refers to plutonium, and then

$$r_{\substack{max \\ min}} = \frac{|0.524 - 0.172|}{0.1812\sqrt{31/32}} = 1.974 < r_{cr} = 1.483 \times 30^{0.187} = 2.773 .$$

Thus, the inequality is satisfied, the set of homogeneous, with a mean value is representative.

In this case, it is also important to justify the need to move the linear dependence of origin, as for physical reasons, it is obvious that for an arbitrarily small boiling heat of the self-diffusion activation energy is close to zero. The condition of the output line from the origin to the processing of any set is satisfied if the same or substantially equal to two methods of averaging

$$\frac{1}{n}\sum_{i=1}^{n} \frac{y_i}{x_i} = \frac{\sum_{i=1}^{n} y_i}{\sum_{i=1}^{n} x_i} . \tag{7}$$

In this case,

$$\frac{1}{32}\sum_{i=1}^{32} \frac{E_{sd,i}}{\Delta H_{ev,i}} = 0.524, \quad \sum_{i=1}^{32} E_{sd,i} \Big/ \sum_{i=1}^{32} \Delta H_{ev,i} = \frac{6336.32}{12261.44} = 0.517 .$$

Rounded up to the second significant digit obtain both embodiments the mean 0.52, helping equation relating the activation energy for self-diffusion in the form of heat of evaporation

$$E_{sd} = 0.52\Delta H_{ev} \tag{8}$$

Nonlinear coefficient of multiple correlation of calculated values with reference data was $R = 0.727$, with significance for its 95% confidence level [11] [12] $t_R = 8.43 > 2$. The coefficient of correlation for R. Shannon [13] determines the degree of determination of any model according to the expression $D = R^2$. In this case, $D = 0.528$, indicating a key role in functional connection compared with the influence of random factors or unaccounted for determining the activation energy of self-diffusion as a function of heat of evaporation. It is possible to also determine the confidence interval of this function, as recommended in [14]

$$\delta = \pm |y_{max} - y_{min}|/t_R , \tag{9}$$

where y_{max} and y_{min}—maximum and minimum values in the experimental data set. For this case

$$\delta = \pm|398 - 53.8|/8.43 = \pm40.8 \text{ kJ/mol},$$

that is $\pm20.6\%$ of the average value of the function. This accuracy is typical for self-diffusion activation energy as the average for the different temperature ranges within the solid state of matter, very different for different metals.

Based on the understanding of the nature liquid-mobile particles that have the potential to leave the crystal lattice sites in excess of their thermal energy equal to ΔH_m, it appears that for self-diffusion, *i.e.* Brownian self-motion of the atoms in the crystal requires additional activation energy. Consequently, these particles must have an energy $\Delta H_m + E_{sd}$ and be only part of the total share liquid-mobile particles. This proportion by analogy with the expression for the total share liquid-mobile particles (3) can be expressed as

$$P_{sd} = \exp\left(-\frac{\Delta H_m + E_{sd}}{RT}\right) - \exp\left(-\frac{\Delta H_{ev}}{RT}\right). \tag{10}$$

For illustration, we present a comparative calculations (3) and (10) for the melting point of the metal to have an idea of the total share liquid-mobile particles and that part which is provided by the self-diffusion, with the greatest development of the formation of these particles in the solid state (**Table 2**).

3. Discussion of Results

Here, first of all draws attention to the comparability of shares liquid-mobile particles for different metals with an average value of this indicator 0,382. Because contribute vapor-mobile the particles by melting temperature (and hence at lower) is negligibly small, not exceeding 10^{-9}, the share of crystal-mobile particles to be the difference value with a unit $P_{crm} = 0.612$. As shown in [8], the proximity of the relationship $P_{crm} : P_{lqm}$ exactly at the point of destruction of the crystal melting to the proportion of the golden ratio (1.618) is not accidental and reveals the nature of this phase transition: above the melting point share of crystal-mobile particles responsible for long-range order due, and thus for the integrity of the crystal is smaller than this critical share and are not

Table 2. Shares liquid-mobile (P_{lam}), self-diffusion (P_{sd}) and vapor-mobile (P_{vm}) particles at the melting point of typical metals.

Me	P_{lqm}	P_{sd}	P_{vm}	Me	P_{lqm}	P_{sd}	P_{vm}
Ag	0.333	2.39×10^{-9}	1.61×10^{-11}	Pb	0.378	2.56×10^{-11}	1.06×10^{-16}
Au	0.320	1.15×10^{-7}	8.16×10^{-14}	Pd	0.333	7.07×10^{-9}	4.58×10^{-11}
Be	0.323	3.72×10^{-6}	6.04×10^{-12}	Pr	0.501	2.22×10^{-8}	2.13×10^{-15}
Cd	0.273	5.27×10^{-7}	1.74×10^{-9}	Pt	0.368	2.85×10^{-8}	1.44×10^{-13}
Ce	0.559	3.70×10^{-5}	6.02×10^{-22}	Pu	0.690	1.19×10^{-4}	1.16×10^{-22}
Co	0.348	7.38×10^{-9}	4.70×10^{-12}	Ra	0.355	4.03×10^{-6}	2.01×10^{-9}
Cr	0.462	4.35×10^{-7}	4.40×10^{-9}	Ta	0.315	7.90×10^{-8}	4.38×10^{-13}
Cu	0.314	3.99×10^{-9}	1.88×10^{-12}	Th	0.390	2.86×10^{-10}	5.04×10^{-16}
Fe	0.365	4.50×10^{-8}	1.52×10^{-10}	Ti	0.345	9.26×10^{-5}	2.40×10^{-13}
Hf	0.351	5.33×10^{-5}	9.95×10^{-13}	Tl	0.406	1.04×10^{-8}	1.84×10^{-15}
In	0.400	6.52×10^{-12}	4.69×10^{-29}	U	0.483	3.51×10^{-5}	2.58×10^{-17}
La	0.535	1.74×10^{-8}	1.36×10^{-19}	V	0.381	3.12×10^{-7}	1.25×10^{-11}
Li	0.465	2.91×10^{-7}	9.19×10^{-18}	W	0.313	6.37×10^{-7}	2.93×10^{-11}
Mo	0.318	3.44×10^{-8}	1.92×10^{-11}	Y	0.316	1.10×10^{-8}	3.56×10^{-12}
Nb	0.299	7.91×10^{-9}	9.45×10^{-14}	Zn	0.287	3.00×10^{-8}	1.95×10^{-9}
Pa	0.386	1.15×10^{-8}	9.61×10^{-14}	Zr	0.321	1.09×10^{-3}	5.57×10^{-15}

capable of holding the crystal in the connected state. Apparently, the golden section relates to such a system-wide criteria that characterize an ideal limit for such a resistance ratio of the inner parts, when one of them provides structural definition and the other—the optimum adaptability, flexibility [15].

However, first need to check the homogeneity of the set P_{lqm}. It is characterized by the greatest deviations again for plutonium, which has $P_{lqm} = 0.690$. Verification by Nalimov's criteria gives the following results:

$$r_{\substack{max \\ min}} = \frac{|0.382 - 0.690|}{9.14 \times 10^{-2} \sqrt{31/32}} = 3.424 > r_{cr} = 1.483 \times 30^{0.187} = 2.773 .$$

Consequently, the inequality is not satisfied, and "pop-up" value should be excluded from the set in question. It should be noted that while the demand for plutonium, it is an artificial radioactive element that is difficult to secure treatment [1].

In plutonium excluded from the set of the average $P_{lqm} = 0.372$, and the most deviating value P_{lqm} refers to cerium, accounting for 0.559. This gives a completely uniform set of:

$$r_{\substack{max \\ min}} = \frac{|0.372 - 0.559|}{7.21 \times 10^{-2} \sqrt{30/31}} = 2.636 < r_{cr} = 1.483 \times 29^{0.187} = 2.784 ,$$

and new average value within the mean-square error

$$\overline{P_{lqm}} = 0.372 \pm 7.21 \times 10^{-2} = 0.300 \div 0.444$$

includes the proportion of the golden section on unstructured component (0.382) with difference from the adjusted average only 2.6%.

The proportion of self-diffusion particles, it is certainly much less than the total share of liquid-mobile particles, which serve as a kind of reservoir for the more energetic particles. Because it is a deviation from the equilibrium solid state during the rapid heating, the elastic and plastic deformation and destruction of the crystal, on the one hand, and restoring the equilibrium state of the other. To initiate these processes is sufficient, only a small proportion of liquid-mobile particles closest to the heat of evaporation, accounting for about half of it. The very same proportion self-diffusion particles exceed the share vapor-mobile into two or more orders of magnitude. This agrees very weak evaporation of solids in relation to their properties, to determine the existence of particles liquid-mobile (solubility, plasticity), which, like other characteristics of the solid, liquid and gaseous states, reveals the same positions within the concept of randomized particles [16] [17].

Given the diversity of bands change in the shares self-diffusion particles for various metals restrict example of the temperature dependence of P_{sd} for iron (**Figure 2**) according to (10).

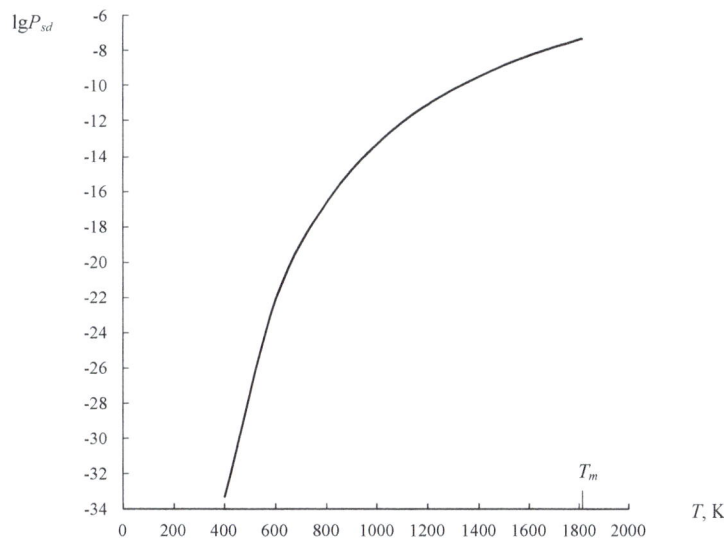

Figure 2. Dependence of the share self-diffusion particles of temperature for iron.

4. Conclusions

1. On the example of typical metals it's found that the activation energy for self-diffusion is above heat of melting and below heat of vaporization. This corresponds to the existence of liquid-mobile particle by classification based on the concept of randomized particles.
2. A formula for estimating the activation energy of self-diffusion by which it is approximately half of the heat of evaporation of the substance was suggested.
3. With the help of the temperature dependence for the share liquid-mobile particles expressed similar pattern to share self-diffusion particles. It is shown that the melting point of this share is typically less than 10^{-5}.
4. Belonging of self-diffusion particles to the class liquid-mobile particles complement representation as the last tank of the equilibrium and non-equilibrium states of a crystalline substance such as a predisposition to dissolve or melt, to the elastic and plastic deformation, in which the role of self-diffusion particle most high energy status is reduced to initiate appropriate changes in these states.

References

[1]   Dritz, M.E. (1985) Element Properties: Refer. In: Dritz, M.E., Ed., Metallurgy, Moscow, 672.

[2]   Boltzmann, L. (1984) Selected Works. Molecular-Kinetic Theory of Gases. Thermodynamics. Statistical Mechanics. The Theory of Radiation. Common Questions of Physics. Nauka, Moscow, 590.

[3]   Leontovich, M. (1983) Introduction to Thermodynamics. Statistical Physics. High School, Moscow, 416.

[4]   Malyshev, V. and Nurmagambetova (Makasheva), A. (2004) United Interpretation of Aggregate Substance Conditions by Degree of Its Chaotization. *Eurasian Physical technicaljournal*, **2**, 10-14.

[5]   Malyshev, V., Bekturganov, N., Turdukozhaeva (Makasheva), A. and Suleimenov, T. (2009) Concepts and According to the Concept of the Chaotical Particles. *Bulletin of the National Academy of Engineering*, **1**, 71-85.

[6]   Malyshev, V., Turdukozhaeva, A. and Suleimenov, T. (2009) Virtuality Solid, Liquid and Gaseous States of Matter. *Encyclopedia of Chemist and Engineer*, **12**, 13-23.

[7]   Malyshev, V. and Turdukozhaeva, A. (2011) Boltzmann Distribution as a Basis of the Concept of Randomized Particles. *Industrial Technology and Engineering*, **1**, 61-76.

[8]   Malyshev, V., Abdrakhmanov, B. and Nurmagambetova, A. (2004) Fusibility and Plasticity of Metals. Scientific World, Moscow, 148.

[9]   Nalimov, V.V. (1977) The Theory of the Experiment. Nauka, Moscow, 207.

[10] Ruzinov, L.P. (1972) Statistical Methods for Optimization of Chemical Processes. Chimiya, Moscow, 486.

[11] Dukarsky, O.M. and Zakurdaev, A.G. (1971) Statistical Analysis and Data on the "Minsk-22". Statistica, Moscow, 179.

[12] Siskov, V.I. (1975) Correlation Analysis in Economic Research. Statistica, Moscow, 168.

[13] Shannon, K.E. (1978) Simulation Modeling Systems—The Art and Science. Mir, Moscow, 418.

[14] Malyshev, V. (2000) The Definition of the Experimental Error, and the Adequacy of the Confidence Interval Approximating Functions. *Bulletin of National Academy of Sciences of Kazakhstan*, **4**, 22-30.

[15] Soroko, E.M. (1985) Managing the Development of Socio-Economic Structures. Nauka I Technika, Minsk, 144.

[16] Malyshev, V., Turdukozhaeva, A. and Ospanov, E. (2010) Evaporation and Boiling of Simple Substances. Scientific World, Moscow, 304.

[17] Malyshev, V., Bekturganov, N. and Turdukozhaeva, A. (2012) Viscosity, Flow and Density of the Substance as a Measure of Chaos. Scientific World, Moscow, 288.

4

A Stereochemically-Bent β-Hairpin: Scrutiny of Folding by Comparing a Heteropolypeptide and Cognate Oligoalanine

Kinshuk Raj Srivastava[1,2*], Susheel Durani[2]

[1]Department of Physics and Astronomy, Michigan State University, East Lansing, USA
[2]Department of Chemistry, Indian Institute of Technology Bombay, Mumbai, India
Email: [*]kinshukraj2@gmail.com

Abstract

A poly-L β-hairpin bent stereochemically as a boat-shaped protein of mixed-L,D structure is scrutinized in basis of ordering as minimum of energy specific for its sequenceand solvent. The model suitable for the scrutiny is accomplished by design. A terminally-blocked oligoalanine is nucleated over $^DPro_6$-Gly_7 and $^DPro_6$-$^LAsp_7$ dipeptide structures as a twelve-residue β-hairpin and bent stereochemically as a boat-shaped fold. The structure is inverse designed with side chains suitable to bind substrate p-nitophenyl phosphate, a surrogate substrate of acetyl choline and CO_2. The designed sequences were proven by spectroscopy and molecular dynamics to order with solvent effects of water and display high binding affinity for the substrate. One of the proteins and a cognate oligoalanine are evolved with molecular dynamics to equilibrium in a solvent bath of water. Molecular dynamics studies establish that heteropolypeptide well ordered as β-hairpin fold and cognate oligoalanine as an ensemble of hairpin-like folds in water. The ordering of cognate oligoalanine as ensembles of hairpin-like folds manifests combined role of water as strong dielectric and weak dipolar solvent of peptides. The roles of stereochemistry and chemical details of sequence in defining polypeptides as energy minima under specific effect of solvent are illuminated and have been discussed.

Keywords

Protein Folding, Protein Stereochemistry, β-Hairpin, Polyalanine Model, Solvent Effects

---

[*]Corresponding author.

1. Introduction

Protein folding presents in determination of the basis of specificity for sequences a formidable challenge [1]-[3]. The challenge is in the size of protein molecule and of the system that orders the structure as minima of energy. The challenge is in characterizing the minima in its defining interactions over main chain, side chains, and solvent. Quantum mechanics is relevant but, being computationally expensive, will only apply to greatly simplified models and typically under vacuum [4]-[14]. Force fields of structures and their interactions developed empirically [15]-[17] will allow equilibria to be computed, provided the models are small enough structures to allow the computation [18]-[20]. Critical in this study are the models that will allow both observing and computing equilibria [21]-[23]. Critical can be the questions posed and the models examined.

Critical in protein folding is the basis for sequence and solvent roles. The basis has been addressed with oligoalanines as main chain models [24]-[31]. The models are scrutinized for effects of structure modifications like extending main chains and mutating side chains [32] [33]. A promising scrutiny, implied in the studies Flory reported in 1967 [34]-[36], was pursued in our lab [37] [38]. The models were modified stereochemically and the effect was scrutinized with specific reporter solvents [37] [39]. It has thus been established that poly-L structure will fold with energetic frustration, hydrogen bonds local in poly-L chain enforce intervening peptides to unfavorable electrostatics of α conformation. Specificity of folding thus is a critical function of screening effects in electrostatics of α conformation. Consequently, folding of protein main chain was proven to manifest two complementary solvent effects, dielectric, to determine if local chain segments fold to α conformation or unfold to β conformation, and dipolar solvation, to determine if peptides hydrogen bond or solvate and thus fold or unfold the chain [39].

Extending the approach, oligoalanines are now targeted for addressing sequence role. The models are scrutinized for effect of mutation as specific heteropolypeptides in water. Equilibria are computed and the folds are scrutinized, over the microstates populated, in the basis of ordering as sequence specific folds. Crucial for the study will be the ability to compute equilibria, which will be easy for "folded" heteropolypeptide but tough for "unfolded" oligoalanine. Thus we implement the studies with the models that are in conformational diversity as oligoalanines restricted with effects of chain length and D residues. β-hairpins are smallest protein models amenable to further minimization and articulation with stereochemistry [40]. With this approach, a twelve-residue model is examined as a stereochemically bent β-hairpin protein in this report. The heteropolypeptides proven to order and display good affinity for the substrate and thus act as receptor though exceptionally simple and small in size. Evolving heteropolypeptides and cognate oligoalanines to equilibrium, the ensembles are resolved to contributing microstates, which are investigated in interactions of main chain and side chains. Contribution of main chain stereochemistry, side chain interactions and peptide-solvent interactions responsible for ordering specific conformation of peptide have been discussed.

2. Results

2.1. Design

The design of the protein examined in this study is illustrated in **Figure 1**. Terminally blocked oligoalanine is nucleated over $^DPro_6$-Gly_7 and $^DPro_6$-$^LAsp_7$ dipeptide structures as Type II' β-hairpin. The sequence positions 3 and 10 were mutated stereochemically and conformationally as D residues; this furnishes the desired boat-shaped fold. Ala residues are mutated to possible other protein residues, except Pro and Gly. Thus minimum energy sequences specific for the desired fold are generated computationally. Sequences were optimized with β-sheet favoring residues compatible with targeted fold. Polar residues were planned over solvent-accessible base of the fold and neutral-aromatic residues were planned in solvent-sequestered molecular cleft. The two sequences H1 and H2 are differ in having Gly_7 or $^LAsp_7$ as the second corner residue in their β-turns. Hydrogen bonds, ion-pairs, hydrophobic contacts, π-π, and cation-π interactions, recruited for possibility of locking the heteropolypeptide as a protein, are listed in **Table 1**. Neutral-aromatic and cation-aromatic residues provided in molecular pocket of the fold are for possible interaction with pNPP (p-nitrophenyl phosphate) as the binding ligand based on π-π, cation-π, and ion-pair interactions. H1 and H2 were synthesized and evaluated for folding, and ligand binding function. A1 is oligoalanine analogous to H1. In addressing $^DPro_6$-$^LAsp_7$ dipeptide for possible role in folding of H1 and A1, the structure is mutated to $^LPro_6$-$^LAla_7$ in the oligoalanine analog A2.

A1: Ac-Ala$_1$-Ala$_2$-^{D}Ala$_3$-Ala$_4$-Ala$_5$-^{D}Pro$_6$-***Asp***$_7$-Ala$_8$-Ala$_9$-^{D}Ala$_{10}$-Ala$_{11}$-Ala$_{12}$-NH$_2$
H1: Ac-Arg$_1$-Glu$_2$-^{D}Ala$_3$-Ser$_4$-Tyr$_5$-^{D}Pro$_6$-***Asp***$_7$-His$_8$-Asn$_9$-^{D}Thr$_{10}$-Lys$_{11}$-Trp$_{12}$-NH$_2$

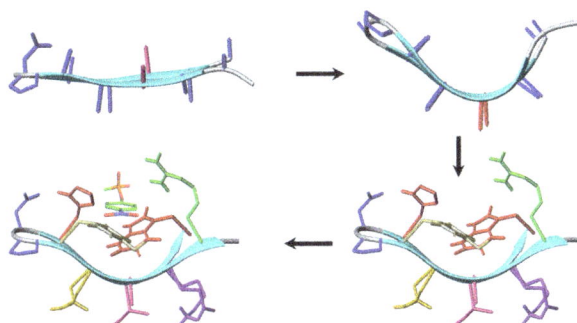

H2: Ac-Arg$_1$-Glu$_2$-^{D}Ala$_3$-Ser$_4$-Tyr$_5$-^{D}Pro$_6$-***Gly***$_7$-His$_8$-Asn$_9$-^{D}Thr$_{10}$-Lys$_{11}$-Trp$_{12}$-NH$_2$
A2: Ac-Ala$_1$-Ala$_2$-^{D}Ala$_3$-Ala$_4$-Ala$_5$-^{L}Pro$_6$-***Ala***$_7$-Ala$_8$-Ala$_9$-^{D}Ala$_{10}$-Ala$_{11}$-Ala$_{12}$-NH$_2$

Figure 1. Stepwise design involving stereochemical mutation of β-hairpin as boat-shaped fold and inverse optimization of H1 as receptor for p-nitrophenyl phosphate.

Table 1. Specific interactions folding H1 as a protein.

Interactions	Residue pairs
H-bond	^{1}R$_{NH}$-^{12}W$_{CO}$, ^{1}R$_{CO}$-^{12}W$_{NH}$, ^{3}A$_{NH}$-^{10}T$_{CO}$, ^{3}A$_{CO}$-^{10}T$_{NH}$, ^{5}Y$_{NH}$-^{8}H$_{CO}$, ^{5}Y$_{CO}$-^{7}D$_{NH}$, ^{7}D$_{NH}$-^{7}D$_O{}^{\delta1}$, ^{7}D$_O{}^{\delta2}$-^{8}H$_{NH}$, ^{7}D$_O{}^{\delta2}$-^{8}H$_N{}^\delta$, ^{4}S$_O{}^\gamma$-^{9}N$_N{}^\delta$, ^{10}T$_O{}^\gamma$-^{3}A$_{CO}$
Salt bridge	^{2}E-^{11}K
π-π	^{12}W-^{5}Y-^{8}H

2.2. Synthesis and Experimental Studies

H1 and H2 were synthesized by solid-phase peptide synthesis. Requisite ion peaks appearing in MALDI mass spectra, as noted in **Figure S1** (Supplementary Material), characterize the structures. The peptides display ^{1}H NMR resonances broadly in accordance with their structures as is noted in **Figure S2** (Supplementary Material). In concentration regime of the NMR experiment (0.25 - 2.5 mM), peptides were soluble and manifested no noteworthy changes in chemical shifts or line widths on tenfold dilution (results not shown). The structures thus appear to be freely soluble contrasted with poly-L hairpins, which tend to aggregate. The stereochemical modification of hairpin by bending the structure appears to have suppressed tendency of natural hairpin to aggregate to a diminished aqueous solubility.

CD spectra recorded for peptide H1 **Figure 2**. A maximum of ellipticity at ~198 nm and a coupled minimum at ~208 nm suggests ordering of the peptides as β-hairpins. Another minimum of ellipticity at ~215 coupled with a maximum at ~228 nm appears and could be coupled exciton due to involvement of aromatic residues in π-π interaction. As is noted in **Figure 1**, the boat-shaped fold has aromatic side chains clustered in its molecular cleft, which may promote aromatic-aromatic interactions within or between strands of the bent hairpin. The bend in artificial hairpin has provided for the interactions that are atypical for canonical poly-L β-hairpin. Ordering of the structures was tested in a thermal unfolding experiment, which was monitored with CD. Ellipticity at 228 nm diminishes with increase of temperature and recovers fully on cooling, as is noted in the inset of **Figure 2**. This suggest that Peptide H1 may fold and unfold like proteins.

H1 and H2 were tested for binding with p-nitrophenyl phosphate (pNPP). On titration with pNPP, both peptides manifest strong quenching of fluorescence as is noted in **Figure 3** and **Figure S3** of Supplementary material. Analysis of the data with Stern-Volmer equation gave the binding constants given in **Table 2**. Peptides were evaluated for ligand binding with AutoDock. The evaluation was undertaken with central member of the largest cluster in H1 populating equilibrium modeled with MD, as we shall discuss. The result of the calculation is in **Table 2**. The calculated binding energy is in reasonable agreement with the experimentally determined value. Aromatic side chains in molecular cleft provide a specific site for ligand binding conforming to quenching of fluorescence on ligand binding. On combined evidence of reversible folding, and ligand binding, H1 and

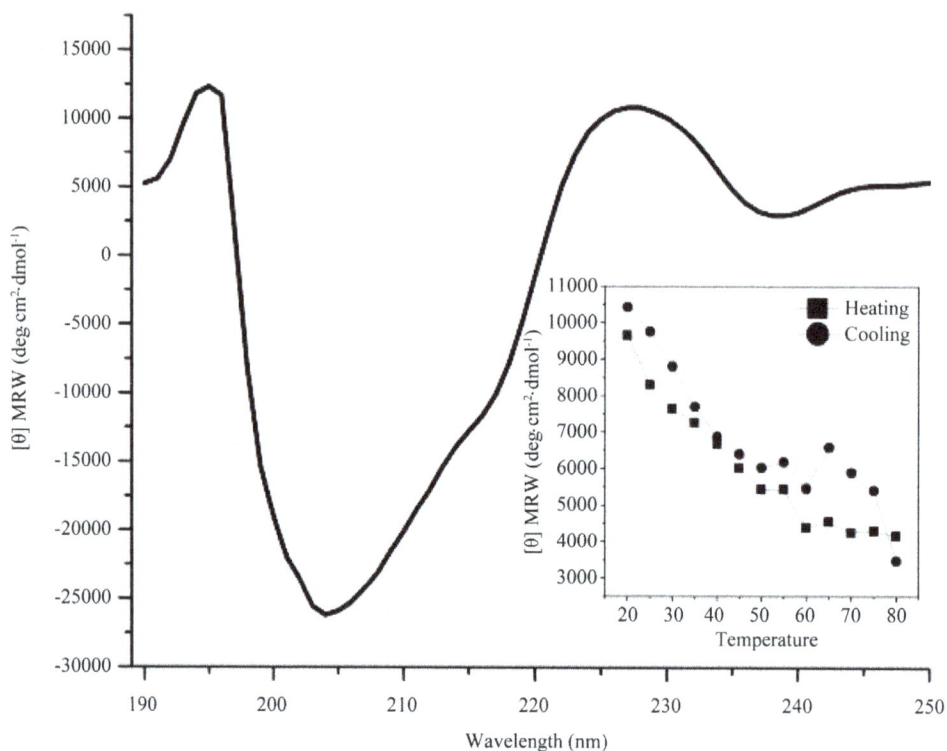

Figure 2. CD spectra of heteropolypeptides H1 in water at 25˚C. CD thermal melting curves of hete-ropolypeptides H1 recorded at 228 nm presented as inset.

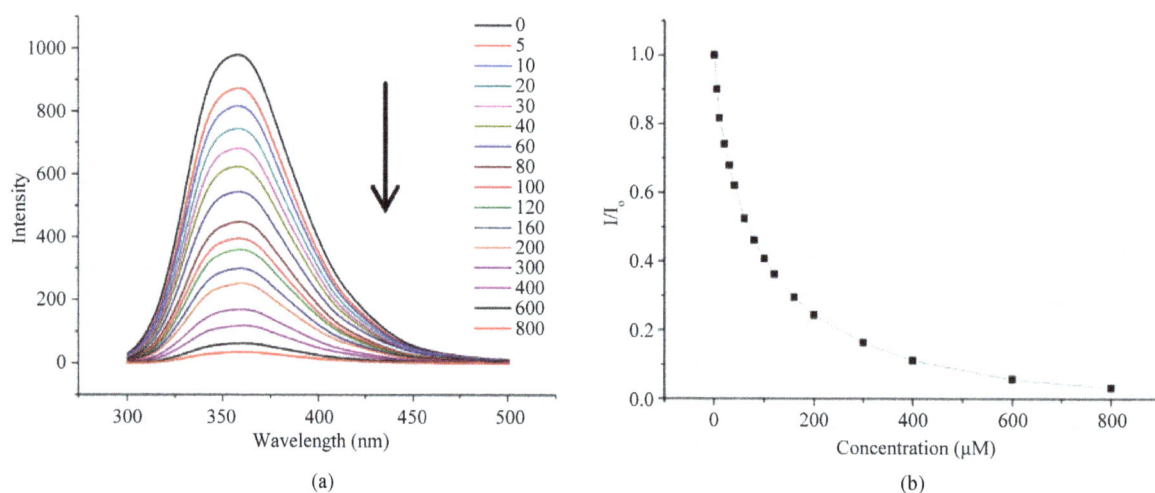

Figure 3. Quenching of tryptophan fluorescence of peptide H1 (20 μM) in 20 mM Tris-HCl buffer at pH 7.5, on progressive titration with increasing titration with pNPP (panel (a)), and plot of relative fluorescence intensity as a function of pNPP concentration (panel (b)).

H2 are justified as protein models.

2.3. Modeling of Folding

H1, A1, and A2 were submitted to molecular dynamics in a solvent bath of water to sample their conformational ensemble. Gromos 96 force field modified for accommodation of D residues is implemented at 298 K using SPC-water model as the solvent. Trajectories were monitored in conformational phase space of polypeptide structure, which was mapped to discrete microstates using 0.15 nm RMSD cut-off over backbone atoms to dis-

Table 2. Observed binding energy of peptide varients for pNPP.

Varients	Binding energy (Kcal/mol)	
	AutoDock	Fluorescence
H1	−4.33	−5.01 ± 0.19
H2	−4.03	−4.34 ± 0.34

tinguish conformationally specific clusters. The clustering was implemented with Duara *et al.* algorithm [41]; the central member of each cluster was taken to model a specific microstate, viz., as a discrete fold populating equilibrium. Time evolution of population in microstates during MD is compared in **Figure 4**. H1 equilibrates early and saturates as an ensemble of 6 microstates. A1 and A2 evolve slowly and do not manifest a robust asymptote in evolution of microstates even after 250 ns of MD. However, assuming equilibria to be reasonably approximated, the simulations were concluded at this point in time. Compared to 6 microstates in H1, A1 and A2 are noted to populate in, respectively, 523 and 983 microstates (see **Table 3**). Clearly, sequence length is critical for the ability to compute equilibria. Side chains and β-turn fold H1. β-turn in ^{D}Pro$_6$-^{L}Asp$_7$ structure also restricts conformational diversity in A1; mutated to ^{L}Pro$_6$-^{L}Ala$_7$ structure, A2 is nearly two-fold greater in density of microstates than A1. According to mole fraction in minima of energy, 0.96 in H1, free energy change in ordering of the structure in its ensemble is −7.8 kJ·M^{-1}. Likewise, 0.13 and 0.21 in mole fraction, A1 and A2 have their energy minima ordered with free energy change of, respectively, 4.7 kJ·M^{-1} and 3.3 kJ·M^{-1}. Relative to A1, the energy minima of H1 manifest net free energy change of −12.5 kJ·M^{-1}; this reflects sequence contribution in ordering of H1.

H1, A1, and A2 were assessed in distribution of Rg over the conformers populating macrostates, in percentage occupancy of specific ϕ, ψ basins, and in percentage occurrence of specific main chain hydrogen bonds, short (SR), medium (MR), and long ranged (LR). The data are summarized in **Table 4** and **Figure S4** of Supplementary Material. Appearnece of Rg maxima of entire macrostate at 0.54 nm for H1 and at 0.52 for A1; and maxima of first microstate at 0.55 nm for H1 and at 0.52 nm for A1, suggest more compaction of A1 than H1. Macrostate in H1 is ~92 % in β + PPII basin and ~6% in α basin, while in A1 it changes to ~80% occupancy in β + PPII basin and ~10% in α basin, as is illustrated in **Figures S5** of Supplementary Material. According to residue-level basin occupancies (**Figure S6** of Supplementary Material), Pro$_6$ and Asp$_7$ are locked in Type II' β-turn and only marginally dispersed in A1. ^{L}Pro$_6$ and ^{L}Ala$_7$ in A2 are considerably dispersed and accordingly, stereochemical effect of D-proline and interactions of Asp side chain contribute in ordering both A1 and H1.

According to the data in **Table 4**, H1 is locked with 8.7 hydrogen bonds per molecule of which 5.0 are with in backbone, 2.8 are with backbone-sidechain (mc-sc), and 0.9 within different side chains. A1 lacks hydrogen bonding groups in Ala side chains and has mc-mc hydrogen bonds diminished to 3.4 from 5.0 in H1. In H1, ~61% of mc-mc hydrogen bonds are LR type (n-n ≥ 6), implying an ordered β-sheet structure, while ~32% are MR type (n-n ± 3), conforming to its β-turn structure. In its conformational dispersal, A1 diminishes in LR hydrogen bonds to 55% from ~61% in H1, and increases in MR hydrogen bonds to ~35% from ~32% in H1.

2.4. Interactions Ordering H1

Interactions ordering H1 were assessed by monitoring distribution of Rg over specific side chain pairs. Specific pairs with Rg distribution peaking at ≤0.3 nm, noted in **Figure 5**, are strongly associated. Specifically, Trp$_{12}$-Tyr$_5$, His$_8$-Tyr$_5$, and Trp$_{12}$-His$_8$ are closely interacting, and could be involved in π-π interaction, and Lys$_{11}$-Glu$_2$ and Ser$_4$-Asn$_9$ are closely interacting, and could be involved in mutual hydrogen bonds. The appearance of peak at 0.25 nm in the Rg distribution of ^{D}Ala$_3$-^{D}Thr$_{10}$ indicate the presence of hydrogen bond between ^{D}Thr$_{10}$ side chain hydroxyl and ^{D}Ala$_3$ main chain carbonyl. Arg$_1$-Tyr$_5$, a case of potential cation-π interaction, and Asp$_9$-His$_8$, a case of potential hydrogen bond, are bimodal in Rg distribution (**Figure 5**), which implies that the side chains only interact transiently. Arg$_1$-Trp$_{12}$ pair, a possible case for π-π and cation-π interaction, has Rg distribution peaking at 0.65 nm; clearly the side chains do not interact.

H1 was assessed in its interactions with solvent. The radial distribution of water oxygen against specific backbone and C$_\beta$ atoms are presented in **Figure 6** and against other specific side chain atoms are presented in **Figure S7** of Supplementary material. The overall spatial distribution plots of water oxygen are presented in **Figure S8** of Supplementary material. As evidenced in reduction of radial distribution function (RDF) magni-

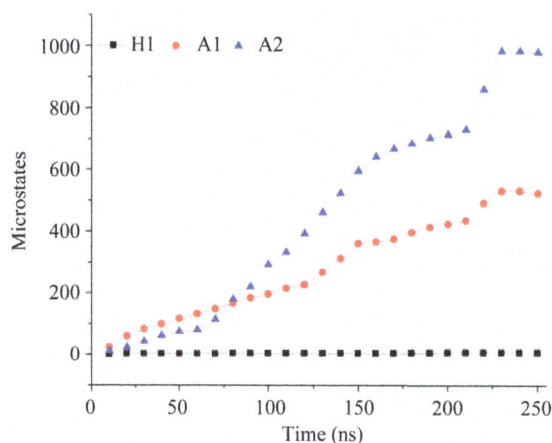

Figure 4. Evolution of microstates during molecular dynamics.

Table 3. Microstate populating equilibria, showing population statistics and radius of gyration distribution.

Model system	No. of microstates	% population in microstate 1	□G	Radius of gyration (nm)	
				Macrostate	Microstate 1
H1	6	96	−7.8	0.54 ± 0.02	0.55 ± 0.02
A1	523	13	4.7	0.52 ± 0.03	0.52 ± 0.02
A2	983	21	3.3	0.54 ± 0.08	0.52 ± 0.01

Table 4. Conformational parameters of microstates.

		[a]Hydrogen bonds statistics				[b]Percent occupancy of ϕ, ψ basins		
		Avg./conf.	% SR	% MR	% LR	α	β	PPII
H1	mc-mc	5.0	6.9	32.3 (32.3, 0.0)	60.8 (0.0, 60.8)	5.9	60.1	32.3
	p-p	8.7	4.20	33.8	49.0			
	sc-sc	0.9	0.0	13.8	50.6			
	mc-sc	2.8	0.9	42.3	28.4			
A1	mc-mc	3.4	9.4	35.1 (29.7, 5.4)	55.4 (3.1, 52.4)	10.1	42.3	39.2
A2	mc-mc	3.5	13.96	15.47 (15.34, 0.13)	70.56 (7.84, 62.72)	14.5	39.7	38.3

[a]Hydrogen bonds are short (SR; i → i ± 2), medium (MR; i → i ± 3 + i → i ± 4) and long ranged (LR; i → i ± 5 + i → i ± 6 according to sequence separation between donor and acceptor residue. [b]Basin definitions are, α: $^{L/D}\phi = -/+ 20$ to $-/+ 100$, $^{L/D}\psi = -/+ 20$ to $-/+ 80$; β: $^{L/D}\phi = -/+ 90$ to $-/+ 170$, $^{L/D}\psi = +/- 80$ to $+/- 180$; PPII: $^{L/D}\phi = -/+ 30$ to $-/+ 90$, $^{L/D}\psi = +/- 80$ to $+/- 170$. mc-mc indicate main chain-main chain; sc-sc indicate side chain-side chain; mc-sc indicate main chain-side chain hydrogen bonds whereas p-p indicate the total hydrogen bonds observed in the entire peptide.

tudes relative to equilibrium solvent density of 1, H1 and A1 have specific backbone atoms sequestered from solvent possibly in intramolecular hydrogen bonds. The greater density of main chain-main chain hydrogen bonds in H1 than in A1 and A2 support the observation of radial distribution. RDF peaks of water oxygen against oxygen and nitrogen atoms of specific side chains display maxima typically at ~0.3 nm, indicating the strong solvation of polar atoms of side chains. RDF maxima of water oxygen are diminished against β carbons of aromatic and aliphatic side chains, conformed to solvent sequestered nature of non polar side chain groups and their participation in π-π interaction and hydrophobic clustering.

2.5. Specific Microstates Populating the Equilibria

Figure 7 shows ribbon representation of top five microstates in each ensemble. The individual populations, shown in parenthesis, add up to ~99% in H1 and ~40% in A1 and A2. The microstates in H1 and A1 have the general appearance of bent hairpins, which are well locked in H1 and floppy in A1. From ϕ, ψ plots, shown alongside, H1 and A1 are noted to have ^{D}Pro and ^{L}Asp locked in Type II' β-turn. Remarkably, many microstates

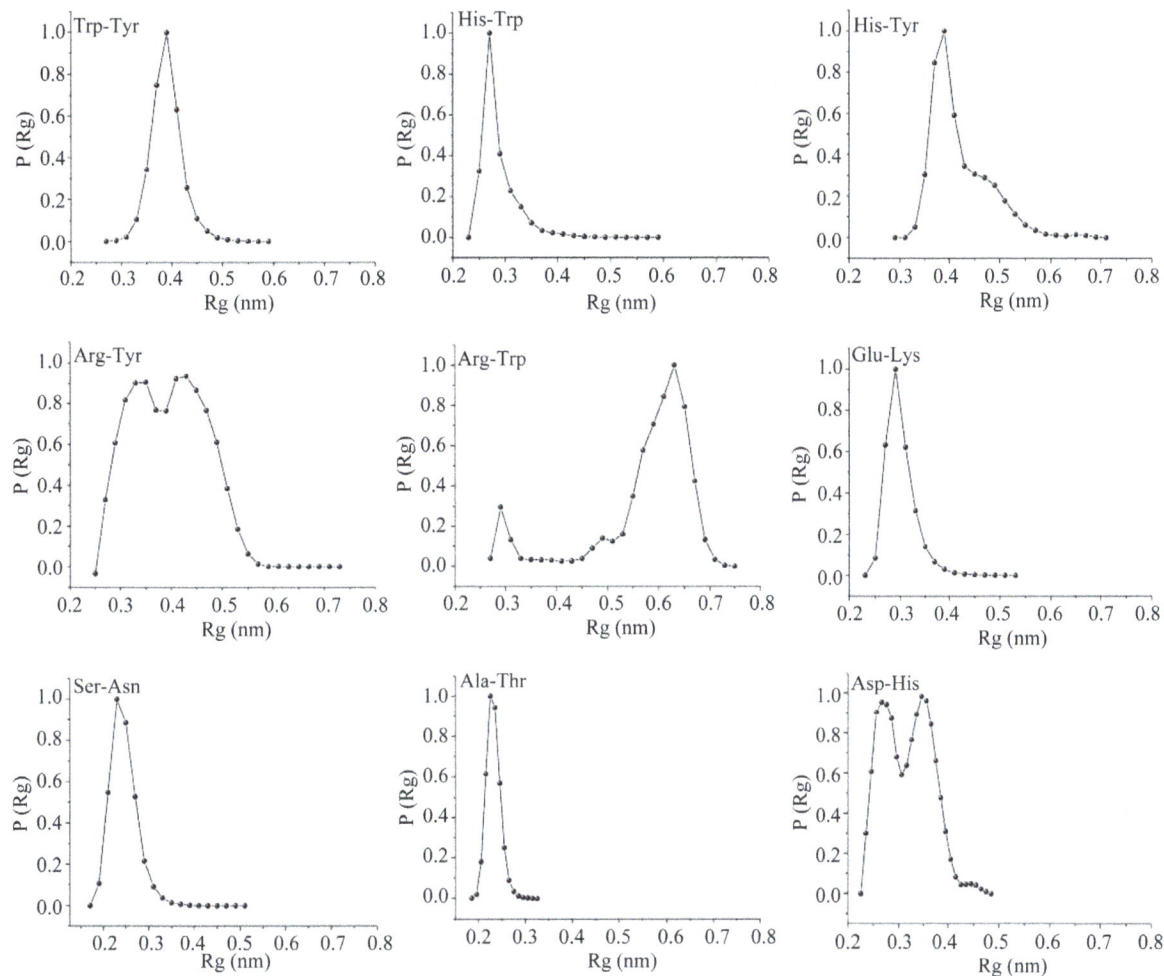

Figure 5. Radius of gyration (Rg) distribution of specific side chain pairs of H1 over its macrostate.

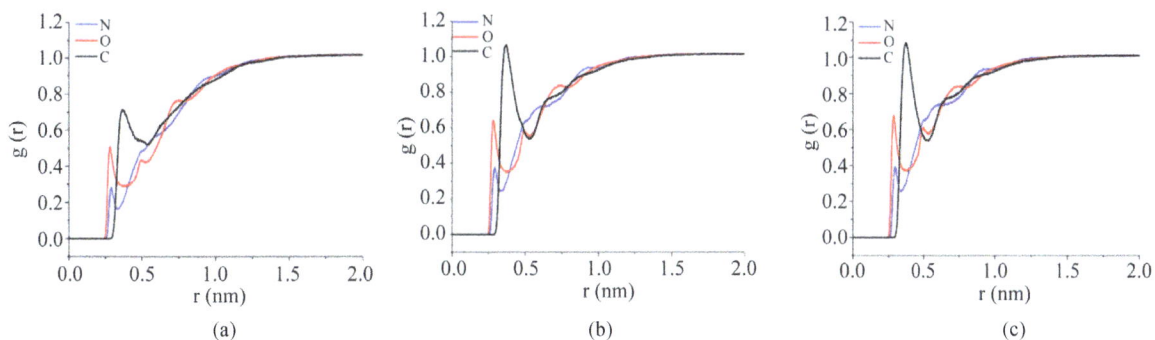

Figure 6. Radial distribution of water oxygen atoms against main-chain nitrogen (blue trace), oxygen (red trace), and side chain Cβ of H1 (panel (a)), A1 (panel (b)) and A2 (panel (c)) over macrostates.

in A2 also appear to be in hairpin-like folds. Presumably ^{L}Pro induces chain reversal to promote the folds. An effect contributing specific folds may involve L, D, L segments ordering to a curved morphology due to $^{L}\beta$, $^{D}\beta$, $^{L}\beta$ conformation, in practically all the microstates, according to ϕ, ψ plots. Identical in curvature, the strand sections are well poised for mutual antiparallel β-sheet hydrogen bonds in curved hairpins. Indeed, A2 is comparable to A1 in the number of backbone hydrogen bonds at ~3.5 (see **Table 4**), and is significantly higher in LR hydrogen bonds, at ~70%, implying more extensive β-sheet structure, compared to 60% in H1 and only

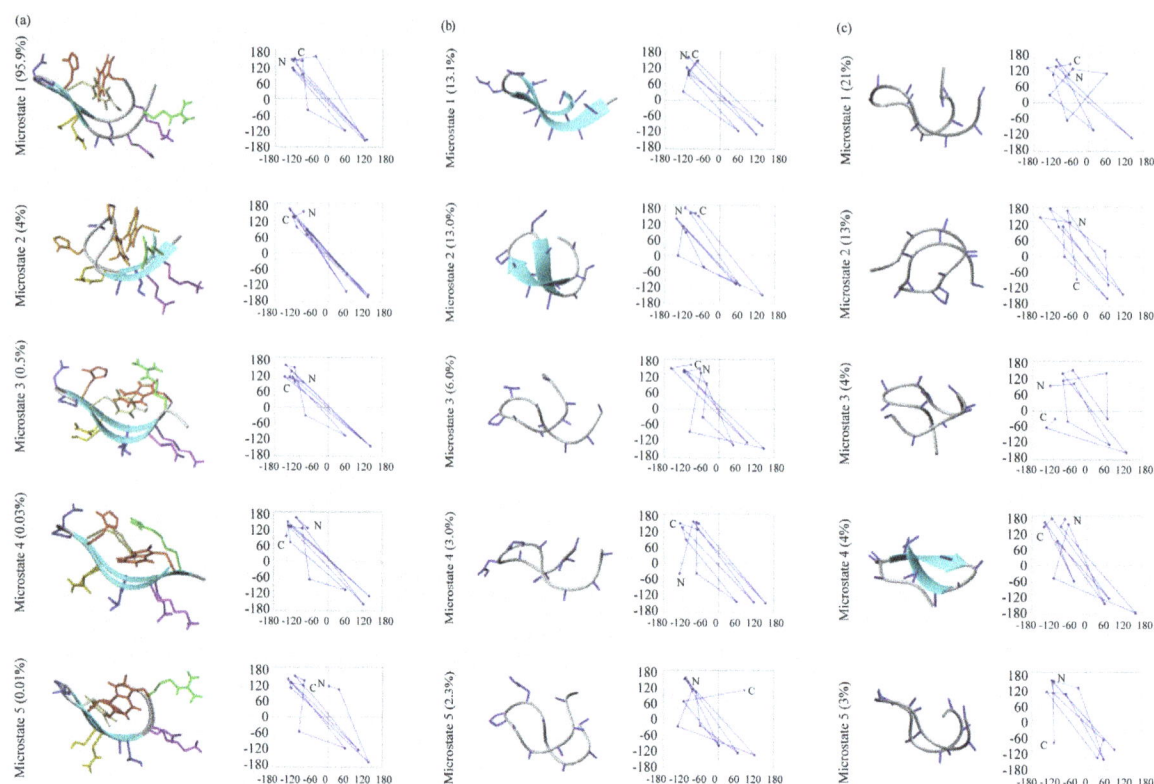

Figure 7. Ribbon representation of central members of top microstates of H1 (panel (a)), A1 (panel (b)) and A2 (panel (c)) showing populations in parenthesis and ϕ, ψ plots underneath.

55% in A1. Thus the identical curved morphology of alternating L, D, L structures may constrain conformational diversity in H1, A1, and A2 due to intramolecular main chain hydrogen bonds between strands facilitated by chain reversal in the models. H1 is well locked in its first microstate with aromatic interactions, interactions of several cross-strand side chains, and in intramolecular main chain hydrogen bonds. Aromatics are clustered in all the microstates of H1; polydispersity of microstates involves considerable fraying of Arg as N-terminal residue. Due to a local twist in main chain, Asp_7 is pushed out of α basin uniquely in microstate 2 of the ensemble.

3. Discussion

Proteins fold over a complex interplay of interactions of main chain, side chains, and solvent [1]-[3]. In addressing the interplay, oligoalanines have been fruitful models [4] [5] [23] [26] [27] [30] [32] [39]. A fruitful enquiry of the models has involved mutating stereochemical structure and probing the effect with specific reporter solvents [37]-[39]. Consequently, stereochemistry has been implicated in critical role: poly-L folds order to local hydrogen bonds of main chain under conflict with unfavorable electrostatics of α conformation. Accordingly, solvents fold main chain with two complementary effects, screening of electrostatics, to allow or disallow α conformation, and dipolar solvation of peptide, to allow or disallow hydrogen bonds [39]. Crucial to unmasking of the folding model has been mutation of stereochemistry and examination of solvent effect of water.

The approach has now been extended to scrutiny of sequence role. Model proteins have been mutated as cognate oligoalanines and the effect examined with reporter solvents. Applying water as solvent, side chain and main chain structures were probed in the critical interactions involved. The fold probed with water in this study was designed and validated as receptor protein against targeted ligand. Applied as reporters against homochiral and heterochiral models, water and DMSO illuminated critical effects of main chain [37] [39]. Comparison of heteropolypeptides and cognate oligoalanines in water have illuminated critical effects of sequences.

The critical effect of stereochemistry was exemplified in this study on the effect of D-proline-locked β-turn structure. ^{D}Pro-^{L}Asp structure restricted conformation in "unfolded" oligoalanine considerably relative to ^{L}Pro-^{L}Ala structure. ^{D}Pro and Asp side chain at turn position locked the structure as an ensemble of hairpin-like folds. The ordering of ologoalanines as an ensemble of hairpin-like folds manifest the role of set ereochemistry in nucleating the turn and ordering specific β-hairpin. The stereochemical effect of a turn in nucleating structure, and main chain hydrogen bonds were noted to be the important factor for locking the structure as hairpin-like folds. Indeed, ordering of oligoalanines to ~3.5 main chain hydrogen bonds, only marginally less than 5 in H1, suggests that water could be a significant fold promoting solvent directly at level of main chain. This observation manifest a role for solvent water as a weak solvent of peptides and thus as a promoter of folds passively by allowing formation of main chain hydrogen bonds. Thus, the present study establishes the folding of protein in water involves combined effect one, dielectric effect in screening of electrostatics of poly-L peptides, and, two, weak dipolar solvation of peptides [39].

With effects of side chains, heteropolypeptide structure were found to fold with solvents effects of water. The folding involved ordering of main chain with increased sampling of peptide hydrogen bonds and increased sampling of β conformation. We found the sequence complement of side chains order the first microstate in A1 to the first microstate in H1 with a free energy gain of modest ~12 kJ·M^{-1}. The magnitude includes entropic cost of ordering ~500 microstates in A1 to only 6 in H1 and the gain in the interactions of specific side chains. Interactions of side chains compensate not only for unfavorable entropy of conformational ordering, but also for unfavorable enthalpy of desolvation of peptides and unfavorable electrostatics of α conformation. Thus heteropolypeptides order with synergy of main chain and side chain effects. Idiosyncrasies of specific folds may be involved but generic effects of main chain and side chain structures are likely to be important. Broadly, similar interactions involving side chains were observed in folding of heteropolypeptides, but the mix of physical effects is expected to be specific for the solvent role as screen of main chain electrostatics and solvation of peptides. Water as a solvent manifested in this study close interactions mainly of His-Trp, Trp-Tyr, Glu-Lys, and Ser-Asn side chains and weaker interaction of several other side chains. In general, hydrogen bonds of polar groups and hydrophobic aggregation of nonpolar groups, could be more critical effects in interaction of the structure with water as solvent.

4. Conclusion

Ordering of proteins as energy minima specific for their sequences was investigated with combination of experiment and computation. A stereochemically bent β-hairpin is designed as receptor protein for scrutiny of the forces ordering proteins. The heteopolypeptide found to order as sequence-specific folds under the influence of position specific interactions over several side chains. The ordering of cognate polyalanine as an ensemble of hairpin-like folds manifests the combined role of water. The water dielectric effect screens the electrostatics of poly-L peptides, and water being weak dipolar solvent passively promoter of the main chain hydrogen bond. The success of achieving high affinity for targeted ligand in exceptionally small peptide illustrates the power of the proposed design principles and affirms stereochemistry as a valuable aid in customizing molecular morphological plans. The proteins, being small in size, facilitated the analysis of its folding and ligand binding with molecular dynamics, which allow scrutinizing the individual roles of backbone and side chain and solvent roles in protein folding.

5. Material and Methods

5.1. Peptide Modeling

Peptides were modeled either with the in-house software package CAPM (Computer Aided Peptide Modeling), capable of handling D-amino acid effectively. In-house program PDB make was used for generation of PDB coordinates of CAPM modeled structure. Sequences were designed with help of in-house sequence optimization program IDeAS, capable of handling D-amino acid effectively.

5.2. Preparation of Equilibrium Ensembles

Molecular dynamics were performed with gromos-96 43A1 force field in GROMACS 3.3.3 [42] [43] in a periodic box of with water as explicit solvent. The simulation was performed under NVT condition [38]. We used 1.4 nm cut off for Non-bonded list with 0.8 nm shift and used 2 fs as integration step. Initial velocities were drawn from Maxwellian distribution. Temperature was coupled to an external bath with relaxation time constant of 0.1 ps. Bond lengths were constrained with SHAKE [44] to geometric accuracy 10^{-4}. Electrostatics were treated with Particle Mesh Ewald [45] [46] for charged system whereas SHIFT was used for neutral peptides. We implemented a 1.4 nm coulomb cutoff, 0.12 nm fourier spacing, and 4 as an interpolation order. Peptides constrained to the center of the periodic cubic box were surrounded by solvent water to 1 atm density at 298 K. We first minimize the energy of solute followed by minimization of solvent energies while restraining solute, and finally both were energy minimized after removing restraint. We started the molecular dynamics simulation and sampled the trajectory at 10 ps intervals. We have discarded initial 3 ns trajectory as pre-equilibration period, before analyzing the data.

5.3. Characterization of Macrostate and Polypeptide Microstates

We used Daura *et al.* Algorithm for clustering peptide conformers in cartesian space with ≤0.15 nm RMSD cutoff over backbone atoms. This out put different microstates in their diminishing population, viz., diminishing thermodynamic stability. We calculated the free energy of first microstate (most populous) using equation $\Delta G = -RT \ln K$, where $K = p_1/p_{total} - p_1$, where is R gas constant, and T is temperature, and p_1 is the population of the first microstate. We considered the most populated first microstate as ordered state and evaluated its stability with respect to the remaining microstate considered as unordered state.

5.4. Solvation Shell Analysis

We calculated the radial distribution and spatial distribution of specific solvent atoms around peptide for first microstate with g_rdf and g_spatial functions in GROMACS.

5.5. Molecular Docking

For docking the ligand to peptide, we used in build flexible docking algorithm of AutoDock 4.0 [47]. The representative structure of first microstate was obtained by clustering the three aromatic residues over entire trajectory. This structure was used as a receptor structure for docing with ligand. Using genetic algorithm with RMSD tolerance of 2 Å, structurally distinct conformational clusters of the ligand were ranked in terms of increasing energy. The observed the lowest energy of peptide-ligand complex were reported as the binding energy.

5.6. Peptide Synthesis

Pepetides were synthesized by solid phase peptide synthesis using Fmoc chemistry on Rink Amide AM resin with HOBt/DIC as coupling reagents [48]. Each coupling, monitored with Kaiser and chloranil tests were periodically performed to check the coupling of each amino acid. 30% (v/v) piperidine-DMF was used for deprotection in each step of synthesis. Acetylation of N-terminus were achieved with Ac_2O:DIPEA:DMF in 1:2:20 ratio. Reagents K (82.5% TFA/5% dry-phenol/5% thioanisole/2.5% ethandithiol/5% water) was used for simultaneous deprotection of side chain and final cleavage of peptide chain from resin. The finally cleaved peptides were precipitated in anhydrous diethyl ether. The precipitated peptides were then lyophilized in 1:4 H_2O:^{t}BuOH solution, resultant peptides were stored in freeze. Peptide purity was assessed with HPLC over RP-C18 (10 μM, 10 mm × 250 mm; Merck) eluting with $CH_3CN\backslash H_2O$ (0.1%TFA) 0% - 100% gradients.

5.7. Mass Spectrometry

Mass spectra were recorded either by MALDI-TOF (Matrix Assisted Laser Desorption Ionization-Time of Flight) mode on AXIMA-CFR Kratos instrument.

5.8. Nuclear Magnetic Resonance Spectroscopy

^{1}H NMR spectra were recorded on 800 MHz Bruker instrument at 298 K in 90% H_2O/10% D_2O in citrate buffer at pH ~3 with 2.5 mM and 0.25 mM concentrations of peptides. Solvent was suppressed with pre-saturation or WATERGATE sequence, as provided in Bruker softwares.

5.9. Circular Dichroism

Far-UV Circular Dichroism (CD) spectrum were recorded on JASCO J-810 CD instrument at 298 K in 0.2 cm path length quartz cell. Using 2 nm bandwidth and scanning speed 100 nm/min with 1.0 s time constant in 1 nm steps, we record five scans and averaged them. Each spectra was corrdcted for solvent absorbance. We finally report the values in molar residue ellipticity [θ_{MRW}] by converting the observed values in millidegrees using well reported equation.

5.10. Spectrofluorometry

Fluorescence spectra were recorded on a Perkin Elmer LS-55 spectrofluorimeter. We collectec the data at 298 K in 1 mL cell by exiting the sample at 280 nm and recording the emission in the wavelength range of 300 - 500 nm range, with 5 nm excitation and emission slits width. A scan rate of 100 mm/min with 1 nm steps were used. We kept the fixed concentrations of peptide 20 μM and varied the substrate (pNPP and pNPA) concentration in the range of 0 - 400 μM. All experiments were performed in 20 mM Tris-HCl buffer at ~7.5 pH. We calculated Stern-Volmer constant (K_{SV}) for the external quencher *i.e.* pNPP using the following biomolecular quenching equation.

$$I_0/I = 1 + K_{sv}[Q] \tag{1}$$

where I_0 = fluorescence intensity in the absence of external quencher, I = fluorescence intensity in the presence of quencher, Q = concentration of the quencher, and K_{SV} = Stern-Volmer constant calculated from the slope of line. The emission maximum intensities of tryptophan were fit as a function of pNPP concentration to the described 1:1 binding isotherm and K_d and hence binding energy were estimated.

Acknowledgements

We acknowledge DST (09DST028), Government of India, for financial support and IIT Bombay for the computing facility "Corona". KRS is recipients of fellowships from Council of Scientific and Industrial Research (CSIR).

References

[1]   Dill, K.A., Ozkan, S.B., Shell, M.S. and Weikl, T.R. (2008) The Protein Folding Problem. *Annual Review of Biophysics*, **37**, 289-316. http://dx.doi.org/10.1146/annurev.biophys.37.092707.153558

[2]   Onuchic, J.N. and Wolynes, P.G. (2004) Theory of Protein Folding. *Current Opinion in Structural Biology*, **14**, 70-75. http://dx.doi.org/10.1016/j.sbi.2004.01.009

[3]   Daggett, V. and Fersht, A. (2003) The Present View of the Mechanism of Protein Folding. *Nature Reviews Molecular Cell Biology*, **4**, 497-502. http://dx.doi.org/10.1038/nrm1126

[4]   Dannenberg, J.J. (2005) The Importance of Cooperative Interactions and a Solid-State Paradigm to Proteins: What Peptide Chemists Can Learn from Molecular Crystals. *Advances in Protein Chemistry*, **72**, 227-273. http://dx.doi.org/10.1016/S0065-3233(05)72009-X

[5]   Hua, S., Xu, L., Li, W. and Li, S. (2011) Cooperativity in Long Alpha- and 3(10)-Helical Polyalanines: Both Electrostatic and van der Waals Interactions Are Essential. *Journal of Physical Chemistry B*, **115**, 11462-11469. http://dx.doi.org/10.1021/jp203423w

[6]   Elstner, M., Jalkanen, K.J., Knapp-Mohammady, M., Frauenheim, T. and Suhai, S. (2000) DFT Studies on Helix Formation in N-Acetyl-(L-Alanyl)n-N'-Methylamide for n = 1 - 20. *Chemical Physics*, **256**, 15-27. http://dx.doi.org/10.1016/S0301-0104(00)00100-2

[7]   Rossi, M., Blum, V., Kupser, P., von Helden, G., Bierau, F., Pagel, K., Meijer, G. and Scheffler, M. (2010) Secondary Structure of Ac-Alan-LysH+ Polyalanine Peptides (n = 5, 10, 15) in Vacuo: Helical or Not? *Journal of Physical Chemistry Letters*, **1**, 3465-3470. http://dx.doi.org/10.1021/jz101394u

[8]   Salvador, P., Asensio, A. and Dannenberg, J.J. (2007) The Effect of Aqueous Solvation upon Alpha-Helix Formation for Polyalanines. *Journal of Physical Chemistry B*, **111**, 7462-7466. http://dx.doi.org/10.1021/jp071899a

[9]   Topol, I.A., Burt, S.K., Deretey, E., Tang, T.H., Perczel, A., Rashin, A. and Csizmadia, I.G. (2001) Alpha- and 3(10)-Helix Interconversion: A Quantum-Chemical Study on Polyalanine Systems in the Gas Phase and in Aqueous Solvent. *Journal of the American Chemical Society*, **123**, 6054-6060. http://dx.doi.org/10.1021/ja0038934

[10]   Tsai, M.I., Xu, Y. and Dannenberg, J.J. (2005) Completely Geometrically Optimized DFT/ONIOM Triple-Helical Collagen-Like Structures Containing the ProProGly, ProProAla, ProProDAla, and ProProDSer Triads. *Journal of the American Chemical Society*, **127**, 14130-14131. http://dx.doi.org/10.1021/ja053768y

[11] Tsai, M.I., Xu, Y. and Dannenberg, J.J. (2009) Ramachandran Revisited. DFT Energy Surfaces of Diastereomeric Tri-alanine Peptides in the Gas Phase and Aqueous Solution. *Journal of Physical Chemistry B*, **113**, 309-318. http://dx.doi.org/10.1021/jp8063646

[12] Tsemekhman, K., Goldschmidt, L., Eisenberg, D. and Baker, D. (2007) Cooperative Hydrogen Bonding in Amyloid Formation. *Protein Science*, **16**, 761-764. http://dx.doi.org/10.1110/ps.062609607

[13] Han, W.G., Jalkanen, K.J., Elstner, M. and Suhai, S. (1998) Theoretical Study of Aqueous N-Acetyl-L-Alanine N'-Methyl amide: Structure and Raman, VCD, and ROA Spectra. *Journal of Physical Chemistry B*, **102**, 2587-2602. http://dx.doi.org/10.1021/jp972299m

[14] Wieczorek, R. and Dannenberg, J.J. (2005) Enthalpies of Hydrogen-Bonds in Alpha-Helical Peptides. An ONIOM DFT/AM1 Study. *Journal of the American Chemical Society*, **127**, 14534-14535. http://dx.doi.org/10.1021/ja053839t

[15] Brooks, B.R., Brooks III, C.L., Mackerell Jr., A.D., Nilsson, L., Petrella, R.J., Roux, B., Won, Y., Archontis, G., Bar-tels, C., Boresch, S., Caflisch, A., Caves, L., Cui, Q., Dinner, A.R., Feig, M., Fischer, S., Gao, J., Hodoscek, M., Im, W., Kuczera, K., Lazaridis, T., Ma, J., Ovchinnikov, V., Paci, E., Pastor, R.W., Post, C.B., Pu, J.Z., Schaefer, M., Ti-dor, B., Venable, R.M., Woodcock, H.L., Wu, X., Yang, W., York, D.M. and Karplus, M. (2009) CHARMM: The Bio-molecular Simulation Program. *Journal of Computational Chemistry*, **30**, 1545-1614. http://dx.doi.org/10.1002/jcc.21287

[16] Guvench, O. and MacKerell Jr., A.D. (2008) Comparison of Protein Force Fields for Molecular Dynamics Simulations. *Molecular Modeling of Proteins*, **443**, 63-88. http://dx.doi.org/10.1007/978-1-59745-177-2_4

[17] van Gunsteren, W.F., Dolenc, J. and Mark, A.E. (2008) Molecular Simulation as an Aid to Experimentalists. *Current Opinion in Structural Biology*, **18**, 149-153. http://dx.doi.org/10.1016/j.sbi.2007.12.007

[18] Wang, D., Friedmann, M., Gattin, Z., Jaun, B. and van Gunsteren, W.F. (2010) The Propensity of α-Aminoisobutyric Acid (=2-Methylalanine; Aib) to Induce Helical Secondary Structure in an α-Heptapeptide: A Computational Study. *Helvetica Chimica Acta*, **93**, 1513-1531. http://dx.doi.org/10.1002/hlca.200900420

[19] Yu, H., Ramseier, M., Burgi, R. and van Gunsteren, W.F. (2004) Comparison of Properties of Aib-Rich Peptides in Crystal and Solution: A Molecular Dynamics Study. *ChemPhysChem*, **5**, 633-641. http://dx.doi.org/10.1002/cphc.200301026

[20] Zagrovic, B., Lipfert, J., Sorin, E.J., Millett, I.S., van Gunsteren, W.F., Doniach, S. and Pande, V.S. (2005) Unusual Compactness of a Polyproline Type II Structure. *Proceedings of the National Academy of Sciences of the United States of America*, **102**, 11698-11703. http://dx.doi.org/10.1073/pnas.0409693102

[21] Brooks III, C.L. (2002) Protein and Peptide Folding Explored with Molecular Simulations. *Accounts of Chemical Research*, **35**, 447-454. http://dx.doi.org/10.1021/ar0100172

[22] Shea, J.E. and Brooks III, C.L. (2001) From Folding Theories to Folding Proteins: A Review and Assessment of Si-mulation Studies of Protein Folding and Unfolding. *Annual Review of Physical Chemistry*, **52**, 499-535. http://dx.doi.org/10.1146/annurev.physchem.52.1.499

[23] Karanicolas, J. and Brooks III, C.L. (2004) An Evolution of Minimalist Models for Protein Folding: From the Behavior of Protein-Like Polymers to Protein Function. *Biosilico*, **2**, 127-133.

[24] Chakrabartty, A. and Baldwin, R.L. (1995) Stability of α-Helices. *Advances in Protein Chemistry*, **46**, 141-176. http://dx.doi.org/10.1016/S0065-3233(08)60334-4

[25] Head-Gordon, T., Stillinger, F.H., Wright, M.H. and Gay, D.M. (1992) Poly(L-Alanine) as a Universal Reference Ma-terial for Understanding Protein Energies and Structures. *Proceedings of the National Academy of Sciences of the Uni-ted States of America*, **89**, 11513-11517. http://dx.doi.org/10.1073/pnas.89.23.11513

[26] Makowska, J., Liwo, A., Żmudzińska, W., Lewandowska, A., Chmurzynski, L. and Scheraga, H.A. (2012) Like-Charged Residues at the Ends of Oligoalanine Sequences Might Induce a Chain Reversal. *Biopolymers*, **97**, 240-249. http://dx.doi.org/10.1002/bip.22013

[27] Makowska, J., Rodziewicz-Motowidlo, S., Baginska, K., Makowski, M., Vila, J.A., Liwo, A., Chmurzynski, L. and Scheraga, H.A. (2007) Further Evidence for the Absence of Polyproline II Stretch in the XAO Peptide. *Biophysical Journal*, **92**, 2904-2917. http://dx.doi.org/10.1529/biophysj.106.097550

[28] Pappu, R.V., Srinivasan, R. and Rose, G.D. (2000) The Flory Isolated-Pair Hypothesis Is Not Valid for Polypeptide Chains: Implications for Protein Folding. *Proceedings of the National Academy of Sciences of the United States of America*, **97**, 12565-12570. http://dx.doi.org/10.1073/pnas.97.23.12565

[29] Ramakrishnan, V., Ranbhor, R. and Durani, S. (2004) Existence of Specific "Folds" in Polyproline II Ensembles of an "Unfolded" Alanine Peptide Detected by Molecular Dynamics. *Journal of the American Chemical Society*, **126**, 16332-16333. http://dx.doi.org/10.1021/ja045787y

[30] Shi, Z., Olson, C.A., Rose, G.D., Baldwin, R.L. and Kallenbach, N.R. (2002) Polyproline II Structure in a Sequence of Seven Alanine Residues. *Proceedings of the National Academy of Sciences of the United States of America*, **99**, 9190-

9195. http://dx.doi.org/10.1073/pnas.112193999

[31]   Shi, Z., Woody, R.W. and Kallenbach, N.R. (2002) Is Polyproline II a Major Backbone Conformation in Unfolded Proteins? *Advances in Protein Chemistry*, **62**, 163-240. http://dx.doi.org/10.1016/S0065-3233(02)62008-X

[32]   Cheng, R.P., Girinath, P., Suzuki, Y., Kuo, H.T., Hsu, H.C., Wang, W.R., Yang, P.A., Gullickson, D., Wu, C.H., Koyack, M.J., Chiu, H.P., Weng, Y.J., Hart, P., Kokona, B., Fairman, R., Lin, T.E. and Barrett, O. (2010) Positional Effects on Helical Ala-Based Peptides. *Biochemistry*, **49**, 9372-9384. http://dx.doi.org/10.1021/bi101156j

[33]   Wieczorek, R. and Dannenberg, J.J. (2003) Hydrogen-Bond Cooperativity, Vibrational Coupling, and Dependence of Helix Stability on Changes in Amino Acid Sequence in Small 3_{10}-Helical Peptides. A Density Functional Theory Study. *Journal of the American Chemical Society*, **125**, 14065-14071. http://dx.doi.org/10.1021/ja034034t

[34]   Brant, D.A., Miller, W.G. and Flory, P.J. (1967) Conformational Energy Estimates for Statistically Coiling Polypeptide Chains. *Journal of Molecular Biology*, **23**, 47-65. http://dx.doi.org/10.1016/S0022-2836(67)80066-4

[35]   Flory, P.J. and Schimmel, P.R. (1967) Dipole Moments in Relation to Configuration of Polypeptide Chains. *Journal of the American Chemical Society*, **89**, 6807-6813. http://dx.doi.org/10.1021/ja01002a001

[36]   Flory, P.J. (1969) Statistical Mechanics of Chain Molecules. InterScience Publishers, New York.

[37]   Kumar, A., Ramakrishnan, V., Ranbhor, R., Patel, K. and Durani, S. (2009) Homochiral Stereochemistry: The Missing Link of Structure to Energetics in Protein Folding. *The Journal of Physical Chemistry B*, **113**, 16435-16442. http://dx.doi.org/10.1021/jp906811k

[38]   Ramakrishnan, V., Ranbhor, R., Kumar, A. and Durani, S. (2006) The Link between Sequence and Conformation in Protein Structures Appears to Be Stereochemically Established. *The Journal of Physical Chemistry B*, **110**, 9314-9323. http://dx.doi.org/10.1021/jp056417e

[39]   Srivastava, K.R., Kumar, A., Goyal, B. and Durani, S. (2011) Stereochemistry and Solvent Role in Protein Folding: Nuclear Magnetic Resonance and Molecular Dynamics Studies of Poly-L and Alternating-L,D Homopolypeptides in Dimethyl Sulfoxide. *The Journal of Physical Chemistry B*, **115**, 6700-6708. http://dx.doi.org/10.1021/jp200743w

[40]   Durani, S. (2008) Protein Design with L- and D-Alpha-Amino Acid Structures as the Alphabet. *Accounts of Chemical Research*, **41**, 1301-1308. http://dx.doi.org/10.1021/ar700265t

[41]   Daura, X., van Gunsteren, W.F. and Mark, A.E. (1999) Folding-Unfolding Thermodynamics of a Beta-Heptapeptide from Equilibrium Simulations. *Proteins: Structure, Function, and Bioinformatics*, **34**, 269-280. http://dx.doi.org/10.1002/(SICI)1097-0134(19990215)34:3<269::AID-PROT1>3.0.CO;2-3

[42]   Lindahl, E., Hess, B. and van der Spoel, D. (2001) GROMACS 3.0: A Package for Molecular Simulation and Trajectory Analysis. *Journal of Molecular Modeling*, **7**, 306-317.

[43]   Van Gunsteren, W.F., Billeter, S.R., Eising, A.A., Hünenberger, P.H., Krüger, P., Mark, A.E., Scott, W.R.P. and Tironi, I.G. (1996) Biomolecular Simulation: The GROMOS96 Manual and User Guide. Hochschulverlag AG an der ETH Zürich, Zürich.

[44]   Ryckaert, J., Ciccotti, G. and Berendsen, H. (1977) Numerical Integration of the Cartesian Equations of Motion of a System with Constraints: Molecular Dynamics of *n*-Alkanes. *Journal of Computational Physics*, **23**, 327-341. http://dx.doi.org/10.1016/0021-9991(77)90098-5

[45]   Darden, T., York, D. and Pedersen, L. (1993) Particle Mesh Ewald: An *N*-Log(*N*) Method for Ewald Sums in Large Systems. *The Journal of Chemical Physics*, **98**, 10089-10092. http://dx.doi.org/10.1063/1.464397

[46]   Essmann, U., Perera, L., Berkowitz, M.L., Darden, T., Lee, H. and Pedersen, L.G. (1995) A Smooth Particle Mesh Ewald Method. *The Journal of Chemical Physics*, **103**, 8577-8593. http://dx.doi.org/10.1063/1.470117

[47]   Morris, G.M., Goodsell, D.S., Halliday, R.S., Huey, R., Hart, W.E., Belew, R.K. and Olson, A.J. (1998) Automated Docking Using a Lamarckian Genetic Algorithm and an Empirical Binding Free Energy Function. *Journal of Computational Chemistry*, **19**, 1639-1662. http://dx.doi.org/10.1002/(SICI)1096-987X(19981115)19:14<1639::AID-JCC10>3.0.CO;2-B

[48]   Chan, W.C. and White, P.D. (1989) Fmoc Solid Phase Peptide Synthesis: A Practical Approach. IRL Press, Oxford.

Supplementary Material

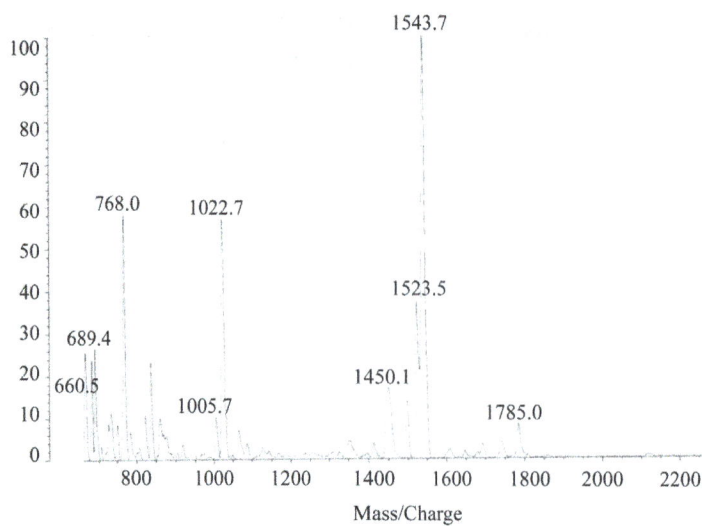

(a)

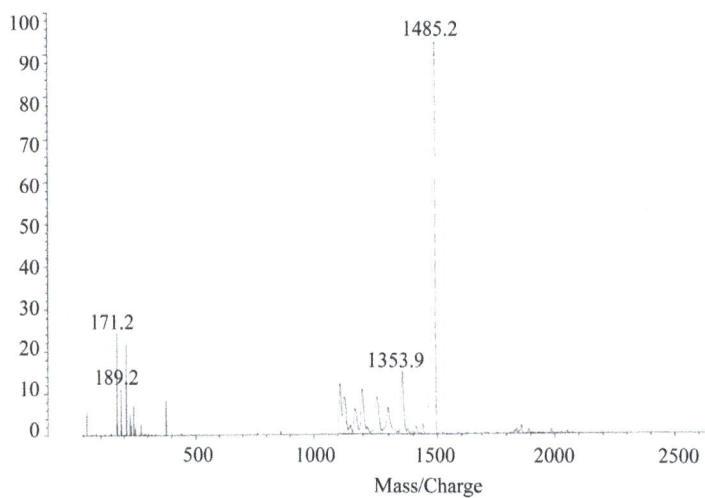

(b)

Figure S1. MALDI-Mass spectra of H1 (panel (a)) and H2 (panel (b)).

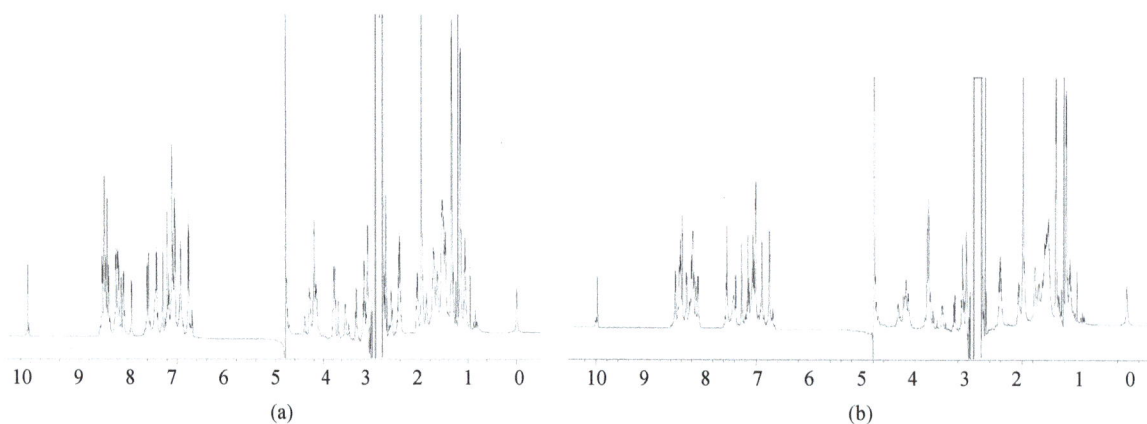

(a)   (b)

Figure S2. (a) ^{1}H NMR spectra of H1 in citrate buffer (pH ~3) and (b) ^{1}H NMR spectra of H2 in citrate buffer (pH ~3).

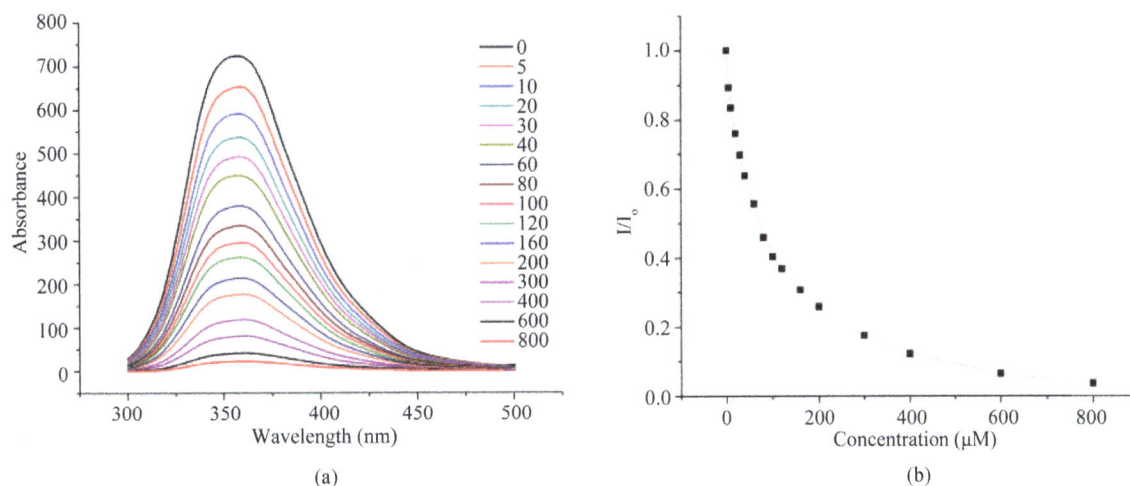

(a)   (b)

Figure S3. Quenching of tryptophan fluorescence of peptide H2 (20 μM) in 20 mMTris-HCl buffer at pH 7.5, on progressive titration with increasing titration with pNPP (panel (a)), and plot of relative fluorescence intensity as a function of pNPP concentration (panel (b)).

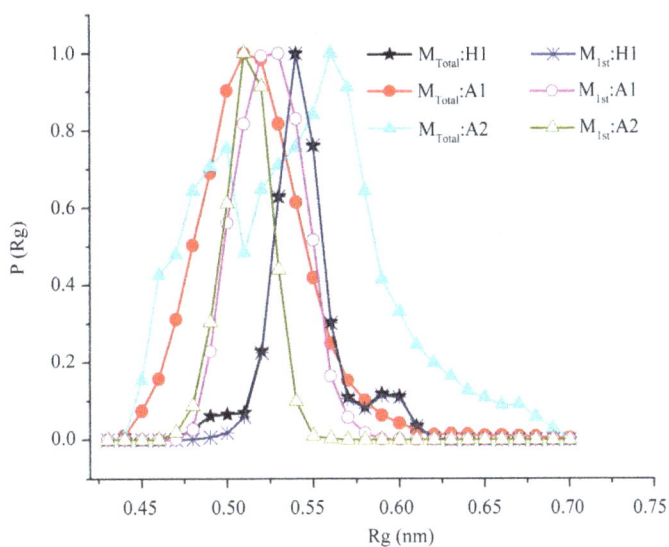

Figure S4. Radius of gyration (Rg) distribution of main chain atoms over conformers populating equilibrium.

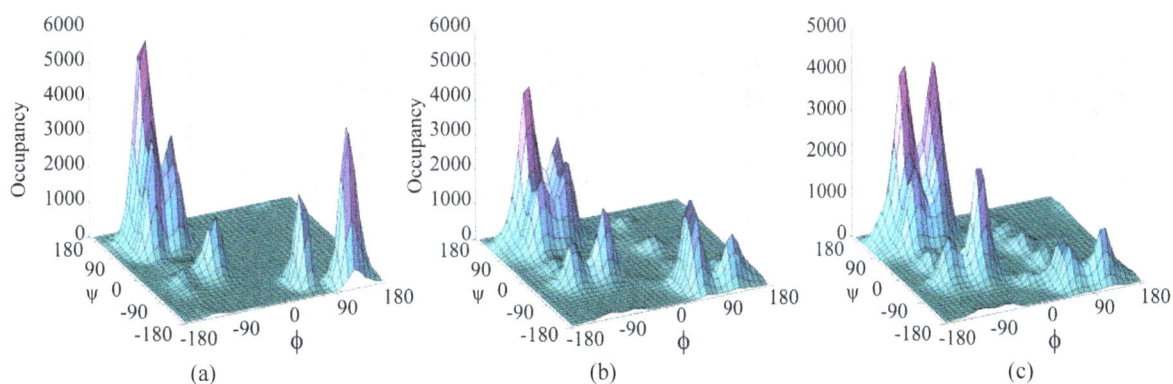

(a)   (b)   (c)

Figure S5. Occupancy of ensembles in H1 (panel (a)), A1 (panel (b)) and A2 (panel (c)) in specific basins of ϕ, ψ.

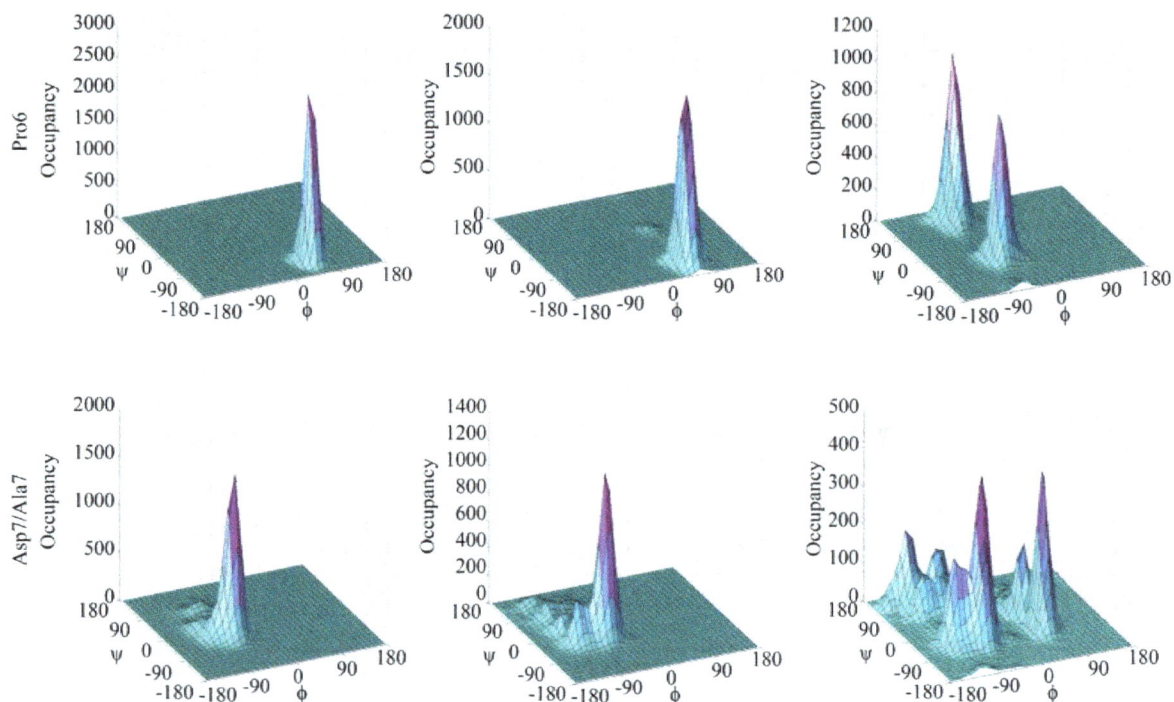

Figure S6. Specific *β*-turn residues in H1 (panel (a)), A1 (panel (b)), and A2 (panel (c)) are varied in *φ*, *ψ* space.

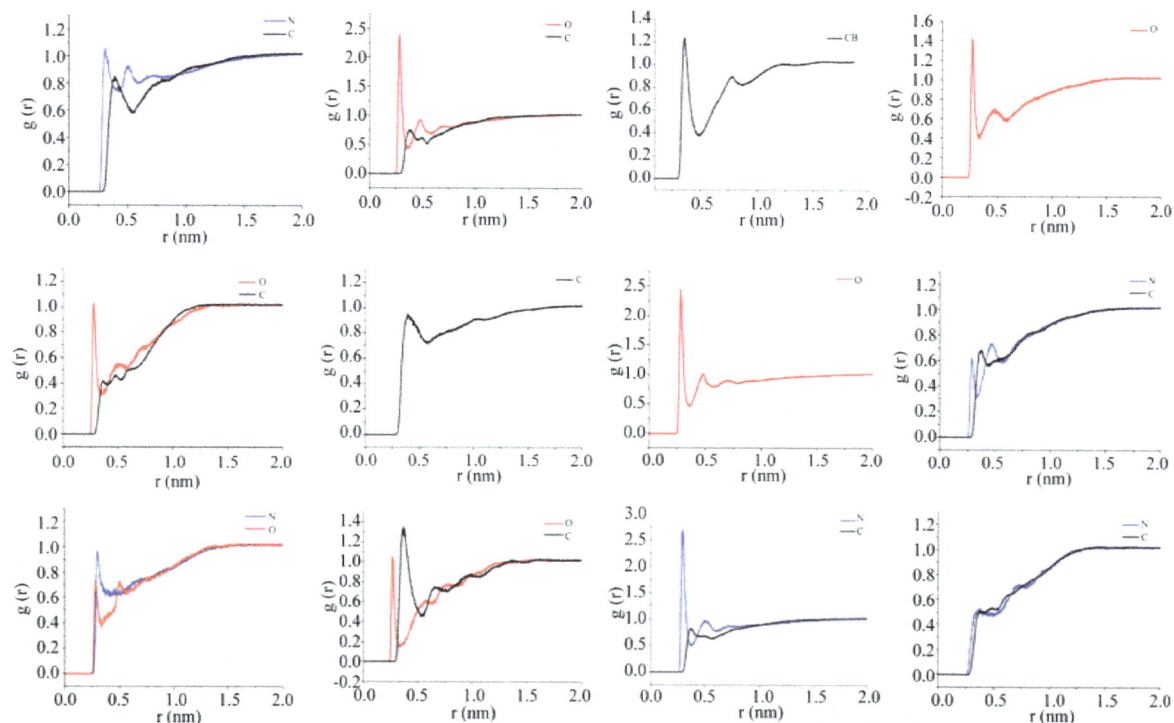

Figure S7. Radial distribution of water oxygen atoms against specific side chain nitrogen (blue trace), oxygen (red trace), and carbon (black trace) atoms of H1 over macrostate.

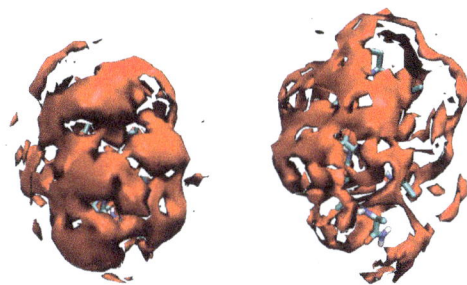

Figure S8. Spatial distribution of water oxygen atom around atoms of H1 (left panel) and A1 (right panel) over macrostate.

5

Synthesis, Characterization and Application of Mg/Al Layered Double Hydroxide for the Degradation of Congo Red in Aqueous Solution

Nimibofa Ayawei[1]*, Seimokumo Samuel Angaye[1], Donbebe Wankasi[1,2], Ezekiel Dixon Dikio[1,2]

[1]Department of Chemical Sciences, Niger Delta University, Wilberforce Island, Nigeria
[2]Applied Chemistry and Nanoscience Laboratory, Department of Chemistry, Vaal University of Technology, Vanderbijlpark, South Africa
Email: *ayawei4acad@gmail.com

Abstract

The adsorption properties of layered double hydroxide (Mg/Al-CO$_3$) for the removal of Congo Red (CR) dye from aqueous solution were studied. The layered double hydroxide was synthesized by co-precipitation method and characterized by X-ray diffraction (XRD), Fourier Transform Infrared spectroscopic (FTIR) and Energy-Dispersive X-ray Spectroscopic (EDX). The effects of various experimental parameters such as contact time, dye concentrations and temperature variation were investigated. The results show that the amount of Congo Red adsorbed increases with increase in temperature but decreases with increase in initial dye concentration and contact time. The data were also fitted to several kinetic models: zero-order kinetic model, first-order kinetic model, second-order kinetic model, pseudo-second-order kinetic model and third-order kinetic model respectively. The adsorption process was best defined by zero-order-kinetic model ($R^2 = 1$). Langmuir, Freundich, Temkin and Dubinin-kaganer-Radushkevich (DPK) adsorption isotherm models were applied to analyze adsorption data with Temkin isotherm being the most applicable to the adsorption process. Thermodynamic parameters e.g. ΔG^o, ΔS^o, ΔH^o and ΔH_x of the adsorption process were found to be endothermic, spontaneous and feasible.

Keywords

Congo Red, Layered Double Hydroxides, Kinetic, Dye, Adsorption, Isotherms, Thermodynamics

---

*Corresponding author.

1. Introduction

The wastewater disposed by textile industries is causing major hazards to the environment and drinking water due to presence of a large number of contaminants like acids, bases, toxic organic, inorganic, dissolved solids and colour [1]. The presence of such compounds in the industrial wastewater may create serious environmental problems due to toxicity to aquatic life and mutagenicity to humans. In spite of resistance to biodegradation under aerobic conditions, dyes (in particular azo dyes) undergo reductive splitting of the azo bond relatively easily under anaerobic conditions releasing corresponding aromatic amines [2]-[4]. Congo Red is investigated as a mutagen and reproductive effector. It is a skin, eye, and gastrointestinal irritant. It may affect blood factors such as clotting, and induce somnolence and respiratory problems [5]. Therefore, an increased interest has been focused on removing of such dyes from the wastewater. Various physical, chemical and biological methods, including adsorption, biosorption, ozonation, coagulation/flocculation, advanced oxidation, membrane filtration and liquid-liquid extraction have been widely used for the treatment of dye-bearing wastewater [6]-[9]. Adsorption is a very effective separation technique and now it is considered to be superior to other techniques for water treatment in terms of initial cost, simplicity of design, ease of operation and insensitive to toxic substances [10]-[13].

Layered double hydroxides also known as hydrotalcite-like compounds or anionic clays, have received much attention in the past decades due to their wide spread applicability. LDHs have positively charged layers of metal hydroxides and the anions and water molecules are located between the layers. The positive charges that are produce from the isomorphous substitution of divalent cations and trivalent cations, are counter balanced by anions located between the layers [14]. LDHs have a general formula of $\left[M^{2+}_{1-x} M^{3+}_{x} (OH)_2 \right]\left[A^{n-}_{x/n} \cdot m H_2 O \right]$, where M^{2+} and M^{3+} are divalent and trivalent metal cations, respectively; A is the anions, and x is ratio $M^{3+}/(M^{2+} + M^{3+})$ [15]. The anions between the layers can be polymers, organic dyes, surfactants and organic acids [16]. Layered double hydroxides (LDHs) have in the past three decades received considerable attention due to their flexible interlayer region. A variety of layered materials have been synthesized by different methods and LDHs have widespread applications as catalysts or catalyst precursors [17], adsorbents [18] [19], anionic exchangers [15], in biochemistry [20], polymer additives [21] and as hybrid pigments [22]. This study aims to replicate hydrotalcites which are clays in a laboratory condition for the degradation of Congo Red in aqueous solution.

2. Experimental

2.1. Synthesis of Mg/Al-CO₃

Carbonate form of Mg-Al LDH was synthesized by co-precipitation method. A 50 ml aqueous solution containing 0.3 M $Mg(NO_3)_2 \cdot 6H_2O$ and 0.1 M $Al(NO_3)_3 \cdot 9H_2O$ with Mg/Al ratios 4:1, was added drop wise into a 50 ml mixed solution of NaOH (2 M) + Na_2CO_3 (1 M) with vigorous stirring and maintaining a pH of greater than 10 at room temperature. After complete addition which last between 2 hours 30 minutes to 3 hours, the slurry formed was aged at 60°C for 18 hours. The products were centrifuged at 5000 rpm for 5 minutes, with distilled water 3 - 4 times and dried by freeze drying.

2.2. Characterization of Layered Double Hydroxide

X-ray diffraction (XRD) pattern of the sample was characterized by using a Shimadzu XRD-6000 diffractometer, with Ni-filtered Cu-Kα radiation (λ = 1.54 Å) at 40 kV and 200 mA. Solid samples were mounted on alumina sample holder and basal spacing (d-spacing) was determined via powder technique. Samples scan were carried out at 10° - 60°, 2θ/min at 0.003° steps.

FTIR spectrum was obtained using a Perkin Elmer 1725X spectrometer where samples will be were finely ground and mixed with KBr and pressed into a disc. Spectrums of samples were scanned at 2 cm^{-1} resolution between 400 and 4000 cm^{-1}.

FESEM/EDX was obtained using Carl Zeiss SMT supra 40 VPFESEM Germany and inca penta FET ×3 EDX, Oxford. It was operated at extra high tension (HT) at 5.0 kV and magnification at 20000×. FESEM uses electron to produce images (morphology) of samples and was attached with EDX for qualitative elemental analysis.

2.3. Preparation of Congo Red Solution

Congo Red (**Figure 1**) was supplied by Merck (Mumbai, India). A stock solution of CR dye was prepare

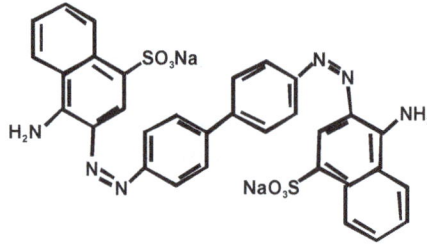

Figure 1. Molecular formula of Congo Red.

(100 mg/L) by dissolving a required amount of dye powder in deionized water. The stock solution was diluted with deionized water to obtain the desired concentrations of 20, 30 and 40 mg/L. The supernatants were analyzed using a UV-vis spectrophotometer (Shimadzu, Kyoto, Japan) at wavelength of 497 nm.

2.4. Adsorption Isotherm Studies

The quantity of Congo Red removed by the layerd double hydroxide in aqueous solution and the percentage were calculated using Equations (1) and (2) below:

$$q_{eql} = \frac{C_{init} - C_{eql}}{m} \tag{1}$$

$$R\% = \frac{C_{init} - C_{eql}}{C_{eql}} \times 100 \tag{2}$$

where C_{init} and C_{eql} are, respectively, the initial and equilibrium concentrations of dye in solution (mmol/l) and m is the layered double hydroxide dosage (g/l).

The data for the uptake of Congo Red at different temperatures has been processed in accordance with the linearised form of the Freundlich and Langmuir isotherm equations.

The Langmuir model linearization (a plot of $1/q_{eql}$ vs $1/C_{eql}$) was expected to give a straight line with intercept of $1/q_{max}$:

$$\frac{1}{q_{eql}} = \frac{1}{K_l q_{mas} C_{eql}} + \frac{1}{q_{eql}} \tag{3}$$

The essential characteristics of the Langmuir isotherm were expressed in terms of a dimensionless separation factor or equilibrium parameter S_f.

$$S_f = \frac{1}{1 + aC_o} \tag{4}$$

With C_o as initial concentration of Congo Red in solution, the magnitude of the parameter S_f provides a measure of the type of adsorption isotherm. If $S_f > 1.0$, the isotherm is unfavourable; $S_f = 1.0$ (linear); $0 < S_f < 1.0$ (favourable) and $S_f = 0$ (irreversible).

For the Freundlich isotherm the *In-In* version was used:

$$Inq_{eql} = InK_f + \frac{1}{n} InC_{eql} \tag{5}$$

The DKR isotherm is reported to be more general than the Langmuir and Freundlich isotherms. It helps to determine the apparent energy of adsorption. The characteristic porosity of adsorbent toward the adsorbate and does not assume a homogenous surface or constant sorption potential [23].

The Dubinin-Kaganer-Radushkevich (DKR) model has the linear form

$$Inq_e = InX_m - \beta\varepsilon^2 \tag{6}$$

where X_m is the maximum sorption capacity, β is the activity coefficient related to mean sorption energy, and ε is the Polanyi potential, which is equal to

$$\varepsilon = RTln\left(1 + \frac{1}{C_e}\right) \qquad (7)$$

where R is the gas constant (kJ/kmol). The slope of the plot of $\ln q_e$ versus ε^2 gives β (mol^2/J^2) and the intercept yields the sorption capacity, X_m (mg/g). The values of β and X_m, as a function of temperature are listed in **Table 1** with their corresponding value of the correlation coefficient, R^2. It can be observed that the values of β increase as temperature increases while the values of X_m decrease with increasing temperature.

The values of the adsorption energy, E, was obtained from the relationship [24]

$$E = \left(2\beta\right)^{-1/2} \qquad (8)$$

The Temkins isotherm model was also applied to the experimental data, unlike the Langmuir and Freundlich isotherm models, this isotherm takes into account the interactions between adsorbents and dye to be adsorbed and is based on the adsorption that the free energy of adsorption is simply a function of surface coverage [25]. The linear form of the Temkins isotherm model equation is given in (9).

$$q_e = B \ln A + B \ln C_e \qquad (9)$$

where $B = [RT/b_T]$ in (J/mol) corresponding to the heat of adsorption, R is the ideal gas constant, T (K) is the absolute temperature, b_T is the Temkins isotherm constant and A (L/g) is the equilibrium binding constant corresponding to the maximum binding energy.

Kinetic Studies

The experimental data were further subjected to certain kinetic parameters.

Zero-order kinetic model,

$$q_t = q_o + K_o t \qquad (10)$$

First-Order Kinetic model,

$$\ln q_t = \ln q_o + K_1 t \qquad (11)$$

Second-Order Kinetic model,

$$\frac{1}{q_t} = \frac{1}{q_o} + K_2 t \qquad (12)$$

Third-order kinetic model

$$\frac{1}{q_t^2} = \frac{1}{q_o^2} + K_3 t \qquad (13)$$

Pseudo-second order model

$$\frac{t}{q_t} = \frac{1}{h_o} + \frac{1}{q_e t} \qquad (14)$$

where q_o (mg/g) and q_t (mg/g) are the adsorbed amounts of CR at equilibrium and time t(min); K_o, K_1, K_2 and K_3 are the adsorption rate constants for the kinetic models.

2.5. Thermodynamic Studies

The thermodynamic parameters such as change in free energy ΔG^o, enthalpy change ΔH^o and entropy change ΔS^o were determined by using the following equations:

$$G^o = -RT \ln K_d \qquad (15)$$

$$\Delta G^* = \Delta H^* - T\Delta S^* \qquad (16)$$

where K_d equals the ratio of C_{solid} and C_{liquid}. C_{solid} is the equilibrium concentration of adsorbate on the adsorbent (mg/L), C_{liquid} is the equilibrium concentration of adsorbate in solution (mg/L), T is temperature (K) and R is the ideal gas constant (8.314 J·mol^{-1}·K^{-1}).

Table 1. Characteristic parameters of the adsorption isotherm models for Congo Red adsorption by layered double hydroxide.

Isotherm model	Isotherm parameter	Results
Freundlich	$1/n$	1.0875
	K_F, mg/L	2.2403
	R^2	0.9936
Langmuir	R_L	0.802
	R^2	0.9925
Dubinin-Kaganer-Radushkevich	E, kJ/mol	0.698
	β_D, mol^2/kJ^2	1.0263
	q_D, mg/g	0.9543
	R^2	0.996
Temkin	A	1.372
	b	1.36×10^3
	B	1.6637
	R^2	1

The differential isosteric heat of adsorption (ΔH_x) at constant surface coverage was calculated using the Clausius-Clapeyron equation:

$$\frac{dIn\left(C_{eql}\right)}{dT} = -\frac{\Delta H_x}{RT^2} \tag{17}$$

Integration gives the following equation [18]:

$$In\left(C_{eql}\right) = \frac{\Delta H_x}{R}\frac{1}{T} + k \tag{18}$$

where K is a constant. The differential isosteric heat of adsorption was calculated from the slope of the plot of $\ln(C_{eql})$ vs $1/T$ and was used for an indication of the adsorbent surface heterogeneity.

The linear form of the modified Arrhenius expression was applied to the experimental data to evaluate the activation energy (E_a) and sticking probability S^* as shown in Equation (19).

$$In\left(1-\theta\right) = S^* + \frac{E_a}{RT} \tag{19}$$

where θ is the degree of surface coverage, T is absolute solution temperature and R is gas constant (8.314 J/mol^{-1}·K^{-1}.

3. Results and Discussion

3.1. Characterization of LDH

1) SEM

Figure 2 clearly show the pre & post adsorption SEM images. The SEM image of post adsorption shows coverage of available pores in relation to pre-adsorption image.

2) XRD

The typical XRD pattern (**Figure 3**) shows a lamellar structure of LDH material. Peaks at 8.6, 23.4 and 34.6 correspond to d-spacing at 1.027 nm, 0.3797 nm and 0.259 nm respectively. This is consistent with layered materials.

3) FT-IR

The pre and post adsorption FT-IR spectra as shown in **Figure 4** resemble those of other hydrotalcite-like phases. The pre-adsorption (a) band at 3408 cm^{-1} could be attributed to the stretching vibration of hydroxyl

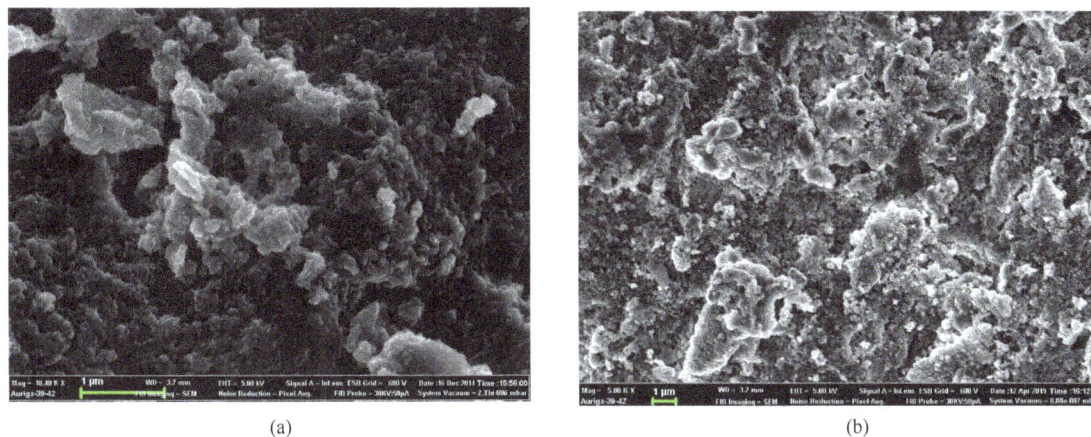

Figure 2. Scanning Electron Microscope (SEM) micrograph of Mg/Al-CO$_3$ before (a) and after (b) adsorption studies.

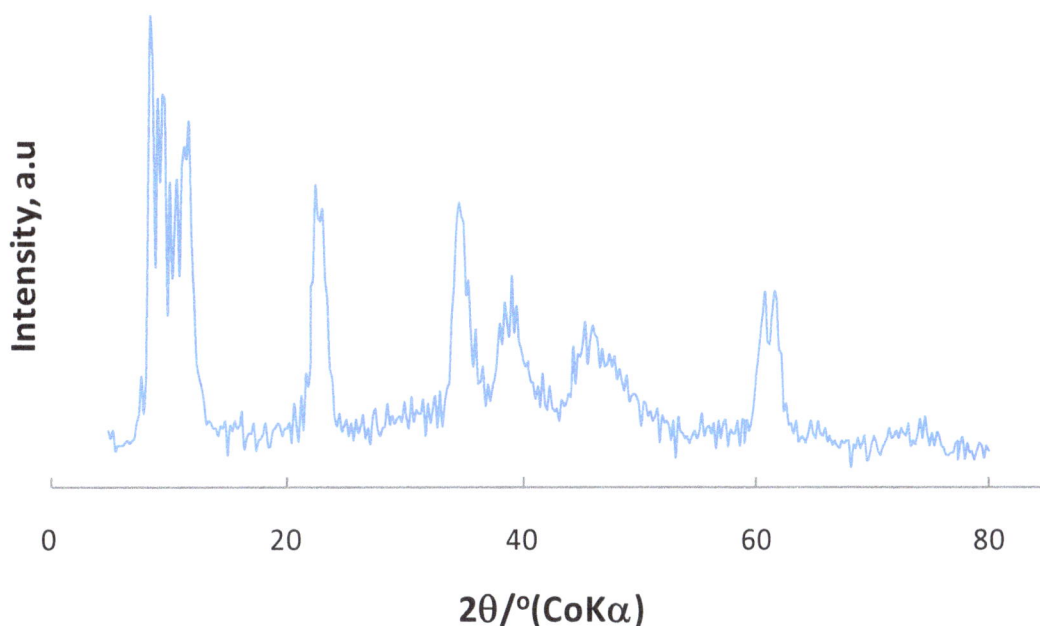

Figure 3. Mg/Al-CO$_3$ X-ray powder diffraction.

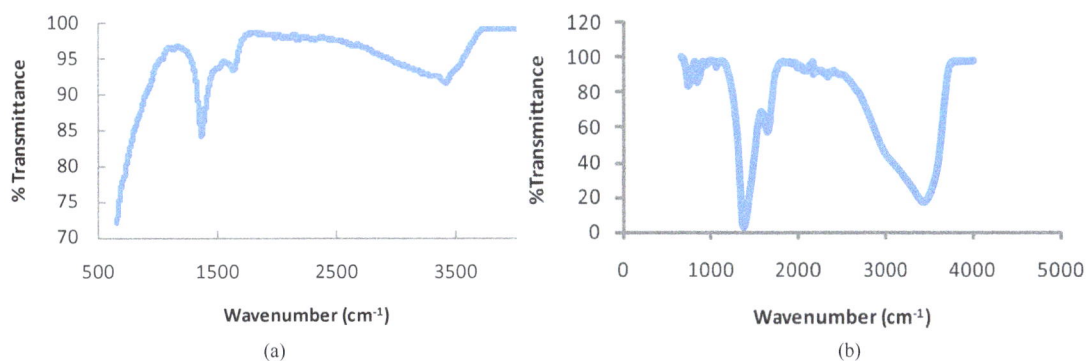

Figure 4. Mg/Al-CO$_3$ Fourier transform infrared spectroscopy, before (a) and after (b) adsorption studies.

group. The low intensity band at 1632 cm^{-1} is assigned to bending vibration of strongly adsorbed water (solvation water for compensating anion vibration). The band at 1363 cm^{-1} is assigned to carbonate vibration $\left(CO_3^{2-} \right)$, the bands at 672 is due to M-O vibration. While the shrinking of the post-adsorption spectra as shown in (b), is attributed to the aromatic ring about 1015 cm^{-1}, the band between 1100 cm^{-1} - 1200 cm^{-1} is due to phosphate stressing and the out-of-plane wagging at 650 cm^{-1} - 900 cm^{-1} are characteristic of 1° amines. The change in the FTIR spectra confirms the formation of complex between the functional groups present in the adsorbent and Congo Red [26] [27].

3.2. Effect of Concentration

Removal efficiency of Congo Red by adsorbents is illustrated in **Figure 5**. It shows that removal efficiency decreased with increasing of initial concentration (47.5%, 46.7% and 43%) respectively, this is probably due to rapid adsorption at all available site and relatively small amount of adsorbent that was used, an increase in the amount of adsorbent may therefore reverse adsorption trend.

3.3. Isotherm Analysis

To investigate an interaction of adsorbate molecules and adsorbent surface, four well-known models, the Langmuir, Freundlich, Dubinin-Kaganer-Radushkevic and Temkin isotherms, were selected to explicate LDH interaction in this study.

The Langmuir plot in **Figure 6** fitted the experimental data with $R^2 = 0.9925$ and therefore, confirm monolayer coverage. The favourability or otherwise of a Langmuir type isotherm is determined by a dimensionless constant separation factor (R_L), given by Equation (4). The calculated value of R_L from **Figure 6** is 0.802, which is within the range of 0 - 1, thus confirms the favourable uptake of the Congo Red by the LDH. The degree of favourability is generally related to the irreversibility of the system, giving a qualitative assessment of the layered double hydroxide interactions.

K_f is a constant describing the adsorption capacity (mg/L) and n is an empirical parameter related to the adsorption intensity, the plot of lnq_e against lnC_e is shown in **Figure 7** gives 2.2403 and 0.92 as values for K_F and n respectively.

The values of k_F and n determine the steepness and curvature of the isotherm. The Freundlich equation frequently gives n adequate description of adsorption data over a restricted range of concentration, even though it is not based on any theoretical background. Apart from a homogeneous surface, the Freundlich equation is also suitable for a highly heterogeneous surface and an adsorption isotherm lacking a plateau, indicating a multilayer adsorption [28]. The values of $1/n$, less than unity is an indication that significant adsorption takes place at low concentration but the increase in the amount adsorbed with concentration becomes less significant at higher concentration and *vice versa* [29]. The magnitude of K_F and n shows easy separation of Congo Red dye from wastewater and high adsorption capacity.

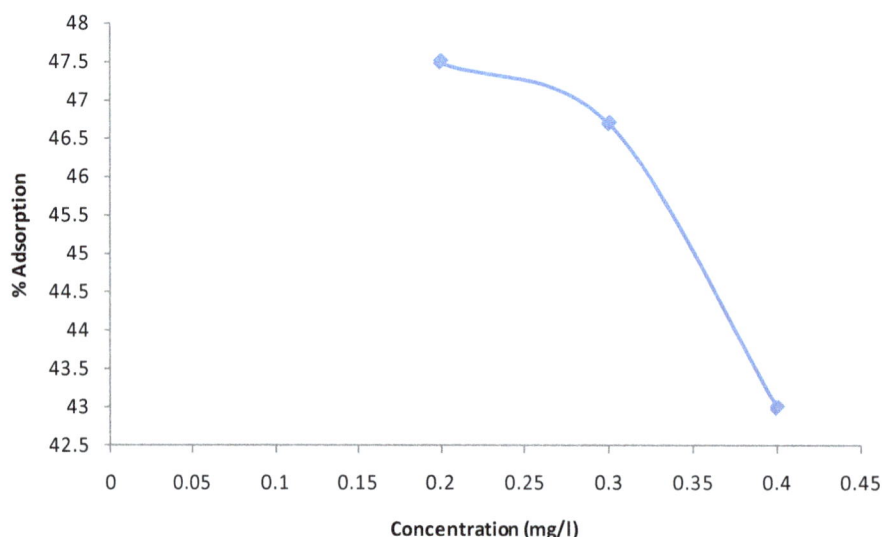

Figure 5. Effect of concentration on adsorption of Congo Red onto layered double hydroxide.

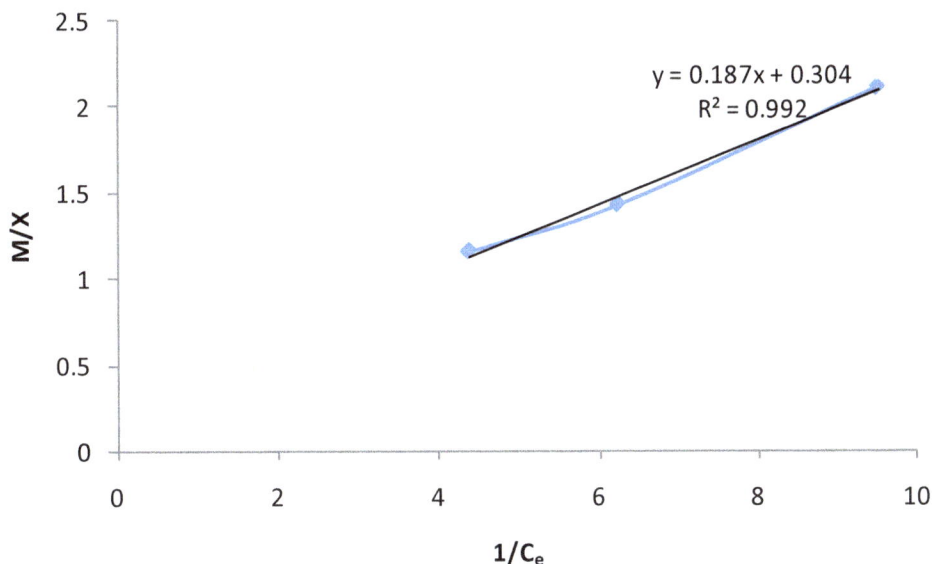

Figure 6. Langmuir isotherm plot for adsorption of Congo Red onto layered double hydroxide.

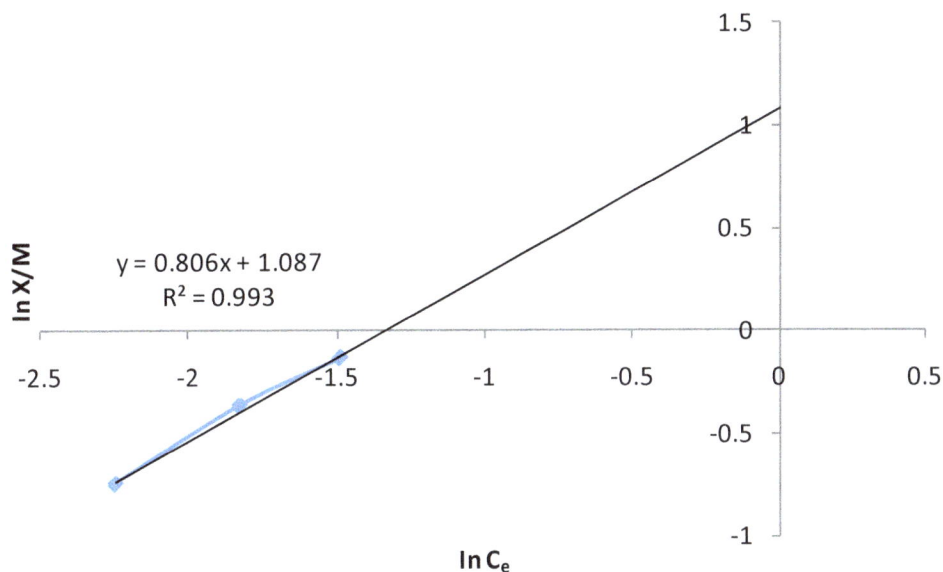

Figure 7. Freundlich isotherm plot for adsorption of Congo Red onto layered double hydroxide.

The fraction of the layered double hydroxide surface covered by the Congo Red is given as 0.47 (**Table 1**). This value indicates that 47% of the pore spaces of the layered double hydroxide surface were covered by the Congo Red which means less than average degree of adsorption.

The plots of Inq_e against ε^2 as shown in **Figure 8** yielded straight lines and indicates a good fit of the isotherm to the experimental data. The values of linear regression R^2, q_D, B_D and apparent energy E are calculated from the intercepts slopes of the plots respectively are shown on **Table 1** and Equation (8). From the linear plot of DRK model, q_D was determined to 0.9543 mg/g, the mean energy, $E = 0.698$ kJ/mol indicating a physiosorption process and the $R^2 = 0.996$.

Temkin adsorption isotherm model is usually chosen to evaluate the adsorption potentials of an adsorbent for the adsorbate from an experimental data. This model gives the mechanism and adsorption capacity of an adsorbate in a sorption process. From the Temkin plot shown in **Figure 9**, the following values were estimated:

$A = 1.372$ L/g, $B = 1.6637$ J/mol which is an indication of the heat of sorption, indicating a physical adsorption process and the $R^2 = 1$.

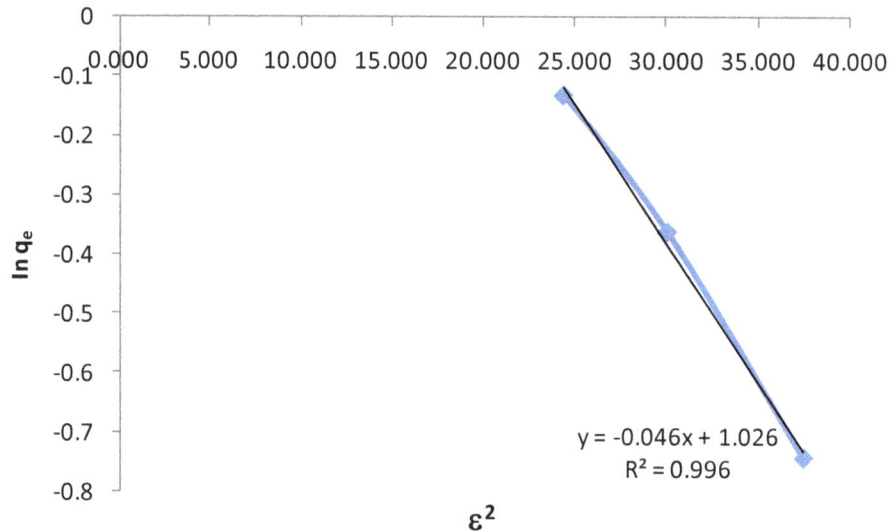

Figure 8. Dubinin-Kaganer-Radushkevic (DKR) isotherm plot for adsorption of Congo Red onto layered double hydroxide.

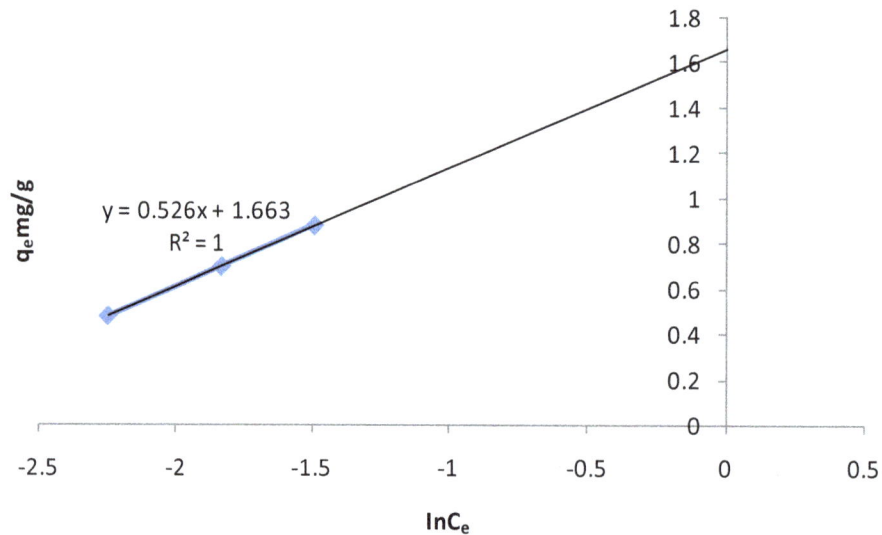

Figure 9. Temkin isotherm model plot for adsorption of Congo Red onto layered double hydroxide.

3.4. Effect of Temperature

As shown in **Figure 10** adsorption was lowest at 313 K (52%), and increased slightly to 333 K (55%) and 353 K (56%). This means that adsorption capacity increase with higher temperature.

The values of the enthalpy change (ΔH^o) and entropy change ΔS^o were calculated from Equation (10) to be 3.67 kJ/mol and 12.5 J/mol·K respectively, as shown in **Figure 11**. A positive ΔH^o suggests that sorption proceeded favourably at a higher temperature and the sorption mechanism was endothermic. A positive value of ΔS^o (12.5 J/mol·K) reflects the affinity of the adsorbent towards the adsorbate species. In addition, positive value of ΔS^o suggests increased randomness at the solid/solution interface with some structural changes in the adsorbate and the adsorbent. The adsorbed solvent molecules, which are displaced by the adsorbate species, gain more translational entropy than is lost by the adsorbate ions/molecules, thus allowing for the prevalence of randomness in the system. The positive ΔS^o value also corresponds to an increase in the degree of freedom of the adsorbed species.

Isosteric heat of adsorption ΔH_x is one of the basic requirements for the characterization and optimization of an adsorption process and is a critical design variable in estimating the performance of an adsorptive separation process. It also gives some indication about the surface energetic heterogeneity. Knowledge of the heats of sorption is very important for equipment and process design. A plot of InC_e against $1/T$ in **Figure 12** gives a slope equal to ΔH_x. The value of ΔH_x derived from Equation (11) was 40.03 kJ/mol which indicates that adsorption mechanism was physical adsorption and in an heterogeneous surface.

The activation energy E_a and the sticking probability S^* were calculated from Equation (12). The values shown in **Table 2** for E_a and S^* are -10.13 kJ/mol and 0.47 respectively, extrapolated from the plot in **Figure 13**. The value of activation energy shows that the sorption process was a physical one less than 4.2 kJ/mol.

3.5. Effect of Time

The adsorption kinetic study is important in predicting the mechanisms (chemical reaction or mass-transport process) that control the rate of the pollutant removal and retention time of adsorbed species at the solid-liquid interface [28] [29]. That information is important in the design of appropriate sorption treatment plants.

The effect of contact time of the phases on removal of Congo Red by the Layered double hydroxide from solutions of initial concentration equal to 400 mg CR/L at three different times (10, 20 and 30 minutes) is presented in **Figure 14**.

The result shows that adsorption was highest at 10 minutes, thereafter, a gradual decrease occurred (10 = 56.6%, 20 = 55% and 30 = 53%).

The experimental data were fitted into different kinetic models as shown in **Figures 15-18** including zero-order-kinetic model, second-order-kinetic model, pseudo-second-order-kinetic model and third-order-kinetic model to ascertain the suitability of the models [30]. The correlation coefficient values of 1, 0.9995, 0.9996 and 0.999 respectively confirming the applicability of all the studied models.

4. Conclusion

Layered double hydroxide (Mg/Al-CO$_3$) was successfully synthesized and characterized for the adsorption of

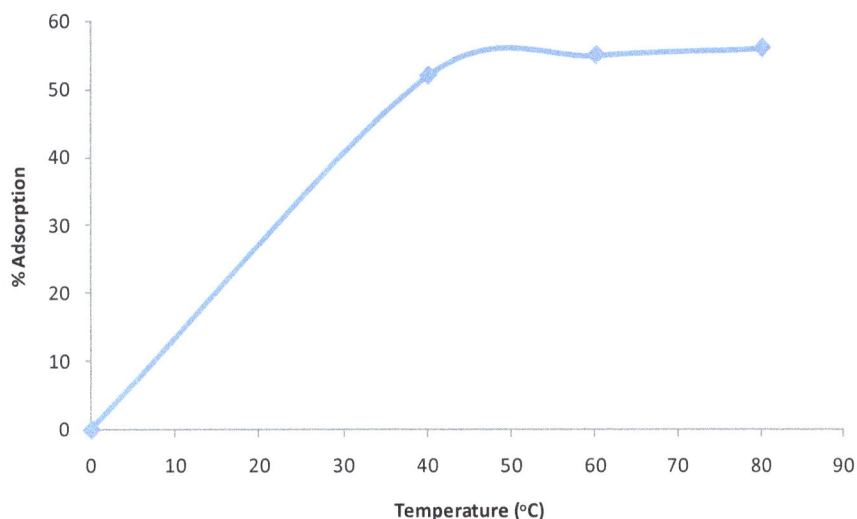

Figure 10. Effect of temperature on adsorption of Congo Red onto layered double hydroxide.

Table 2. Thermodynamic parameters of the adsorption of Congo Red onto layered double hydroxide.

T, K	ΔG^o, kJ/mol	ΔH^o, kJ/mol	ΔS^o, J/mol·K	E_a, kJ/mol	ΔH_x, kJ/mol
313	−0.208				
333	−0.550	3.67	12.5	−10.13	40.03
353	−0.707				

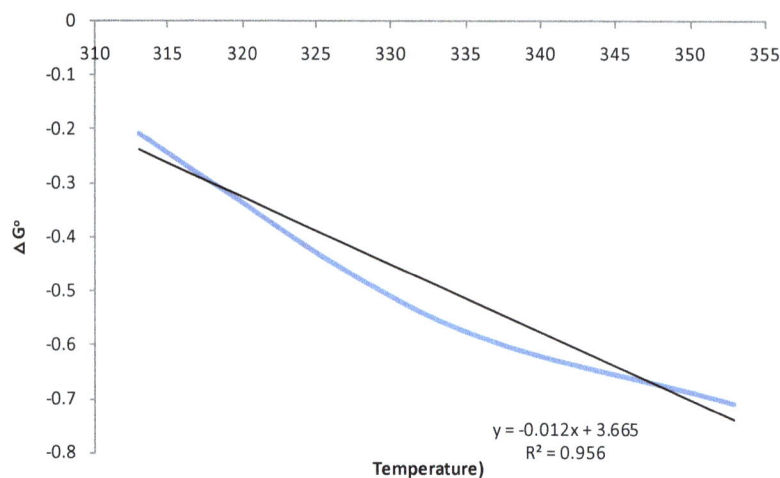

$$y = -0.012x + 3.665$$
$$R^2 = 0.956$$

Figure 11. Plot of ΔG^o vs. temperature for the adsorption of Congo Red onto layered double hydroxide.

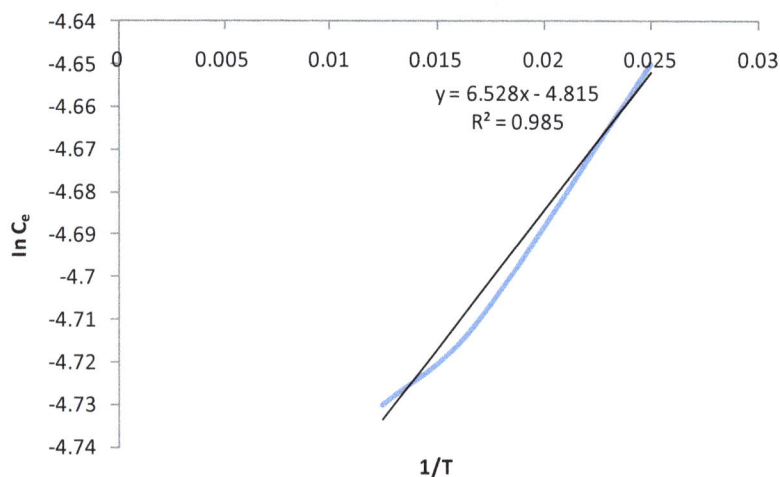

$$y = 6.528x - 4.815$$
$$R^2 = 0.985$$

Figure 12. Plot of $\ln C_e$ vs. $1/T$ for the adsorption of Congo Red onto layered double hydroxide.

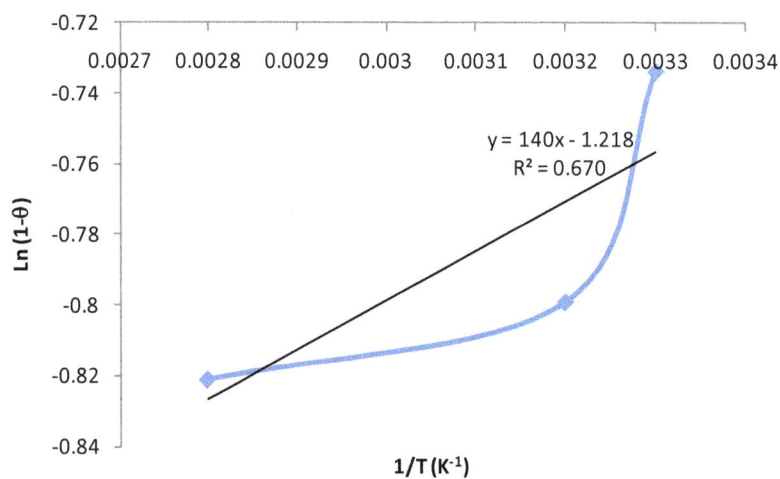

$$y = 140x - 1.218$$
$$R^2 = 0.670$$

Figure 13. Plot of $\mathrm{Ln}(1 - \theta)$ vs. $1/T$ (K^{-1}) for the adsorption of Congo Red onto layered double hydroxide.

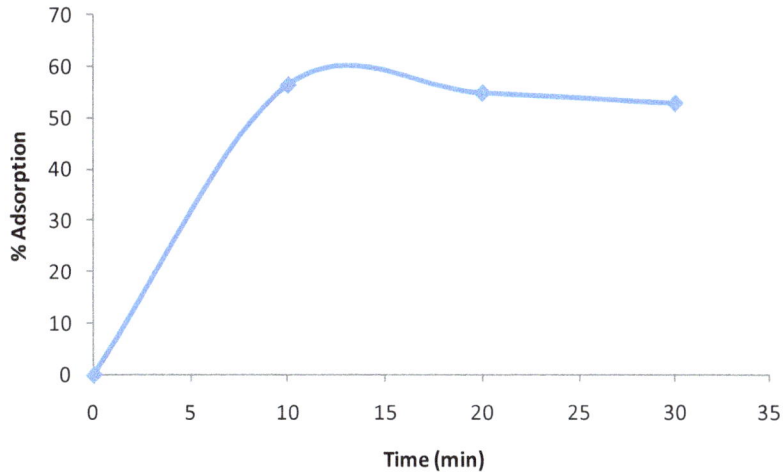

Figure 14. Effect of contact time on adsorption of Congo Red onto layered double hydroxide.

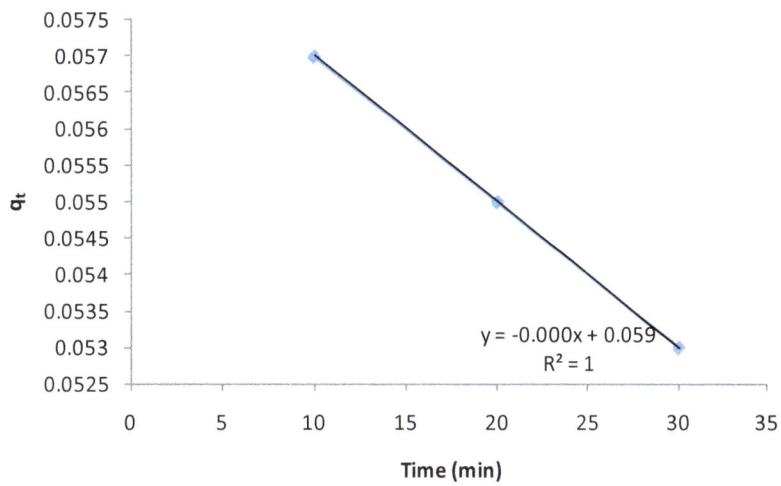

Figure 15. Plot of q_t vs. t for the adsorption of Congo Red onto layered double hydroxide.

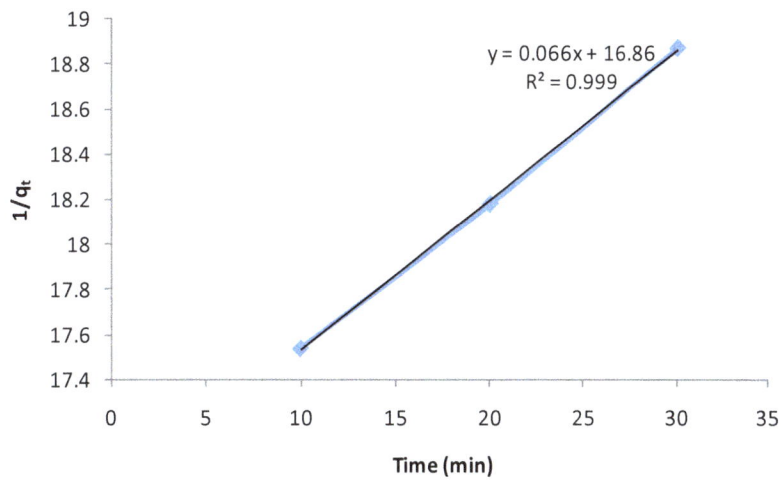

Figure 16. Plot of $1/q_t$ vs. t for the adsorption of Congo Red onto layered double hydroxide.

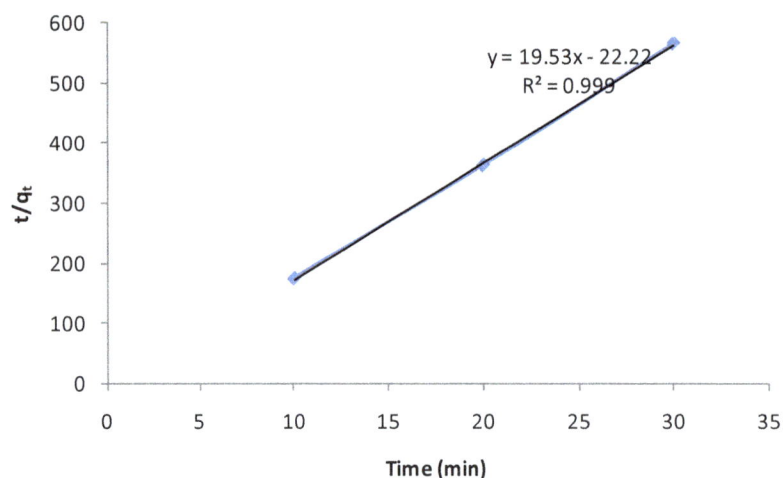

Figure 17. Plot of $1/q_t$ vs. t for the adsorption of Congo Red onto layered double hydroxide.

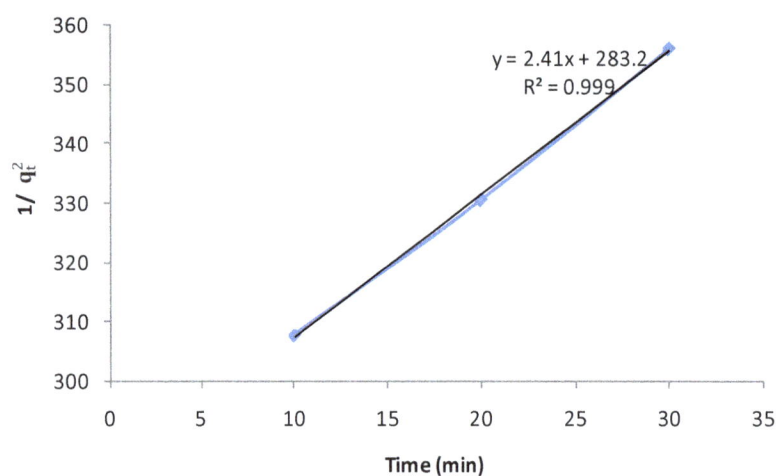

Figure 18. Plot of $1/q_t^2$ vs. t for the adsorption of Congo Red onto layered double hydroxide.

Congo Red dye in aqueous solution. The experimental data were best defined by Temkin isotherm ($R^2 = 1$) and zero-order kinetic model ($R^2 = 1$). The values of ΔH^o and ΔS^o indicated that the adsorption process was endothermic and process was dependent on increase in temperature, thereby increasing the randomness of the solid/liquid phase of the reaction system.

References

[1]   Gupta, V.K., Mittal, A., Krishnan, L. and Gajbe, V. (2004) Adsorption Kinetics and Column Operations for the Removal and Recovery of Malachite Green from Wastewater Using Bottom Ash. *Separation and Purification Technology*, **40**, 87-96. http://dx.doi.org/10.1016/j.seppur.2004.01.008

[2]   Singh, K. and Arora, S. (2011) Removal of Synthetic Textile Dyes from Wastewater: A Critical Review on Present Treatment Technologies. *Critical Reviews in Environmental Science and Technology*, **41**, 807-878. http://dx.doi.org/10.1080/10643380903218376

[3]   Anjaneyulu, Y., Chary, N.S. and Raj, D.S.S. (2005) Decolourization of Industrial Effluents-Available Methods and Emerging Technologies. *Reviews in Environmental Science and Biotechnology*, **4**, 245-273. http://dx.doi.org/10.1007/s11157-005-1246-z

[4]   Prasad, A.L. and Santhi, T. (2012) Adsorption of Hazardous Cationic Dyes from Aqueous Solution onto *Acacia nilotica* Leaves as an Eco Friendly Adsorbent. *Sustainable Environment Research*, **22**, 113-122.

[5]   Gupta, V.K., Pathania, D., Singh, P., Kumar, A. and Rathore, B.S. (2014) Adsorptional Removal of Methylene Blue by Guar Gum-Cerium (IV) Tungstate Hybrid Cationic Exchanger. *Carbohydrate Polymers*, **101**, 684-691. http://dx.doi.org/10.1016/j.carbpol.2013.09.092

[6]   Ghaedi, M., Hassanzadeh, A. and Nasiri Kokhdan, S. (2011) Multiwalled Carbon Nanotubes as Adsorbents for the Kinetic and Equilibrium Study of the Removal of Alizarin Red S and Morin. *Journal of Chemical and Engineering*, **56**, 2511-2520.

[7]   Yao, Z., Wang, L. and Qi, J. (2009) Biosorption of Methylene Blue from Aqueous Solution Using a Bioenergy Forest Waste: Xanthocerassorbifolia Seed Coat. *Clean*, **37**, 642-648.

[8]   Abd Ei-Latif, M.M., Ibrahim, A.M. and EI-Kady, M.F. (2010) Adsorption Equilibrium, Kinetics and Thermodynamics of Methylene Blue from Aqueous Solutions Using Biopolymer Oak Sawdust Composite. *Journal of American Science*, **6**, 267-283.

[9]   Vimonses, V., Lei, S., Jin, B., Chow, C.W.K. and Saint, C. (2009) Kinetic Study and Equilibrium Isotherm Analysis of Congo Red Adsorption by Clay Materials. *Chemical Engineering Journal*, **148**, 354-364. http://dx.doi.org/10.1016/j.cej.2008.09.009

[10]   Sen, T.K., Afroze, S. and Ang, H.M. (2011) Equilibrium, Kinetics and Mechanism of Removal of Methylene Blue from Aqueous Solution by Adsorption onto Pine Cone Biomass of Pinus Radiate. *Water, Air, Soil Pollution*, **218**, 499-515. http://dx.doi.org/10.1007/s11270-010-0663-y

[11]   Zenasni, M.A., Meroufel, B., Amrouche, A., Naâr, F., Merzouka, F. and Difallah, F. (2012) Phytoremédiation de Zn (II) par les racines de Calotropis Procera (Bechar, Algérie). *ScienceLib*, **4**, 1-15.

[12]   Dawood, S. and Kanti Sen, T. (2012) Removal of Anionic Dye Congo Red from Aqueous Solution by Raw Pine and Acid-Treated Pine Cone Powder as Adsorbent: Equilibrium, Thermodynamic, Kinetics, Mechanism and Process Design. *Water Research*, **46**, 1933-1946. http://dx.doi.org/10.1016/j.watres.2012.01.009

[13]   Zenasni, M.A., Benfarhi, S., Mansri, A., Benmehdi, H., Meroufel, B., Desbrieres, J. and Dedriveres, R. (2011) Influence of pH on the Uptake of Toluene from Water by the Composite Poly(4-vinylpyridinium)-Maghnite. *African Journal of Pure and Applied Chemistry*, **5**, 486-493.

[14]   Hsu, L.C., Wang, S.L., Tzou, Y.M., Lin, C.F. and Chen, J.H. (2007) The Removal and Recovery of Cr(VI) by Li/Al Layered Double Hydroxide (LDH). *Journal of Hazardous Materials*, **142**, 242-249. http://dx.doi.org/10.1016/j.jhazmat.2006.08.024

[15]   Hu, Q.H., Xu, Z.P., Qiao, S.Z., Haghseresht, F., Wilson, M., Lu, G.Q., *et al.* (2007) A Novel Color Removal Adsorbent from Heterocoagulation of Cationic and Anionic Clays. *Journal of Colloid and Interface Science*, **308**, 191-199. http://dx.doi.org/10.1016/j.jcis.2006.12.052

[16]   Roto, R., Nindiyasari, F. and Tahir, I. (2009) Removal of Hexacyanoferrate (II) Using Zn-Al-OA Hydrotalcite as an Anion Exchanger. *Journal of Physical Science*, **20**, 73-84.

[17]   Iyi, N., Ebina, Y. and Sasaki, T. (2011) Synthesis and Characterization of Water-Swellable LDH (Layered Double Hydroxide) Hybrids Containing Sulfonate-Type Intercalant. *Journal of Materials Chemistry*, **21**, 8085-8095. http://dx.doi.org/10.1039/c1jm10733j

[18]   Ayawei, N., Ekubo, A.T., Wankasi, D. and Dikio, E.D. (2015) Synthesis and Application of Layered Double Hydroxide for the Removal of Copper in Wastewater. *International Journal of Chemistry*, **7**, 122-132.

[19]   Ayawei, N., Ekubo, A.T., Wankasi, D. and Dikio, E.D. (2015) Equilibrium, Thermodynamic and Kinetic Studies of the Adsorption of Lead (II) on Ni/Fe Layered Double Hydroxide. *Asian Journal of Applied Sciences*, **3**, 207-217.

[20]   Vadi, M. and Rahimi, M. (2014) Langmuir, Freundlich and Temkin Adsorption Isotherms of Propranolol on Multi-Wall Carbon Nanotube. *Journal of Modern Drug Discovery and Drug Delivery Research*, V1I3.

[21]   Yasin, Y., Malek, A.H.A. and Sumari, S.M. (2010) Adsorption of Eriochrome Black Dye from Aqueous Solution onto Anionic Layered Double Hydroxides. *Oriental Journal of Chemistry*, **26**, 1293-1298.

[22]   Badreddine, M., Legrouri, A., Barroug, A., De Roy, A. and Besse, J.P. (1999) Ion Exchange of Different Phosphate Ions into the Zinc-Aluminium-Chloride Layered Double Hydroxide. *Materials Letters*, **38**, 391-395. http://dx.doi.org/10.1016/S0167-577X(98)00195-5

[23]   Arivoli, S., Venkatraman, B.R., Rajachandrasekar, T. and Hema, M. (2007) Adsorption of Ferrous Ion from Aqueous Solution by Low Cost Activated Carbon Obtained from Natural Plant Material. *Research Journal of Chemistry and Environment*, **17**, 70-78.

[24]   Dawodu, F.A., Akpomie, G.K. and Ogbu, I.C. (2012) Isotherm Modeling on the Equilibrium Sorption of Cadmium (II) from Solution by Agbani Clay. *International Journal of Multidisciplinary Sciences and Engineering*, **3**, 9-14.

[25]   Choy, K.K.H., Mckay, G. and Porter, J.F. (1999) Sorption of Acidic Dyes from Effluents Using Activated Carbons. *Resources, Conservation and Recycling*, **27**, 57-71. http://dx.doi.org/10.1016/S0921-3449(98)00085-8

[26] Flavio, A.P., Silvio, L.P.D., Ede, C.L. and Edilson, V.B. (2008) Removal of Congo Red from Aqueous Solution by Anilinepropylsilica Xerogel. *Dyes and Pigments*, **76**, 64-69. http://dx.doi.org/10.1016/j.dyepig.2006.08.027

[27] Ahmet, G. (2009) Synthesis and Characterization of Hybrid Congo Red from Chloro-Functionalized Silsesquioxanes. *Turkish Journal of Chemistry*, **34**, 437-445.

[28] Bulut, Y. and Aidin, H. (2006) A Kinetic and Thermodynamics Study of Methylene Blue Adsorption on Wheat Shells. *Desalination*, **194**, 259-267. http://dx.doi.org/10.1016/j.desal.2005.10.032

[29] Ayawei, N., Ekubo, A.T., Wankasi, D. and Dikio, E.D. (2015) Mg/Fe Layered Double Hydroxide for Removal of Lead (Ii): Thermodynamic, Equilibrium and Kinetic Studies. *European Journal of Science and Engineering*, **3**, 1-17.

[30] Ayawei, N., Ekubo, A.T., Wankasi, D. and Dikio, E.D. (2015) Adsorption Dynamics of Copper Adsorption by Zn/Al-CO$_3$. *IJACSA*, **3**, 57-64.

6

Products and Kinetics of the Reaction of Monomeric Target Bis-(Acetylacetonato) Copper(II) with Transmetalator Bis-(Diethoxydithiophosphato) Zinc(II) in Methylene Chloride

Hisham A. Abo-Eldahab[1,2]

[1]Chemistry Department, Faculty of Science, Alexandria, Egypt
[2]Umm Al-Qura University, University College, Makkah, KSA
Email: hdahab-41@hotmail.com

Abstract

Reaction of bis-(acetylacetonato) copper(II) (A) with transmetalator $Zn((EtO)_2PS_2)_2$ (B_2; Et = ethyl) in methylene chloride is a simple irreversible second-order process over a wide temperature range which is the first example of a second-order reaction of mononuclear target A with a transmetalator. The plots of k_{obsd} vs [A] are linear, meaning that there is one A and one B_2 in the activated complex of the slowest reaction step. The slowest step is precursor formation on the basis that B_2 is an exceptionally weak complex. The product of the A/B_2 reaction is the strong successor complex $Zn(acac)_2 \cdot Cu(ps)_2$. The data are compared with those for reactions of the same target (A) with S-methyle isopropylidenehydrazinecarbodithioate-carbodithioato-metal(II) complexes $M(SN)_2$ (M=Ni (C_1) and Zn (C_2)). The reaction is not like that of A with Ni $((MeO)_2PS_2)_2$ (B_1; Me=methyl), because it is irreversible and also consistent with the measured lower relative thermodynamic stability of B_2 compared to B_1.

Keywords

Kinetics, Mechanism, Transmetalation, Transmetallators, Thermodynamics

1. Introduction

Transmetalation is the stoichiometric replacement of the metals in a polymetallic target with other metals from reagents called transmetalotors [1]. It is a source of many new heteropolymetallic molecules that cannot be obtained by other means [2]. The transmetalation phenomenon has mostly been applied to polynuclear copper(I) [3]-[5] and copper(II) [1] [5]-[13] targets, although it is also applicable to targets containing other transition metals [10]. The best known transmetalators are S-methyl hydrazinecarbodithioate complexes M(NS)₂, where M is Co, Ni, Cu and Zn and NS is monoanionic S-methyl isopropylidenehydrazinecarbodithioate and S-methyl benzyllidenehydrazinecarbodithioate in reagents a and b, respectively. Transmetalation reactions proceed under mild conditions in aprotic solvents and the heteropolymetallic products are easily separated [1] [2].

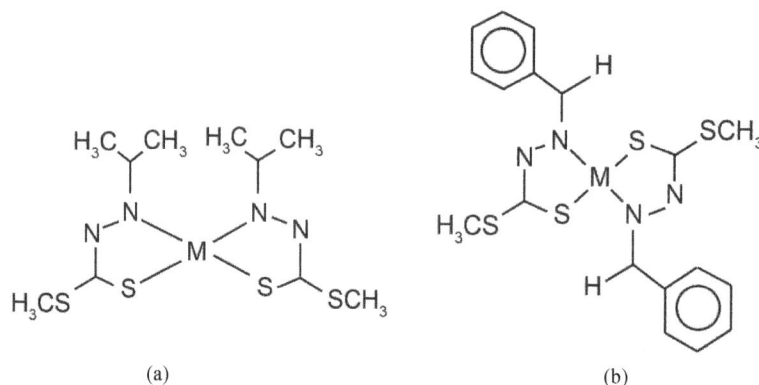

(a)   (b)

The major driving force for stoichiometric copper replacement is the formation of highly stable co-products $Cu(NS)_{(S)}$ and $Cu(NS)_2$ as typified by Equations (1) [5] and (2) [7], where the transmetalator is A; L is an N,N,N',N'-tetraalkyldiamine; N is a mohodentate pyridine ligand and X is Cl or Br.

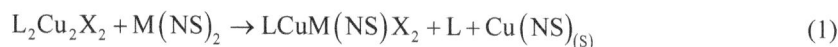

$$L_2Cu_2X_2 + M(NS)_2 \rightarrow LCuM(NS)X_2 + L + Cu(NS)_{(S)} \tag{1}$$

$$N_4Cu_4Cl_6O + xM(NS)_2 \rightarrow N_4Cu_4 \cdot xM_xCl_6O + xCu(NS)_2 \tag{2}$$

Extensive studies have shown that the patterns and specificity of progressive transmetalation reactions strongly depend on the target core structure [2].

Practical transmetalation reactions have the following characteristics:

1. The targets are easy to make and purify.

2. The reactions proceed at high rates under mild conditions in common solvents.

3. The products are in stoichiometric steps, as in Equation (3) with x = 1 - 4.

$$\left(\mu_4\text{-O}\right)N_4Cu_4Cl_6 + xM(NS)_2 \rightarrow \left(\mu_4\text{-O}\right)N_4Cu_4 \cdot xM_xCl_6 + xCu(NS)_2 \tag{3}$$

4. The desired heteropolymetallic products are easy to isolate.

5. They are characterizable solids that exist as members of families containing different metals M in different proportions.

Transmetalation chemistry is being developed and applied in four major areas:

1. The synthesis of heteropolymetallic molecule families as a means of understanding the structures of catalytic and non-catalytic homopolymetallic targets that decomposes on attempted crystallization. We have shown that transmetalation alters the rates and rate laws of copper-catalyzed reactions [1] [14] and gives valuable information about catalyst structures [15].

2. Application of the transmetalation phenomenon to unexplored elements. This requires new labile polymetallic targets and/or transmetalators containing the elements of interest. Until recently, transmetatation chemistry was restricted to the elements Fe [5], Co [8], Ni [1] [11], Cu [10], Zn [13] [16], Cd [14], Hg [14] and Sn [14] [17], because only these elements form useful transmetalators with excellent transmetalator ligand S-methyl isopropylidenehydrazinecarbodithioate (NS) [2]. However, we have now discovered that bis-(dialkoxydithiophosphato)-metal complexes $((RO)_2PS_2)_nM$ (abbreviated M(PS)ₙ) are transmetalators of copper targets. Application of transmetalation to new elements is likely to be successful because most of the metallic elements form dithiophosphato-complexes [18].

3. Thermal and electrochemical [17] [18] conversion of heteropolymetallic transmetalation products to bulk and supported metals, alloys and mixed metal oxides. The development of transmetalation chemistry thus goes hand in hand with new ways of making useful materials and catalysts.

4. Studies of transmetalation mechanisms. Transmetalation is remarkable because (a) direct transmetalation as in Equation (1) leaves the rest of the target unchanged [2] [3] [6] [7]; (b) targets react selectively with mixtures of transmetalators [8$_b$]; and (c) different metals in a heteropolymetallic target are specifically replaced: for example, Zn(NS) specifically replaces copper in targets $(\mu_4-O)N_4Cu_{4-x}M_XCl_6$ (x = 1 - 4) even though replacement of M also is thermodynamically favorable [9] [19]. This specifically has been traced kinetically to specific interaction of the transmetalator with the metal center that is replaced [20]. The interaction occurs in precursors $TM \cdot T_n$ through Equation (4), where TM is the transmetalator; T is the target, n is 1 or 2 and β_n can range from very small [22] to moderate and measurable [22] to very large [21] [22]. It also has been found that different forms of precursors with the same stoichiometry $TM \cdot T$ can exist in a given system at different temperatures [20]. However, the interaction is complicated by apparent involvement of more than one metal center of a polymetallic target with the transmetalator [20].

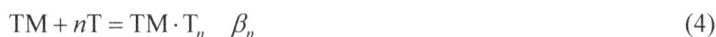

$$TM + nT = TM \cdot T_n \quad \beta_n \tag{4}$$

That is why we turned to bis-(acetylacetonato) copper(II) (A) as a model for the metal centers in polymetallic targets [22] [23]. Our work showed that:

1) A is capable of forming both moderately strong and very strong precursors ((n = 1) Equation (4) with different $M(NS)_n$ transmetalators) [22].

2) Different forms of TM·T precursors can be detected at different temperatures [22] [23] and the thermodynamic data can be correlated with data for precursor formation in irreversible polymetallic transmetalation systems [20]. This comparison shows that specific transmetalation occurs because:

Precursor formation is cooperative in the sense that no single property of T or TM determines the extent of precursor interaction or the rate of metal exchange [23].

3) The $A/M(NS)_2$ systems are reversible, which enable us to compare the thermodynamics and other properties of precursor and successor complexes [22] [23].

4) Comparison of reaction profiles for irreversible and reversible $A/M(NS)_2$ systems enables sub-classification of the systems according to the slow step (precursor formation thermodynamics, metal exchange kinetics and successor complex dissociation) [23].

With the promise of much new transmetalation chemistry based on $M(PS)_n$ tansmetalators on the horizon [18], though to provide a mechanistic comparison with established $M(NS)_n$ transmetalation systems by studying the products and kinetics of reaction of target A with transmetalator $(MeO)_2PS_2)_2Ni$ (B$_1$) [24]. The results show that:

1) the A/B$_1$ system is reversible and involves more stable reaction precursors A·B$_1$ than the A/Ni(NS)$_2$ (A/C) system [22] [25];

2) different precursors A·B$_1$ and different successor complexes $Ni(acac)_2 \cdot Cu(PS)_2$ exist at different temperatures;

3) the first thermodynamic data for precursors B$_1 \cdot$A$_2$ (n = 2 in Equation (2)) can be obtained;

4) precursors, activated complex and successor complex formation involve much lower entropy changes in the A/B$_1$ system than in the A/C system.

2. Experimental

2.1. Materials

2.1.1. Target, Solvent and Reaction Environment
Bis-(acetylacetonato) copper(II) (A) (Alfa) was recrystallized from methylene chloride/diethylether. Methylene chloride was dried with H_2SO_4, stirred with anhydrous sodium carbonate, freshly distilled from P_2O_5 and stored in the dark over anhydrous sodium carbonate. high purity dinitrogen was deoxygenated by passage through a freshly activated column of Alfa DE-OX catalyst.

2.1.2. Synthesis of Transmetalator, B$_2$
Bis(diethoxydithiophosphato) zinc(II) (Zn(PS)$_2$, B$_2$) was synthesized by a modified form of the literature method [2]. B$_2$ thermally unstable and very air sensitive. It was therefore made at low temperature from reactions

Equations (5) and (6).

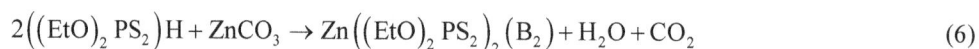

$$P_4S_{10} + 8EtOH \rightarrow 4\left(\left(EtO\right)_2 PS_2\right)H + 2H_2S \tag{5}$$

$$2\left(\left(EtO\right)_2 PS_2\right)H + ZnCO_3 \rightarrow Zn\left(\left(EtO\right)_2 PS_2\right)_2\left(B_2\right) + H_2O + CO_2 \tag{6}$$

In a typical experiment, 12.7 gm P_4S_{10} (57.0 mmol) was placed in a 200 mL round-bottom flask fitted with a dinitrogen inlet, magnetic stirrer and reflux condenser. Deoxygenated anhydrous methanol (75 mL) was added and the mixture was refluxed under flowing dinitrogen until H_2S could no longer be detected in the effluent with lead acetate on damp filter paper (ca 45 min). At this point, $ZnCO_3$ (120 mmol) was added and the mixture was stirred under dinitrogen in the ice bath for 20 min. Pentane (50 mL) was then added and, after filtration, the filtrate was then pumped to dryness in a vacuum rotary evaporator. The white solid $Zn((EtO)_2PS_2)_2$ (B_2) obtained melts sharply at 74°C and has a broad resonance in its ^{31}P NMR spectrum at 93.0 ppm.

2.1.3. Physical Measurements

Details of our procedures for product separation and analysis, kinetic measurements and data analysis can be found in the previous work [20]-[23] [26]. The relative thermodynamic stabilities of $Ni(PS)_2$ (B_1), $Zn(PS)_2$ (B_2), $Ni(NS)_2$ (C_1) and $Zn(NS)_2$ (C_2) in ethanol were established by visual inspection and spectrophotometric measurements following established procedures at room temperature.

The kinetic measurements were made with a DEC PRO380 computer-assisted Hi-Tech SFL41 stopped-flow spectrophotometer over the temperature range −25.0°C to 30.0°C controlled to ±0.05°C. The reaction of A with B_2 was monitored at 659 nm. The concentration ranges were [A] = 2.50 to 50.0 mM and [B_2] = 0.50 mM. All experiments with transmetalator $Zn(PS)_2$ (B_2) were conducted under dinitrogen. The concentration of A was always sufficient to ensure pseudo-first-order conditions. The pseudo-first-order rate constant k_{obsd} at fixed [A], [B], wavelength and temperature was obtained from the slope of a plot of $\ln(A_\infty - A_t)$, where A_t is the absorbance at time t. Each run was repeated at least three times to give a maximum error in each reported rate constant of ±4%.

3. Results and Discussion

3.1. General Observation

Practical transmetalation reactions are irreversible and stoichiometric, proceed rapidly under mild conditions and give easily separated heteropolymetallic products. These considerations require that useful transmetalation systems have a large driving force and that the reactants be kinetically labile [2]. Polynuclear copper transmetalation targets that satisfy these requirements can contain bidentate N,N,N',N'-tetraalkylamine ligands L, Equation (7) [5] [21]

$$L_2Cu_2X_2 + M\left(NS\right)_2 \rightarrow LCuM\left(NS\right)X_2 + L + Cu\left(NS\right)_{(S)} \tag{7}$$

and even an anionic oxo-pyridine ligand that bridges the target metal centers [3] [4] [20]. The relative stabilities of transmetalators $M(NS)_2$ are useful guide to their relative reactivity's with a given target. Previous measurements show that $Cu(NS)_2$ is much more thermodynamically stable than transmetalation product $Ni(acac)_2$ [27].

3.2. Reactant Structures

Target A is a flat, neutral molecule [25] with only 0-donor atoms. Reactant $Ni(PS)_2$ (B_1) is a diamagnetic, flat molecule with the CH_3O groups in a plane perpendicular to the N_iS_4 plane and a center of symmetry [25]. Reactant $Zn(PS)_2$ (B_2) is a diamagnetic, flat molecule with the C_2H_5O groups in a plane perpendicular to the N_iS_4 plane and a center of symmetry [25]. Transmetalator $Ni(NS)_2$ (C) is a flattened tetrahedron (the dihedral angle is 27^0) with a *cis*-geometry [28]. The reactant core structures are shown in (**Figure 1**).

The essential differences between reactants B and C are as follows:

1) B contains four-membered rings while C contains five-membered rings. Electron delocalization in the rings of B is indicated by high sensitivity of the ^{31}P spectrum to the identity of M in $M(PS)_2$ [28].

2) Transmetalation requires the transfer of PS or NS ligands from the transmetalator metal to the target metal (see Equation (7)). Only M-S bonds have to be broken for this purpose in B, but the NS rings in C can, in prin-

(a)

(b)

(c)

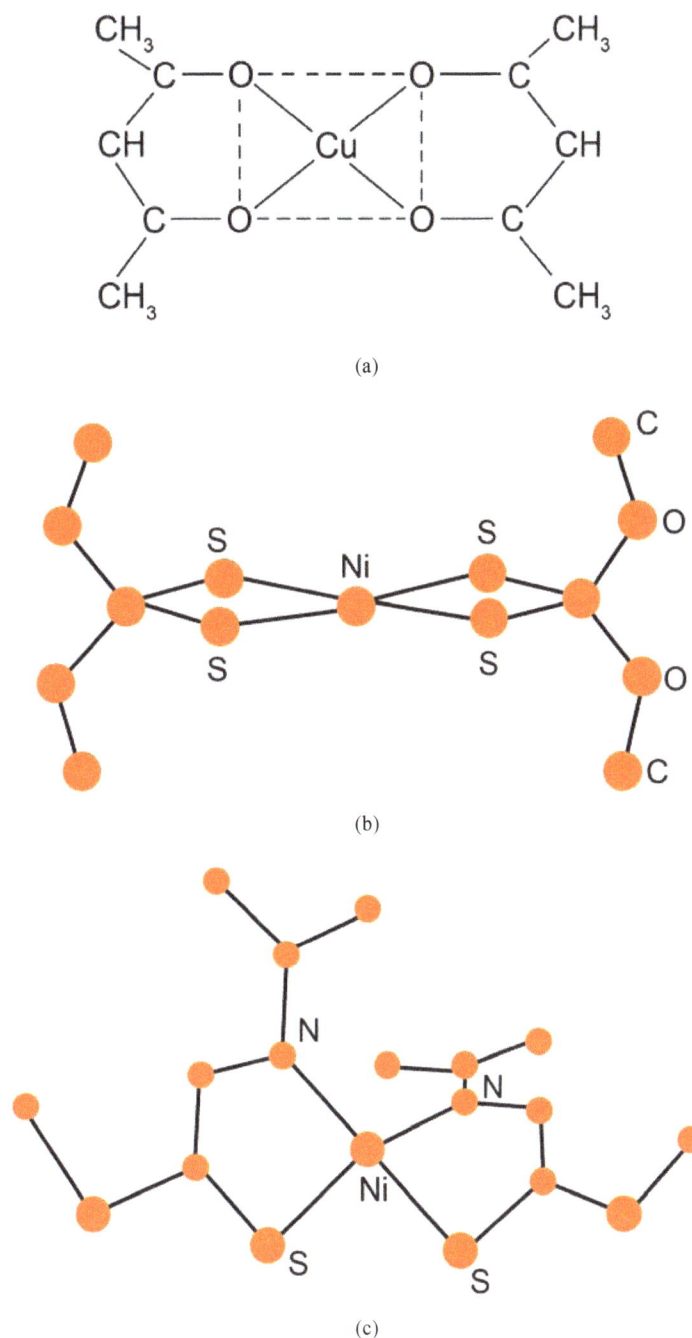

Figure 1. Core structures of target (a); reagent (b); and reagent (c).

ciple, open by Ni-N or Ni-S bond breaking.

3) Reactants B has much lower thermodynamic stability than reactant C [28].

3.3. Transmetalation Steps

Our current view of the sequence of steps in transmetalation reactions is based on a great deal of kinetic information [22] [23] supplemented by very recent structural insights [18]. It is summarized in **Scheme 1**, where X are the outermost framework atoms of the copper target, M is the transmetalator metal and NS is the transmetalator ligand.

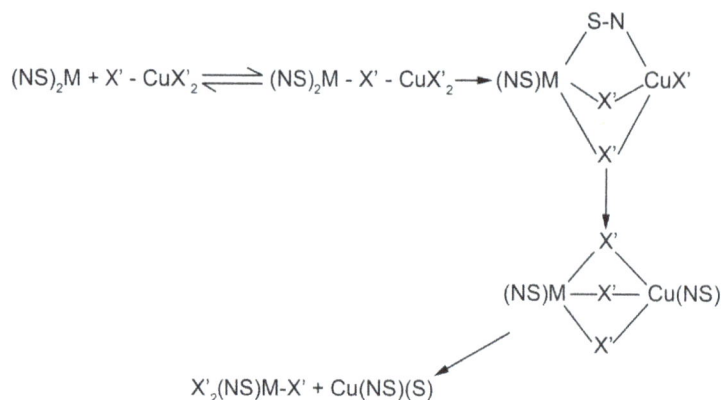

$$(NS)_2M + X' - CuX'_2 \rightleftharpoons (NS)_2M - X' - CuX'_2 \longrightarrow$$

Scheme 1. The sequence of steps in transmetalation reactions.

The transmetalator metal M first interacts with the target atoms X. This increases the coordination number of M, whose NS chelate rings broken in this process and the freed N atoms are coordinated by the target metal center. The extent of this sharing of NS between M and the target metal determines the stoichiometry, strength and character of the precursors formed in Equation (2). The strength of M bonds to target X has now increased. The M-S bonds slowly break and the target metal chelates the transferring NS ligand. The final step is loss of $Cu(NS)_2$ from the original copper(II) target. The net result of transmetalation is the replacement of target metal-X bonds with M-X bonds.

This picture of transmetalation is supported by proof of the existence of precursor structures I and II in the reaction of $Sn(NS)_2Cl_2$ with target $[NCuCl]_4$ (N is N,N-diethylnicotinamide) and the isolation of product III from reaction of equimolar $Sn(NS)_4$ with the same target [19].

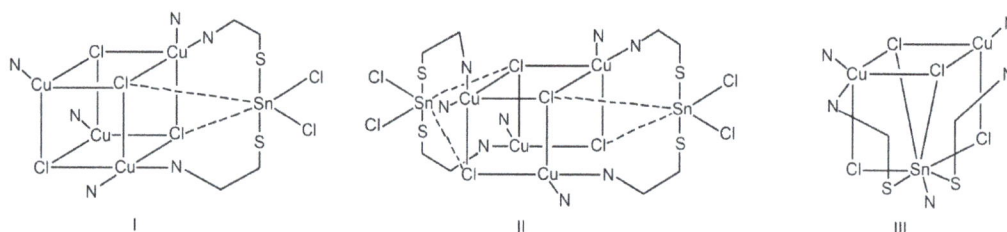

3.4. Kinetics of the Reaction of Cu(acac)₂, (A) with Zn((EtO)₂PS₂)₂, (B₂) in Methylene Chloride

Zinc transmetalator, B_2 is much less thermodynamically stable than nickel transmetalator, B_1 so we should expect B_2 to react more rapidly than B_1 with a target like A. This is borne out by the experimental data, which are collected in **Table 1**. The reaction of A with B_2 is a second-order, irreversible process, as demonstrated in **Figure 2**. The order is the same over a wide temperature range, as demonstrated by the linear plot in **Figure 3**.

3.5. Interpretation of the Data and Comparison with Other Second-Order Transmetalation Systems

Although many polymetallic target transmetalation reactions are second-order [20] [21], this is the first example of a second-order reaction of mononuclear target A with a transmetalator. The other systems either saturate or proceed at rates which are independent of target concentration [A] [22] [23].

3.6. There Are Two Important Characteristics of the A/B₂ System

First, the plots of k_{obsd} vs [A] are linear, which means that there is one A and one B_2 in the activated complex for the slowest reaction step. At the same time, this linearity indicates that the slow step does not involve significant

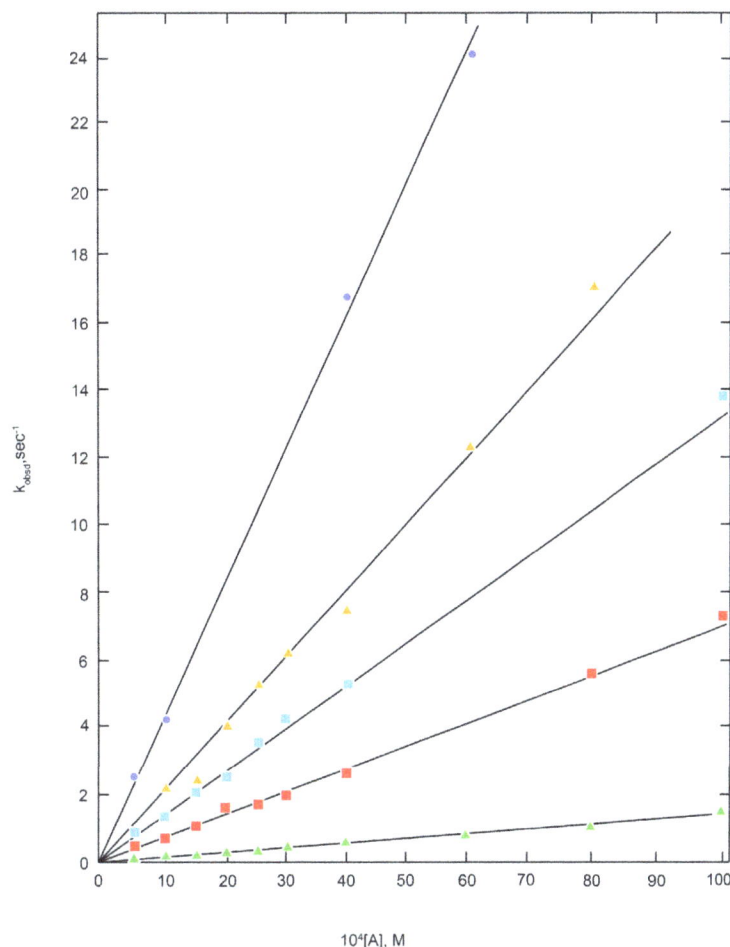

10⁴[A], M

Figure 2. Plots of k_{obsd} vs [A] for the reaction of A with $Zn(PS)_2$ B_2 in methylene chloride at the following temperatures (˚C).

Table 1. Kinetic data for the reaction of target A with $Zn(PS)_2$ (B_2) in methylene chloride.

Temperature	$10^{-2}k_2$, $M^{-1}sec^{-1}$	$\Delta H_2^{\#a}$	$\Delta S_2^{\#b}$
−25.0	1.4		
−9.0	6.9		
0.0	12.6	10.7 ± 0.4	5 ± 5
10.0	21.4		
20.0	39.8		

[a]Units are kcalmol^{-1}; [b]Units are caldeg^{-1}mol^{-1} at 25.0˚C.

proportions of reaction precursors A·B_2. Thus, the slow step is either precursor formation followed by very rapid metal exchange or slow metal exchange proceeding through a very weak precursor.

Second, the reaction is not like that of A with B_1 because it is irreversible. $Zn(PS)_2$ (B_2) is one of the very least stable $M(PS)_2$ complexes, so this is not too surprising because the product of the A/B_2 reaction is the strong successor complex $Zn(acac)_2$·$Cu(PS)_2$.

It is worth recalling 1) that $Zn(NS)_2$ also is a very weak complex compared To $Ni(NS)_2$ and $Cu(NS)_2$ [9] and 2) that it reacts with polynuclear target $(\mu_4$-O)$N_4Cu(Ni(H_2O))_3Cl_6$ via precursors with moderate and measurable equilibrium constants B_1. By contrast, the reaction of the same target $(\mu_4$-O)$N_4Cu(Ni(H_2O))_3Cl_6$ with $Ni(NS)_2$ is

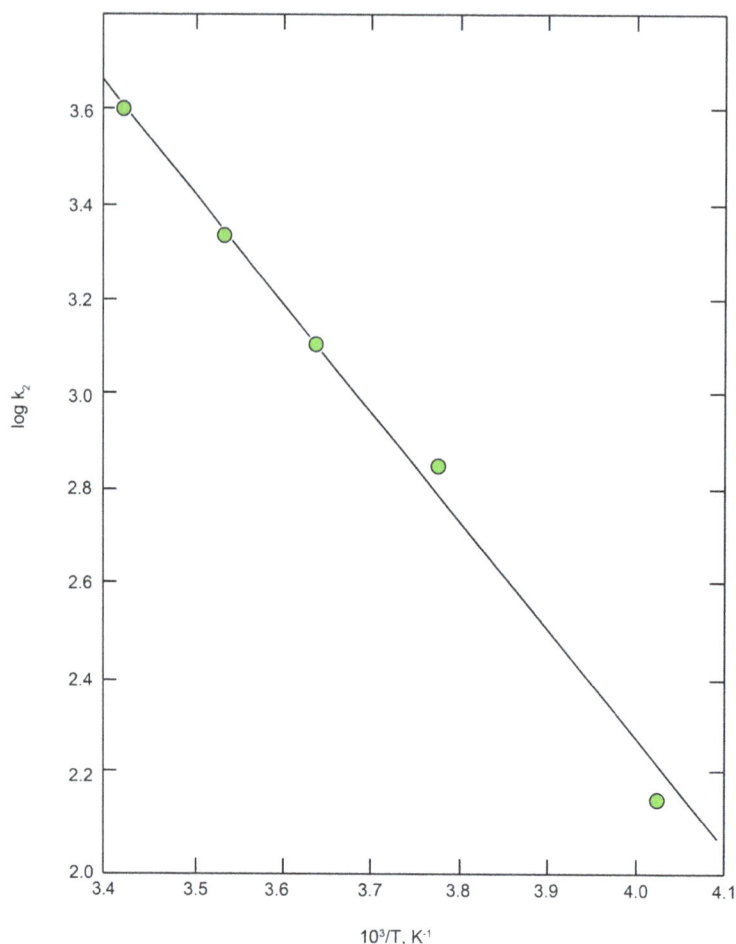

Figure 3. Plot of log k_2 vs 1/T for the reaction of A with Zn(PS)$_2$ B$_2$ in methylene chloride.

third-order and must therefore involve very weak precursors [20].

An ability of zinc transmetalators to ring-open to form more stable precursors than nickel reagents would lead us to expect either the same kind of behavior as in the reaction of A with B$_2$ or a situation where the precursor was so stable that the reaction rate is independent of [A] [22] [23]. However, the second-order rate law for the reaction of A with zinc reactant B$_2$ leads us strongly to suspect that the slow step is precursor formation followed by very rapid, irreversible metal exchange.

Figure 6 of Ref. [21] shows activation parameter correlations for second-order transmetalations of polynuclear copper(II) targets with reactants C and the corresponding S-methylbenzylidenehydrazinecarbodithioate transmetalators D. The latter are generally more thermodynamically stable than C with the same M [10]. The lowest line in that figure M [21] correlates six kinetic data pairs, four of which are for reactants of D (M = Zn) with targets (μ_4-O)N$_4$CuX6 (X = Cl or Br) in methylene chloride or nitrobenzene. The data pair ($\Delta H_2^{\#}$ = 10.7 kcalmol^{-1}, $\Delta S_2^{\#}$ = 5 cal deg^{-1}mol^{-1}) for the reaction of A with Zn(PS)$_2$ (B$_2$) fits this lower line. This suggests that the seven correlated data sets refer to rate-determining precursor formation in these transmetalation systems. All are irreversible and the key to this assignment is the apparent slow step in the A/B$_2$ system.

4. Conclusion

M(PS)$_2$ complexes B is less thermodynamically stable than the corresponding M(NS)$_2$ complexes C. On this basis we might expect them to be faster transmetalators than reagents M(NS)$_2$ (M common in B or C) for a given target. Indeed, Zn(PS)$_2$ (B$_2$) seems to transmetalate A via rate-determining precursor formation. However, my work shows that Ni(PS)$_2$ (B$_1$) is a stronger precursor former than C$_1$ with A and that this results in lower metal exchange rates. Nevertheless, $M(PS)_2$ transmetalation systems might have even greater specificity than that having been found with S-methyl isopropylidenehydrazinecarbodithiothioate ligands [9] [20] [21] [29]. This possibility and the reactions of dithiophosphate transmetalators with other monomeric and polymetallic targets should be explored in the future work.

Acknowledgements

I would like to thank Chemistry Department Faculty of Science, Alexandria University, for support and Professor Mohamed A. El-Sayed for valuable discussions and comments.

References

[1] El-Toukhy, A., Cal, G.Z., Davies, G., Gilbert, T.R., Onan, K.D. and Veidis, M. (1984) Transmetalation Reactions of Tetranuclear Copper(II) Complexes. 2. Stoichiometry and Products of Reaction of [(DENC)CuCl]$_4$O$_2$, [(DENC)CuCl]$_4$(CO$_3$)$_2$, [(DENC)CuCl]$_4$Cl$_4$, and (DENC)$_4$Cu$_4$Cl$_6$O (DENC = N,N-Diethylnicotinamide) with Ni(NS)$_2$, the Kinetics of Product Isomerization in Aprotic Solvents, and Inhibition of Copper-Catalyzed Phenolic Oxidative Coupling by Dioxygen through Transmetalation. *Journal of the American Chemical Society*, **106**, 4596-4605. http://dx.doi.org/10.1021/ja00328a050

[2] Davies, G., El-Sayed, M.A. and El-Toukhy, A. (1989) Transmetalation: A New Route to Heteropolymetallic Molecules and Materials. *Comments on Inorganic Chemistry*, **8**, 203-220. http://dx.doi.org/10.1080/02603598908035795

[3] Cai, G.Z., Davies, G., El-Sayed, M.A., El-Toukhy, A., Gilbert, T.R., Karlin, K.D. and Zubieta, J., Eds. (1986) Biological and Inorganic Copper Chemistry, Vol. 2. Adenne, Guilderland, 237.

[4] Davies, G., El-Sayed, M.A., El-Toukhy, A., Gilbert, T.R. and Nabih, K. (1986) Transmetalation of Tetranuclear Copper Complexes. 5. Transmetalation of Copper(I) Complexes and Stoichiometry and Kinetics of Oxidation of Neutral Tetranuclear (DENC)$_3$Cu$_3$M(NS)X$_4$ (DENC = N,N-Diethylnicotinamide) Complexes by Dioxygen in Aprotic Solvents. *Inorganic Chemistry*, **25**, 1929-1934. http://dx.doi.org/10.1021/ic00232a004

[5] Davies, G., El-Kady, N., El-Sayed, M.A., El-Toukhy, A. and Schure, M.R. (1988) The Kinetics of Primary Events in the Reaction of L$_2$Cu$_2$X$_2$ Complexes (L is an *N,N,N'N'*-Tetraalkyldiamine, X is Cl or Br) with M(NS)$_2$ Reagents. *Inorganica Chimica Acta*, **149**, 31-43. http://dx.doi.org/10.1016/S0020-1693(00)90565-6

[6] Davies, G., El-Sayed, M.A., El-Toukhy, A., Henary, M. and Gilbert, T.R. (1986) Transmetalation of Tetranuclear Copper Complexes. 8. Transmetalation of Tetranuclear Copper(I) Complexes with a Co(NS)$_3$ Reagent. *Inorganic Chemistry*, **25**, 2373-2377. http://dx.doi.org/10.1021/ic00234a019

[7] Davies, G., El-Sayed, M.A. and El-Toukhy, A. (1986) Transmetalation of Tetranuclear Copper Complexes. 7. Spectral Evidence for the Substoichiometric Transmetalation of (.mu.4-O)[(DENC)Cu]$_4$X$_6$ Complexes (DENC = N,N-Diethylnicotinamide; X = Cl or Br) by a Bis(Acetone S-Methyl Hydrazonecarbodithioato)Nickel Reagent. *Inorganic Chemistry*, **25**, 2269-2271. http://dx.doi.org/10.1021/ic00233a036

[8] Henary, M., Davies, G., Abu-Raqabah, A., EI-Sayed, M.A. and El-Toukhy, A. (1988) Progressive Transmetalation of Tetranucleardioxocopper (II) Complexes with Cobalt Reagents. *Inorganic Chemistry*, **27**, 1872.

[9] Davies, G., El-Sayed, M.A., El-Toukhy, A., Henary, M. and Martin, C.A. (1986) Distinguishable Sites in Tetranuclear Oxocopper(II) Complexes (py)$_3$Cu$_4$Cl$_4$O$_2$ and (DENC)$_3$Cu$_3$M(H$_2$O)Cl$_4$O$_2$ (M = Co, Ni, Cu, Zn). *Inorganic Chemistry*, **25**, 4479-4487. http://dx.doi.org/10.1021/ic00245a007

[10] Davies, G., El-Sayed, M.A., El-Toukhy, A., Henary, M., Kassem, T.S. and Martin, C.A. (1986) Selective Transmetalation and Demetalation of Heteropolynuclear Metal Complexes. *Inorganic Chemistry*, **25**, 3904-3909. http://dx.doi.org/10.1021/ic00245a007

[11] Davies, G., El-Sayed, M.A. and El-Toukhy, A. (1986) Transmetalation of Tetranuclear Copper Complexes. 4. Structural Implications of the Kinetics of Direct Transmetalation of Tetranuclear Copper(II) Complexes by Ni(NS)$_2$ Reagents. *Inorganic Chemistry*, **25**, 1925-1928. http://dx.doi.org/10.1021/ic00232a003

[12] Davies, G., El-Sayed, M.A. and El-Toukhy, A. (1986) Transmetalation of Tetranuclear Copper Complexes. 9. Stoichiometry and Kinetics of Transmetalation of (μ_4-O)[NCu]$_4$X$_6$ Complexes by M(NS)$_2$ Reagents in Aprotic Solvents. *Inorganic Chemistry*, **25**, 3899-3903. http://dx.doi.org/10.1021/ic00242a015

[13] Abu-Raqabah, A., Davies, G., El-Sayed, M.A., El-Toukhy, A. and Henary, M. (1989) Limits of Direct Transmetalation of Polynuclear Copper(II) Complexes with M(NS)$_n$ Reagents. Scissor Transmetalators. Synthesis and Properties of the Trimers (μ_3-O)(N,py)$_3$Cu$_3$X$_4$ (N = *N,N*-Diethylnicotinamide; py = Pyridine; X = Cl, Br). *Inorganic Chemistry*, **28**, 1156-1166. http://dx.doi.org/10.1021/ic00305a027

[14] El-Sayed, M.A., El-Wakil, H., Ismail, K.Z., El-Zayat, T.A. and Davies, G. (1998) Stiochiometry, Product and Kinetics of Catalytic Oxidation of 2,6-Dimethylphenol by Bromo(N,N'-Diethylenediamine)Copper Complexes in Methylene

Chloride. *Transition Metal Chemistry*, **23**, 795-800. http://dx.doi.org/10.1023/A:1006970209841

[15] Davies, G., El-Sayed, M.A., El-Toukhy, A. and Henary, M. (1990) Transmetalation of Tetranuclear Copper (I) Complexes with an Fe(NS)₃ Reagent. *Inorganica Chimica Acta*, **168**, 65-76.

[16] Davies, G., El-Toukhy, A., Veidis, M. and Onan, K.D. (1984) Transmetalation Reactions of Tetranuclear Copper(II) Complexes. I. Crystal and Molecular Structures of an Intermediate and a Final Product of Reaction of Di-*µ*-oxo-tetra [chloro(DENC)copper(II)], (DENC = *N*,*N*-Diethylnicotinamide) with [Zn(N₂S₂)], (N₂S₂ = Diacetylbis(hydrazonato-S-methylcarbodithioate) in Aprotic Solvents. *Inorganica Chimica Acta*, **84**, 41-50.

[17] El-Sayed, M.A., Abu-Raqabah, A., Davies, G. and El-Toukhy, A. (1989) Kinetic Proof That the Tetranuclear Oxocopper(II) Complex (py)₃Cu₄Cl₄O₂ Initiates and Catalyzes the Oxidative Coupling of 2,6-Dimethylphenol by Dioxygen in Nitrobenzene. *Inorganic Chemistry*, **28**, 1909-1914. http://dx.doi.org/10.1021/ic00309a028

[18] Mazik, J.V., Carriera, L.G. and Davies, G. (1988) Cu-Ni Alloy Formation by Reduction in Hydrogen of a Polyheterometallic Complex. *Journal of Materials Science Letters*, **7**, 833-835. http://dx.doi.org/10.1007/BF00723777

[19] Davies, G., Giessen, B.C. and Shao, H.L. (1990) Single-Phase Cu₀.₅₀Ni₀.₅₀ Alloy Preparation by Thermolysis of a Simple Heteropolymetallic Precursor. *Materials Letters*, **9**, 231-234. http://dx.doi.org/10.1016/0167-577X(90)90051-M

[20] Al-Shehri, S., Davies, G., El-Sayed, M.A. and El-Toukhy, A. (1990) Rate Law Variations in the Specific Monotransmetalation of (*µ*₄-O)(N,py)₄Cu₄₋ₓMₓX₆ Complexes with Zn(NS)₂ in Nitrobenzene. *Inorganic Chemistry*, **29**, 1206-1210.

[21] El-Sayed, M.A. and Davies, G. (1990) Stoichiometry, Products and Kinetics of Monotransmetalation and Complexation of Dimeric Complexes [N₂CuCl₂]₂ and [N₂NiCl₂]₂ (N Is *N*,*N*-Diethylnicotinamide) with M(NS)₂ Reagents in Nitrobenzene. *Inorganica Chimica Acta*, **173**, 163-173. http://dx.doi.org/10.1016/S0020-1693(00)80209-1

[22] Ali, A. and Davies, G. (1990) Products and Kinetics of the Reactions of Bis-(acetylacetonato)copper(II) with Ni(NS)₂ and Cu(NS)₂ Reagents in Methylene Chloride. *Inorganica Chimica Acta*, **177**, 167-178. http://dx.doi.org/10.1016/S0020-1693(00)85973-3

[23] Ali, A. and Davies, G. (1991) Cooperativity in Metal Exchange Reactions of Bis(acetylacetonato)copper(II) with Co(NS)₂ and Zn(NS)₂ Reagents in Methylene Chloride. *Inorganica Chimica Acta*, **179**, 245-254. http://dx.doi.org/10.1016/S0020-1693(00)85884-3

[24] Zhang, C.X., Kaderli, S., Costas, M., Kim, E., Neuhold, Y.M., Karlin, K.D. and Zuberbuhler, A.D. (2003) Copper(I)-Dioxygen Reactivity of [(L)Cuᴵ]⁺ (L = Tris(2-pyridylmethyl)amine): Kinetic/Thermodynamic and Spectroscopic Studies Concerning the Formation of Cu-O₂ and Cu₂-O₂ Adducts as a Function of Solvent Medium and 4-Pyridyl Ligand Substituent Variations. *Inorganic Chemistry*, **42**, 1807-1824. http://dx.doi.org/10.1021/ic0205684

[25] Kastalsky, V. and MaConnel, J.F. (1969) The Crystal and Molecular Structure of Bis(dimethyldithiophosphato)nickel(II), Ni[(CH₃O)₂PS₂]₂. *Acta Crystallographica Section B*, **25**, 909-915. http://dx.doi.org/10.1107/S0567740869003207

[26] El-Sayed, M.A., Abdel-Hamid, I.A., El-Zayat, T.A. and Abdel-Salam, A.H. (2006) Oxidation of Bis(*µ*-halo)-Bis[(diamine)copper(I)] Complexes [LCuX]₂; L = TMED, X = Cl, Br or I, L = TEED or TMPD, X = Cl, with the Two Electron Organo Oxidizing Agent, Tetrachloro-1,2-Benzoquinone(TClBQ) in Aprotic Media: Models for Intermediates in Catalytic Catechol Oxidase. *Transition Metal Chemistry*, **31**, 776-781. http://dx.doi.org/10.1007/s11243-006-0067-4

[27] Gultnokov, G. and Freiser, H. (1968) Heats and Entropies of Formation of Metal Chelates of Certain 8-Quinolonols, Quinoline-8-Thiols, and 2,4-Pentanedione. *Analytical Chemistry*, **40**, 39-44. http://dx.doi.org/10.1021/ac60257a020

[28] Glowiak, T. and Cizsewska, T. (1978) The Crystal Structure of Bis(S-methyl-N-isopropylidendithiocarbazate)nickel(II). *Inorganica Chimica Acta*, **27**, 27-30. http://dx.doi.org/10.1016/S0020-1693(00)87256-4

[29] Abo-El-Dahab, H.A. (2009) Kinetics and Mechanism of the Reaction of Bis-(Acetylacetonato) Copper(II) with Bis-(Dithiophosphato) Nickel(II) in Methylene Chloride. *Bulletin of the Faculty of Science, Alexandria University*, **46**, 36-55.

7

Interconversion between Planar-Triangle, Trigonal-Pyramid and Tetrahedral Configurations of Boron $\left(B(OH)_3 \text{-} B(OH)_4^-\right)$, Carbon $\left(CH_3^+ \text{-} CH_3X\right)$ and for the Group 15 Elements as Nitrogen $\left(NH_3 \text{-} NH_4^+\right)$. A Modelling Description with *ab Initio* Results and Pressure-Induced Experimental Evidence

Henk M. Buck

Kasteel Twikkelerf 94, Tilburg, The Netherlands
Email: h.m.buck@ziggo.nl

Abstract

Recently a mechanistic understanding of the pressure- and/or temperature-induced coordination change of boron in a borosilicate glass has been demonstrated by Edwards *et al.* *In situ* high-pressure ^{11}B solid-state NMR spectroscopy has been used in combination with *ab initio* calculations in order to obtain insight in the molecular geometry for the pressure-induced conversion. The results indicate a deformation of the $B(OH)_3$ planar triangle, under isotropic stress, into a trigonal pyramid that serves as a precursor for the formation of a tetrahedral boron configuration. From our point of view, the deformation controlling the out-of-plane transition of boron accompanied with a D_{3h} into C_{3v} geometric change is an interesting transformation because it matches with

our molecular description based on Van't Hoff modelling for the tetrahedral change of carbon in CH₃X by substitution of X with nucleophiles via a trigonal bipyramid state in which the transferred carbon is present as a methyl planar triangle "cation". Van't Hoff modelling and *ab initio* calculations have been also applied on the dynamics of the out-of-plane geometry of a transient positively charged carbon in a trigonal pyramidal configuration into a planar trivalent carbon cation. Finally the same model is also used for the C_{3v} trigonal pyramidal configurations as NH₃ of the group 15 elements in their nucleophilic abilities.

Keywords

Pressure-Induced Coordination, Van't Hoff Modelling, *Ab Initio* Calculations, Reaction Dynamics

1. Introduction

We introduced several groups as localized sites in organic molecular systems for accommodation of fundamental modes of bonding via *intra-* and *inter-*molecular reactions. Most of these interactions were focused on well-known reaction types, theoretically described with experimental evidence for complex models. In these studies, three-center four-, three-, and two-electron systems based on carbon-, boron-, hydrogen-, and halogen exchange reactions were studied along the lines of *ab initio* and Van't Hoff (*vtH*) modelling studies [1]-[7]. In the case of a linear three-center model, it was possible with the dynamics of a regular tetrahedron to change its geometry into a trigonal bipyramidal transition- or intermediate complex, to predict the various bond lengths at specific locations on the reaction profile. Based on the numbers of electrons (4, 3 and 2, respectively) in the transient complex, all transfer or exchange reactions share the same ratio of the apical bond distances compared with the corresponding bonding distances in the initial state. These ratio numbers are then 1.333, 1.250, and 1.167, respectively. As illustration for this general concept, we describe the three-center four-electron transition state of the identity S_N2 reaction $X^- + CH_3\text{-}X \rightarrow [X\text{-}CH_3\text{-}X]^- \rightarrow X\text{-}CH_3 + X^-$ with X is a halogen. In the dynamics of this process, the methyl "cation" migrates between the partially negatively charged halogens in the trigonal bipyramid. This is illustrated in **Figure 1**.

In the *vtH* model, the tetrahedral angle of 109.47° results in a ratio number of $1 - \cos 109.47° = 1.333$ (ratio between the apical bond length and the corresponding tetrahedral length). For the reaction shown in **Figure 1**, a good correspondence has been obtained between the ratio numbers defined as $R(d)$ and $R(\theta)$:

$$R(d) = d\left(C(V) - X\right) / d\left(C(IV) - X\right) \text{ and } R(\theta) = 1 - \cos\theta.$$

The results are given in **Table 1** in combination with *ab initio* results [1] [3] [4]. The computations using relativistic DFT at the ZORA-OL0.YP/TZ2P level are obtained from Bento *et al.* [8] [9], the other data from Glukhovtsev *et al.* [10].

Within the scope of the *vtH* model, we used the experimental values of the bond distances and angles of the corresponding tetrahedral configuration for calculating $d\left(C(V) - X\right)$.

As mentioned before, this model has been used to describe the pressure-induced conversion of a BO₃ triangle in planar B(OH)₃ into a trigonal pyramid [11]. The latter structure may be considered as transition state in the formation of a tetrahedral configuration effectuated by local coordination environments via atomic oxygen sites

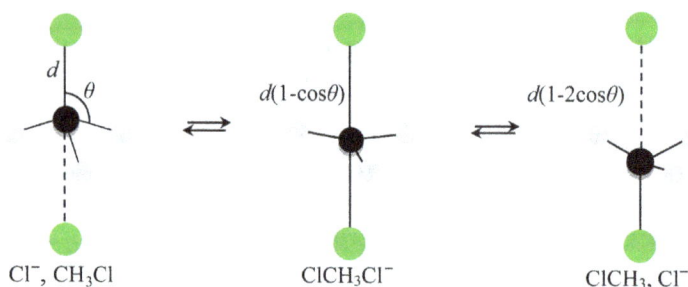

Figure 1. The identity substitution reaction of Cl⁻ + CH₃Cl via a pentavalent carbon.

Table 1. A comparison between $R(d)$ and $R(\theta)$ values in combination with the pentavalent carbon state $[XCH_3X]^-$. The (*ab initio*) distances (d) are given in Å and the corresponding $R(\theta)$ values from the experimental angle data of their tetrahedral configurations.

Halogen	Method	*Ab initio*			Van't Hoff		
X	Level	$d(C(V) - X)$	$d(C(IV) - X)$	$R(d)$	$R(\theta)$	$d(C(IV) - X)$	$d(C(V) - X)$
F	MP2/6-31+G(d)	1.837	1.407	1.306	1.322	1.383	1.828
	MP2/6-31++G(d, p)	1.832	1.405	1.304			
	ZORA-OLYP/TZ2P	1.860	1.396	1.332			
Cl	MP2/6-31+G(d)	2.317	1.780	1.302	1.319	1.776	2.343
	ZORA-OLYP/TZ2P	2.360	1.791	1.318			
Br	MP2/6-31+G(d)-AE	2.466	1.949	1.265	1.304	1.934	2.522
	MP2/6-31+G(d)-ECP	2.480	1.954	1.269			
	ZORA-OLYP/TZ2P	2.510	1.959	1.281			
I	MP2/6-31+G(d)-ECP	2.673	2.140	1.249	1.319	2.132	2.812
	ZORA-OLYP/TZ2P	2.720	2.157	1.261			

in the crystalline lattice.

In our opinion, this type of reaction deserves a more fundamental appreciation because it demonstrates an electrophilic agent, generating an intrinsic conformational change without the formation of an intermediate bonding necessary for the ultimately tetrahedral configuration.

2. Results and Discussion

2.1. Combined Analysis of the Molecular Transformation for the Pressure-Induced B(OH)₃ into (HO)₃B-O-Lattice Conversion in the Crystal Material by *ab Initio* and Van't Hoff Modelling

The intramolecular transformation of planar $B(OH)_3$ (D_{3h} symmetry) into a trigonal pyramid (C_{3v} symmetry) following the *vtH* modelling under formation of the tetrahedral configuration as illustrated in **Figure 2**.

The corresponding geometric values for the various bond distances (in Å) and bond angles (in degrees) are given in **Table 2**. The site selected for the tetrahedral configuration of boron may be considered as an oxygen site. An interesting aspect of this specific isomerization by an intrinsic conformational change is that in the trigonal pyramid, boron chirality can be introduced through differentiation in the three B-O groups. Such an intermediate structure will be strongly dependent on the local orientation of surrounding sites. In this way a more precise information is obtained of the molecular bonding state on the principal reaction coordinate. The question remains whether this approach also gives a more definite answer for the mechanism of a "normal" encounter complex as e.g. between the electrophile $B(OH)_3$ and the nucleophile OH^-, *vide infra*.

Combination of the *ab initio* results and the *vtH* model results in a linear relationship between the O-B-axis angle (y) and the out-of-plane boron distance (x):

$$y = -41.81x + 90.00 \text{ or } y = 41.81x + 90.00.$$

The latter expression corresponds with the O-B-axis angle values in parentheses.

For the (virtual) tetrahedral configuration a B-O distance of 1.397 Å is calculated with the critical value of 0.466 Å for the out-of-plane displacement. Both values correspond with 1.407 Å of the B-OH distance in the cage-like complex anion formed by sodium borate and 1,1,1-tris(hydroxymethyl) ethane: $[Na(H_2O)_3]^+[CH_3C(CH_2O)_3B(OH)]^-$ [12]. The other $B-OCH_2-$ bond lengths are 1.492 Å. The Na^+ ion is octahedrally surrounded by six oxygens with one of the hydroxyl group on boron.

In the report of Edwards *et al.* [11] it was established that the out-of-plane displacement of the boron atom is consistent with the *in situ* high-pressure NMR spectra. This mode of deformation for the $^{11}B\delta_{iso}$ value is given in

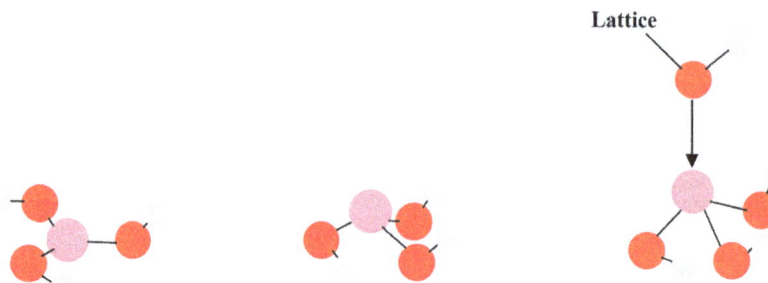

Lattice

Figure 2. Out-of-plane displacement of boron in B(OH)$_3$ following the characteristics of the Van't Hoff (*vtH*) modelling under subsequent formation of a tetrahedral configuration.

Table 2. Geometry calculations of the vertical displacement of the boron out-of-plane orientation along the C_{3v} symmetry axis starting from B(OH)$_3$ with D_{3h} symmetry. The bond lengths are given in Å, the angles in degrees.

Out-of-plane B	Symmetry	B-O	O-B-axis	O-B-O
0.000	D_{3h}	1.370	90.00 (90.00)	120.00
0.196	C_{3v}	1.374	81.80 (98.20)	118.00
0.279	C_{3v}	1.379	78.33 (101.67)	116.01
0.345	C_{3v}	1.385	75.58 (104.42)	114.02
0.466[a]	C_{3v}	1.397[a]	70.53 (109.47)[a]	109.47[a]

[a]These values correspond with the ultimate "out-of-plane" tetrahedral configuration as derived from the *vtH* model. The other values are derived from Ref. [11].

Figure 3. In this figure a linear plot is given in combination with a curved line. The latter line has relevance because it demonstrates clearly the asymptotic character based on the maximum value of the out-of-plane orientation. A linear plot has been shown for the relation between [11]Bδ_{iso} and the variation in bond length of one of the B-O bonds in the planar B(OH)$_3$. This representation conflicts with the absence of significant changes in the quadrupole parameters. The out-of-plane mode is in correspondence with the latter notification, demonstrating the effect of hybridization change in the dynamics with the corresponding symmetry change. The unique character of this intramolecular change will be demonstrated with relatively old descriptions as an illustration for the transition state in bimolecular reactions.

From the foregoing it is evident that simple addition reactions as $\mathrm{B(OH)}_3 + \mathrm{OH}^- \rightarrow \mathrm{B(OH)}_4^-$, the planar B(OH)$_3$ changes its planar geometry into a tetrahedral configuration for collapse with the hydroxyl anion. Here it is experimentally demonstrated that the hypothetic model of Bell-Evans-Polanyi (BEP) is a fundamental principle in the construction of the progress of addition-substitution reactions [13]. The BEP model is shown in **Figure 4** in a primitive way. Generally known, the crossing points correspond with the energy of the transient complexes on the reaction coordinate showing a shift from an early to a late transition state for that specific conversion.

2.2. *Ab Initio* and Van't Hoff Modelling for a Planar Trivalent Carbon Cation into a Trigonal Pyramidal Configuration. Experimental Evidence with a Diamond-Like Model as the *in Situ* 1-Adamantyl Cation

Recently, Fitzgibbons *et al.* published a paper on a high-pressure solid-state reaction of benzene [14]. There is strong experimental evidence based on various spectroscopic measurements that close-packed bundles of sub-nanometre-diameter sp^3-bonded carbon threads (1.52 Å) are formed. The change in hybridization from a planar carbon into a tetrahedral configuration looks similar to the foregoing displacement. However, this conversion is radical-like that promotes an intermediate trigonal pyramidal conformational change.

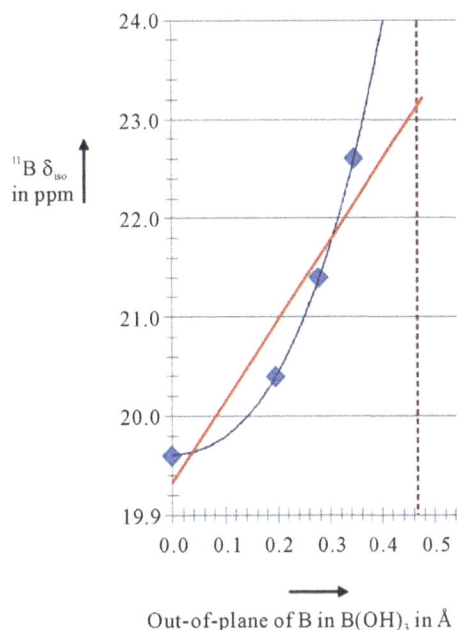

Figure 3. Relation between ^{11}B isotropic chemical shift in ppm versus the out-of-plane displacement of boron in B(OH)$_3$ in Å (red: linear relation, blue: curved line following the data points). The 0.466 Å value is the maximum for the out-of-plane, indicated by the vertical line.

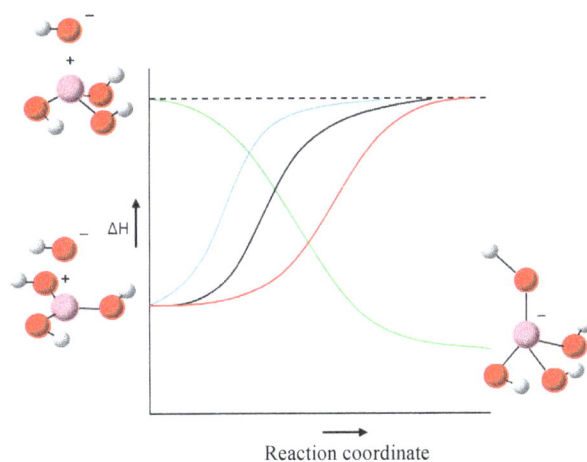

Figure 4. The Bell-Evans-Polanyi energy profile for the reaction dynamics of $B(OH)_3 + OH^- \rightarrow B(OH)_4^-$.

With the *vtH* model the relationship between the C-C-axis angle (y) and the out-of-plane carbon distance (x) is given by:

$$y = \pm 37.93x + 90.00.$$

The ± sign corresponds with (0.513 Å, 109.47°) and (0.513 Å, 70.53°), respectively, for a C-C distance of 1.540 Å. Both (x, y) functions can be given a more general character via y = mx + 90:

$$90/m = d_{\text{C-C}} \times d_{\text{X-Y}}$$

with $d_{\text{C-C}}$ = 1.540 Å and $d_{\text{X-Y}}$ is the bond distance in the tetrahedral configuration as can be seen by substitution of the corresponding *m* values.

The 1-adamantyl cation related to the diamond-like structure adamantane may be considered as an excellent cation for demonstrating a trigonal pyramidal configuration via *in situ* preparation. This aspect has been obtained attention by Jung *et al.* in their synthesis of highly substituted adamantanones from bicyclo [3.3.1] nonanes [15]. The *in situ* prepared 1-adamantyl cation without substituents necessary for the corresponding substituted one is given in **Figure 5**.

In the relaxed form this cation has a near planar geometry for the C1 as has been established by Harding *et al.* on the basis of 13C NMR data combined with sophisticated calculations [16]. Geometry calculations have been performed by Rasul *et al.* for this cation [17]. The bond distance of the planar C1 cation with the surrounding carbons has a value of 1.458 Å. This stringent reduction of the normal C-C distance (1.54 Å) of the *in situ* geometry promotes the increase of the three vertical bond lengths of 1.629 Å. The impact of the introduced stress shows some correspondence with the work of Schreiner *et al.* [18]. That study was directed on steric effects. They mentioned the importance of Van der Waals forces, including London dispersion forces, for the understanding of the stabilizing sterical interactions. X-ray evidence was obtained for the coupling products as e.g. diamantine-triamantane with a C-C distance of 1.704 Å. It seems realistic to assume that both aspects can be explained by (small) changes in the hybridization resulting in (small) angle variation concerning the crucial carbon locations [19].

The *in situ* preparation of a corresponding adamantyl cation was performed by the reaction of trifluoromethanesulphonic acid and 1,5-dimethyl-3,7-dimethylenebicyclo [3.3.1] nonan-9-one. The cation formed is directly trapped by nucleophiles. The main point of the reaction may be considered as a cyclic three-center cyclopropenyl cation with a geometry corresponding with the intermediate transition structure of the unsymmetrically bridged 2-norpinyl cation. For details of the reaction see Ref. [6].

2.3. Van't Hoff Modelling of the C_{3v} Trigonal Pyramidal Configurations as NH$_3$ for the Group 15 Elements in Their Nucleophilic Abilities

The conceptualization as given in the aforementioned sections afford various possibilities to describe the nucleophilic abilities of the [N, P, As, ⋯, Bi]H$_3$ molecules of the group 15 elements. Going from nitrogen to the lower placed elements, the H-X-H angle (X = N, P, As, ⋯, Bi) decreases from 107.8° to 90.48° for BiH$_3$. This means that the out-of-plane location of X with regard to the triangle formed by the three hydrogens increases drastically from nitrogen towards phosphorus from $0.36d$ to $0.54d$ ($0.56d$, $0.56d$ and $0.57d$) in which d is the XH bond length. Focused on nitrogen and phosphorus the difference between the C_{3v} and the T_d symmetry after protonation, the change in the out-of-plane location differs substantially for PH$_4^+$ from 0.767 Å to 0.473 Å compared with the transition for NH$_3$ to NH$_4^+$ from 0.366 Å to 0.368 Å[1]. This approach based on the *vtH* modelling gives an unique visualization of the drop in gas-phase proton affinity from NH$_3$ to PH$_3$. The influence of methyl substitution is significant. The exclusive difference in proton affinity between the methylated nitrogen and phosphorus compounds decreases under simultaneous increase of the proton affinity of the individual compounds. These changes in proton affinity correspond with the bond angles. Usually these effects are explained by hyperconjugation of the methyl group. From the foregoing model study it seems acceptable to assume that sterical effects play an important role. The role of these simple compounds is of importance specifically as an essential part of biological systems [20] [21].

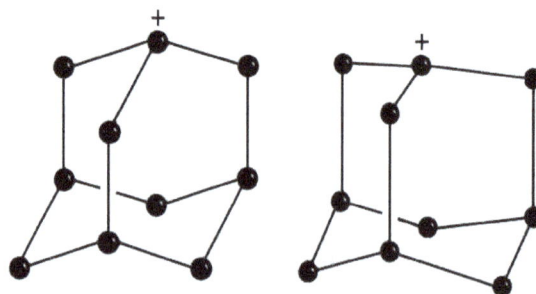

Figure 5. *In situ* geometric configuration of 1-adamantyl cation (left) and the corresponding relaxed cation (right).

[1]Bond angels and bond distances are taken from textbooks and literature.

3. Conclusions

It has been established by Edwards *et al.*, that *in situ* high-pressure ^{11}B solid-state NMR spectroscopy for a borosilicate glass in combination with *ab initio* calculations provides insight in the molecular geometry for the pressure-induced conversion. The results indicate a deformation of the $B(OH)_3$ planar triangle, under isotropic stress, into a trigonal pyramid that serves as a precursor for the formation of a tetrahedral boron configuration. The out-of-plane transition of boron with a D_{3h} into C_{3v} geometric change is an interesting transformation because it matches with the Van't Hoff modelling originally based on the molecular description of the tetrahedral change of carbon by nucleophilic substitution reactions as CH_3X (X is a leaving group) with Cl^- via a trigonal bipyramid in which the transferred carbon is present as a methyl planar triangle "cation" linked to the two partially negatively charged apical positions Cl and X. Although the ultimately tetrahedral configuration for boron is a hypothetical case under the high-pressure conditions, this virtual state can be specified with *ab initio* calculations in combination with molecular modelling based on Van't Hoff's tetrahedron. It is our working model to give a more general support for this type of simple addition reactions which are of fundamental significance. Van't Hoff modelling and *ab initio* calculations are also applied on the dynamics between a planar trivalent carbon cation and its trigonal pyramidal configuration. Experimental evidence is obtained with a diamond-like model as the *in situ* 1-adamantyl cation. Finally the Van't Hoff modelling is also used for the C_{3v} trigonal pyramidal configurations as NH_3 for the group 15 elements in their nucleophilic abilities.

The combinations of sophisticated calculations and on the other hand modelling studies based on the Van't Hoff tetrahedron model clearly demonstrate the fundamental significance for the understanding of the organization of bimolecular interactions via a specific intermediate state by extrapolation of the experimental results of Edwards *et al.* Reactions of this type are focused on symmetry changes as $D_{3h} \rightarrow C_{3v} \rightarrow T_d$.

Acknowledgements

I thank my grandchildren Robin van Dorrestein and Martin Buck BSc for their valuable technical assistance.

References

[1]   Buck, H.M. (2008) A Combined Experimental, Theoretical, and Van't Hoff Model Study for Identity Methyl, Proton, Hydrogen Atom, and Hydride Exchange Reactions. Correlation with Three-Center Four-, Three-, and Two-Electron Systems. *International Journal of Quantum Chemistry*, **108**, 1601-1614. http://dx.doi.org/10.1002/qua.21683

[2]   Buck, H.M. (2010) A Linear Three-Center Four Electron Bonding Identity Nucleophilic Substitution at Carbon, Boron, and Phosphorus. A Theoretical Study in Combination with Van't Hoff Modeling. *International Journal of Quantum Chemistry*, **110**, 1412-1424. http://dx.doi.org/10.1002/qua.22252

[3]   Buck, H.M. (2011) A Model Investigation of *ab Initio* Geometries for Identity and Nonidentity Substitution with Three-Center Four- and Three-Electron Transition States. *International Journal of Quantum Chemistry*, **111**, 2242-2250. http://dx.doi.org/10.1002/qua.22529

[4]   Buck, H.M. (2012) Mechanistic Models for the Intramolecular Hydroxycarbene-Formaldehyde Conversion and Their Intermolecular Interactions: Theory and Chemistry of Radicals, Mono-, and Dications of Hydroxycarbene and Related Configurations. *International Journal of Quantum Chemistry*, **112**, 3711-3719. http://dx.doi.org/10.1002/qua.24127

[5]   Buck, H.M. (2013) An Adjusted Model for Simple 1,2-Dyotropic Reactions. *Ab Initio* MO and VB Considerations. *Open Journal of Physical Chemistry*, **3**, 119-125. http://dx.doi.org/10.4236/ojpc.2013.33015

[6]   Buck, H.M. (2014) Three-Center Configuration with Four, Three, and Two Electrons for Carbon, Hydrogen, and Halogen Exchange. A Model and Theoretical Study with Experimental Evidence. *Open Journal of Physical Chemistry*, **4**, 33-43. http://dx.doi.org/10.4236/ojpc.2014.42006

[7]   Yamashita, M., Yamamoto, Y., Akiba, K., Hashizume, D., Iwasaki, F., Takagi, N. and Nagase, S. (2005) Syntheses and Structures of Hypervalent Pentacoordinate Carbon and Boron Compounds Bearing an Anthracene Skeleton. Elucidation of Hypervalent Interaction Based on X-Ray Analysis and DFT Calculation. *Journal of the American Chemical Society*, **127**, 4354-4371. http://dx.doi.org/10.1021/ja0438011

[8]   Bento, A.P. and Bickelhaupt, F.M. (2008) Nucleophilicity and Leaving Group Ability in Frontside and Backside S_N2 Reactions. *Journal of Organic Chemistry*, **73**, 7290-7299. http://dx.doi.org/10.1021/jo801215z

[9]   Bento, A.P. and Bickelhaupt, F.M. (2008) Frontside versus Backside S_N2 Substitution of Group 14 Atoms. Origin of Reaction Barriers and Reasons for Their Absence. *Chemistry—An Asian Journal*, **3**, 1783-1792. http://dx.doi.org/10.1002/asia.200800065

[10] Glukhovtsev, M.N., Pross, A. and Radom, L. (1995) Gas-Phase Identity S_N2 Reactions of Halide Anions with Methyl Halides: A High-Level Computational Study. *Journal of the American Chemical Society*, **117**, 2024-2032. http://dx.doi.org/10.1021/ja00112a016

[11] Edwards, T., Endo, T., Walton, J.H. and Sen, S. (2014) Observation of the Transition State for Pressure-Induced $BO_3 \rightarrow BO_4$ Conversion in Glass. *Science*, **345**, 1027-1029. http://dx.doi.org/10.1126/science.1256224

[12] Taylor, M.J., Grigg, J.A. and Rickard, C.E.F. (1992) The Structure of the Cage-Like Complex Anion Formed by Sodium Borate and 1,1,1-Tris(Hydroxymethyl)Ethane. *Polyhedron*, **11**, 889-892. http://dx.doi.org/10.1016/S0277-5387(00)83337-9

[13] Dewar, M.J.S. (1969) The Molecular Orbital Theory of Organic Chemistry. Chapter 8, McGraw Hill Book Company, New York.

[14] Fitzgibbons, T.C., Guthrie, M., Xu, E.S., Crespi, V.H., Davidowski, S.K., Cody, G.D., Alem, N. and Badding, J.V. (2014) Benzene-Derived Carbon Nanothreads. *Nature Materials*, **14**, 43-47. http://dx.doi.org/10.1038/nmat4088

[15] Jung, M.E. and Lee, G.S. (2014) Synthesis of Highly Substituted Adamantanones from Bicyclo[3.3.1]Nonanes. *Journal of Organic Chemistry*, **79**, 10547-10552. http://dx.doi.org/10.1021/jo501368d

[16] Harding, M.E., Gauss, J. and von Ragué Schleyer, P. (2011) Why Benchmark-Quality Computations Are Needed to Reproduce 1-Adamantyl Cation NMR Chemical Shifts Accurately. *Journal of Physical Chemistry A*, **115**, 2340-2344. http://dx.doi.org/10.1021/jp1103356

[17] Rasul, G., Olah, G.A. and Prakash, G.K.S. (2010) Density Functional Theory Study of Adamantanediyl Dications $C_{10}H_{14}^{2+}$ and *Protio*-Adamantyl Dications $C_{10}H_{16}^{2+}$. *Proceedings of the National Academy of Sciences of the United States of America*, **101**, 10868-10871. http://dx.doi.org/10.1073/pnas.0404137101

[18] Schreiner, P.R., Chernish, L.V., Gunchenko, P.A., Tikhonchuk, E.Y., Hausmann, H., Serafin, M., Schlecht, S., Dahl, J.E.P., Carlson, R.M.K. and Fokin, A.A. (2011) Overcoming Lability of Extremely Long Alkane Carbon-Carbon Bonds through Dispersion Forces. *Nature*, **477**, 308-311. http://dx.doi.org/10.1038/nature10367

[19] Buck, H.M. (2000) Symmetry Restrictions as Starting Point for the Determination of Geometric Representations and the Dynamics of Cyclic π Systems. *International Journal of Quantum Chemistry*, **77**, 641-650. http://dx.doi.org/10.1002/(SICI)1097-461X(2000)77:3<641::AID-QUA5>3.0.CO;2-R

[20] Buck, H.M. (2011) DNA Systems for B-Z Transition and Their Significance as Epigenetic Model: The Fundamental Role of the Methyl Group. *Nucleosides, Nucleotides and Nucleic Acids*, **30**, 918-944. http://dx.doi.org/10.1080/15257770.2011.620580

[21] Buck, H.M. (2013) A Conformational B-Z DNA Study Monitored with Phosphatemethylated DNA as a Model for Epigenetic Dynamics Focused on 5-(Hydroxy)Methylcytosine. *Journal of Biophysical Chemistry*, **4**, 37-46. http://dx.doi.org/10.4236/jbpc.2013.42005

8

The Even-Odd Rule on Single Covalent-Bonded Structural Formulas as a Modification of Classical Structural Formulas of Multiple-Bonded Ions and Molecules

Geoffroy Auvert

CEA-Leti, Grenoble, France
Email: Geoffroy.auvert@grenoble-inp.org

Abstract

In organic chemistry, as defined by Abegg, Kossel, Lewis and Langmuir, compounds are normally represented using structural formulas called Lewis structures. In these structures, the octet rule is used to define the number of covalent bonds that each atom forms with its neighbors and multiple bonds are frequent. Lewis' octet rule has unfortunately shown limitations very early when applied to non-organic compounds: most of them remain incompatible with the "rule of eight" and location of charges is uncertain. In an attempt to unify structural formulas of octet and non-octet molecules or single-charge ions, an even-odd rule was recently proposed, together with a procedure to locate charge precisely. This even-odd rule has introduced a charge-dependent effective-valence number calculated for each atom. With this number and the number of covalent bonds of each element, two even numbers are calculated. These numbers are both used to understand and draw structural formulas of single-covalent-bonded compounds. In the present paper, a procedure is proposed to adjust structural formulas of compounds that are commonly represented with multiple bonds. In order to keep them compatible with the even-odd rule, they will be represented using only single covalent bonds. The procedure will then describe the consequences of bond simplification on charges locations. The newly obtained representations are compared to their conventional structural formulas, *i.e.* single-bond representation vs. multiple-bond structures. Throughout the comparison process, charges are precisely located and assigned to specific atoms. After discussion of particular cases of compounds, the paper finally concludes that a rule limiting representations of multiple covalent bonds to single covalent bonds, seems to be suitable for numerous known compounds.

Keywords

Multiple Bond, Single Bond, Covalent Bond, Molecule, Ion, Even-Odd, Rule, Structural Formula

1. Introduction

In classical 2D structural formulas of ions and molecules, single bonds are represented by drawing one line between next neighbor atoms, like in di-hydrogen [1]. Other molecules are represented with multiple bonds, *i.e.* several lines between connected atoms: For instance in di-nitrogen where nitrogen atoms are connected with three lines [2]. These different types of bonds have been in combination with the octet rule to represent molecules for about one century [3]-[5]. Lewis' octet rule, at first successful in representing organic molecules, has later shown limitations, mainly for non-organic molecules [6] [7]. In an attempt to address both octet and non-octet molecules, a new rule, named the "even-odd" rule, was recently proposed and shown to apply to molecules like Li2 [8]. This rule was subsequently checked against ions involving only single-bonded interconnections between neighbors [9]. The even-odd rule seems to be generally capable of representing many single-bonded compounds and well-known octet or non-octet compounds [8] [9].

The aim of the present paper is to widen the scope of the even-odd rule by addressing multiple-bonded compounds. In the process, it proposes an adjustment to the representation of multiple-bonded ions and molecules. The proposed modification does not affect the position of atoms in ions or molecules, but only impacts the representation of interconnections between atoms.

This paper thus begins by describing a method to convert multiple bonds into single bonds in ions and molecules. Then a definition of the even-odd rule is given. Many types of multiple-bonded compounds are then redrawn by applying the conversion method and the even-odd rule. This includes octet and non-octet compounds in order to illustrate the versatility of the even-odd rule. At last, a discussion is proposed to point out important cases like the unambiguous determination of charge position in both charge-neutralized molecules and multiple-charged ions.

2. Redefining Multiple Bonds as Single Bonds

In this paper, a modification in the drawing of multiple-bonded connections between atoms in molecules or ions is proposed as follows:

- A single covalent bond between two first neighbors classically drawn with one line, is not modified.
- A classical representation of a double-bond is replaced by a single bond and by adding one positive charge on one connected atoms and a negative charge to the other one.
- A triple bond is replaced by a single covalent bond without any other impact.
- A quadruple bond is transformed into a single bond using the same procedure as a double bond.

Since Lewis never proposed any other bonding type in organic structural formulas, there is no other type of bond to address.

3. The Even-Odd Rule

The even-odd rule is a procedure to draw chemical structural formulas of molecules and ions. The structure is composed of one or several atoms of the periodic table.

As a reminder [9], the rule is as follows:

- Each atom:
- Is an element with one or several electron shells;
- Possesses an outer-shell filled with one to eight electrons;
- Has a number of electrons in the outer-shell, also called valence number, indicated in the periodic table;
- Has a valence number of the element giving the highest number of covalent bonds that the element can form.
- A structure meets the criteria below:
- When it is composed of only one atom, it forms no covalent bond;
- When it is constituted of several atoms, each atom forms a single covalent-bond with each of its first neighboring atoms. This covalent bond involves two electrons, one from each interconnected atom;

- A covalent bond is represented by one line between both connected atoms;
- An atom may have zero, one or more than one line around it;
- In the 2D structure, each atom is represented by the letters from its element as in the periodic table;
- Two numbers have to be evaluated and written on each side of the atom.
- The left side number and the effective valence number:
- The left side number is the valence number as in the periodic table. It ranges from one for elements like sodium (Na) up to eight for noble gas like Argon (Ar);
- The effective valence number has to be evaluated: For a neutral atom, *i.e.* without charge, it is equal to the valence number; for a negatively charged atom, *i.e.* that possesses an extra-electron, it is the valence number increased by one; for a positively charged atom, it is the valence number decreased by one.
- The right side number of an atom:
- The right-side number, the "Lewis number", is equal to the sum of the effective valence number and the number of covalent bonds of the atom. It can also be expressed as the sum of the number of electrons left in the outer-shell and twice the number of covalent bonds.
- The Lewis number must be an even number. This is only possible when the number of bonds and the effective valence number are both odd or both even;
- The smallest value the Lewis number can take is zero: The atom has lost electrons from the outer-shell so it is empty and the atom has no bond;
- The Lewis number can range up to twice the effective valence number: this is twice the maximum number of covalent bonds for this element. This number is charge dependent through the effective valence number;
- If all atoms of a compound have Lewis numbers equal to eight, the compound is compatible with Lewis' octet rule.
- Electron pairs in the outer-shell of an interconnected atom:
- The number of electrons in the outer-shell is calculated by subtracting the effective valence-number and the number of covalent bonds. It is an even number;
- As a consequence, the outer-shell contains electron-pairs not involved in any covalent bond;
- This electron-pair number ranges from 0 to 4 whatever is the charge of the element;
- When this electron-pair number is 0, no additional covalent bond can be formed by the element.

The even value of the right side number *i.e.* the Lewis number, and the even value of the number of electrons in the outer-shell are important keys to the validity of the even-odd rule. With these even values, molecules and ions belong to a group of electron-paired compounds [8] [9].

This rule and the modification of multiple-bonded structure will now be applied to redraw ions and molecules classically represented with multiple bonds between atoms.

4. Application to Multiple-Bonded Compounds

In the following, we purposely reduce the scope of the comparison to compounds containing elements of the main group of the periodic table. The proposed change of multiple-bonded compounds is illustrated and compared to classical structure. **Table 1** lists neutral molecules, composed of charged or uncharged atoms, and **Table 2** ions, also composed of charged or uncharged atoms. For each compound, the available classical structures and the modified even-odd structure are shown.

Classical structures of compounds, ions or molecules, are drawn first. The even-odd drawing follows with both associated numbers. In **Table 2**, classical structures of ions are specifically surrounded by square brackets.

When several classical drawings are available for the same compound, they are shown and specifically referenced.

In **Table 1**, 2D structural formulas of neutral molecules composed of charged and uncharged atoms are shown. The objective is to show that the overall neutrality of each molecule is conserved in both: classical structural formulas and the even-odd structural formulas.

The first example is the structure of nitrogen molecule, conventionally represented with a triple bond. With the even-odd rule, as proposed above, the triple bond is replaced by a single covalent-bond. The new structure is shown underneath with both numbers of the even-odd rule. The right side number is 6, which means that this atom does not follow the octet rule. The difference between the effective valence number and the number of bonds

Table 1. Comparison of structures drawn classically or following the even-odd rule: Application to neutral molecules. Molecules are ordered by increasing number of atoms. For each molecule, the classical representation is in the upper position. They can be easily identified thanks to the multiple bonds. Underneath, new designs are composed of only single bonds, no multiple bonds and two associated numbers: the left one is the valence number and the right one is calculated using the proposed even-odd rule.

Table 2. Comparison of structures drawn classically or following the even-odd rule: Application to ions. For each well-known ion, classical structures are drawn first, between square brackets. Multiple bonds are clearly identified. Underneath, new designs are composed of only single bonds between next neighbor atoms. Each element is associated with its valence number on the left side and an even right-side number calculated with the proposed even-odd rule. The charge position is clearly defined.

Two atoms		
N2 — Nitrogen [10]-[12] Sol Liq Gas [12]	N≡N ; N—N 5 6 5 6	[10] [11]
BN — Boron nitride [10] [11] Sol Gas [12]	B≡N ; B—N 3 4 5 6	[10] [11]
B2 — Boron gas Diboryne [10]	B≡B ; B⁺=B⁻ ; B—B 3 4 3 4	[11] ; [10]
O2 — Oxygen [10]-[12] Sol Liq Gas [12] Different charge position	O=O ; O⁻—O⁺ 6 8 6 6	[10] [11]
SiC — Silicon carbide [10] Sol Liq [12] Same charge position	Si⁺≡C⁻ ; Si⁺—C⁻ 4 4 4 6	[10] [11]
C2 — Diatomic carbon [12] Dicarbon [11] [12] Stellar gas [12] Same charge position	:C=C: ; C⁺≡C⁻ ; C⁺—C⁻ 4 4 4 6	[12] ; [11]

Two atoms		
CN(−) [10] — Cyanide anion [10]-[12] Sol Liq Gas [12] From hydrocyanic acid	[C⁻≡N]⁻ ; C⁻—N 4 6 5 6	[10] [11] [13]
CN(+) — Monocyanogen [11] Nitrilomethylium [10] From cyanide chloride	[C⁺≡N]⁺ ; C⁺—N 4 4 5 6	[10] [11]
IO(+/−) — Iodosyl radical [10] [11] Gas [12] from Hypoiodous acid	[I⁺/⁻=O]⁺/⁻ ; I—O⁺/⁻ 7 8 6 6/8	[10]-[12]
NO(+) — Nitrosonium Ion [10]-[12] In solution	[N≡O⁺]⁺ ; N—O⁺ 5 6 6 6	[10]-[12]
NO(+/−) — Nitric oxide [10]-[12] From HNO Nitroxyl [10] [12] Or ClNO Nitrosyl Chloride [12]	[N≣O]⁺/⁻ ; [N⁺/⁻=O]⁺/⁻ ; [N⁻=O]⁻ ; N—O⁺/⁻ 5 6 6 6/8	[12] ; [10] ; [11]
BO(−) — Boron oxide [11]	[B=O]⁻ ; B—O⁻ 3 4 6 8	[11]

Continued

CS Carbon monosulfide [10]-[12] Gas [12] Same charge position	$:C \equiv S:$ $^-$ $^+$ [10]-[12] $_4C^+_4 — _6S^-_8$

Three atoms		**Three atoms**	
HCN Hydrogen cyanide [10]-[12] Sol Liq Gas [12] Same charge position	$H — C \equiv N$ [10] [12] [13] $_1H_2 — _4C_6 — _5N_6$	CNO(−) Cyanate ion [11] [12] NCO radical [11] Oxomethylenamino Radical [10] In solution	$\left[N \equiv C — O \right]^-$ [10]-[13] $\left[N^{+/-} = C = O \right]^{+/-}$ [10] [11] $_5N_6 — _4C_6 — _6O_8^-$
NCCl Cyanogen chloride [10]-[12] Sol Liq Gas [12] Same charge position	$N \equiv C — Cl$ [10] [12] $_5N_6 — _4C_6 — _7Cl_8$	CNO(−) Fulminate [11] [12]	$\left[C^- \equiv N^+ — O^- \right]^-$ [11] [12] $_4C_6^- — _5N_6^+ — _6O_8^-$
N2O Nitrous oxide [10]-[12] Sol Liq Gas [12] Same and different charge position	$N^- = N^+ = O$ [12] $N \equiv N^+ — O^-$ [10] [11] $_5N_6 — _5N_6^+ — _6O_8^-$	HN2(+) Diazenyl radical [10] Diazynium cation [10]	$\left[H — N = N^+ \right]^+$ [10] $\left[H — N^+ \equiv N \right]^+$ [10] $_1H_2 — _5N_6^+ — _5N_6$
HNO Nitroxyl [10]-[12] Gas [12] different charge posi- tion	$H — N = O$ [10]-[12] $_1H_2 — _5N_6^+ — _6O_8^-$	N3(−) Azide anion [10]-[12]	$\left[N^- = N^+ = N^- \right]^-$ [10]-[12] $_5N_6 — _5N_8^- — _5N_6$
C3 Tricarbon [11] [12] Stellar gas [12]	$:C = C = C:$ [11] [12] $_4C_6^- — _4C_6 — _4C_4^+$	OPO(−)or PO2(−) Phosphinate [10] [11]	$\left[O = P — O^- \right]^-$ [10] [11] $_6O_8^- — _5P_6^+ — _6O_8^-$
CaC2 Calcium carbide [10] [12] Sol Liq Gas [12] Different charge position	$[:C \equiv C:]^{2-}$ Ca^{2+} [10] [12] $_4C\backslash_6 — _4C/_6$ $\backslash Ca/$ $_2\ \ _4$	HCO(−) or (+) Formyl Radical [10] [11] Solution [10]	$\left[H — C^- = O \right]^-$ [10] [11] $_1H_2 — _4C_6 — _6O_8^-$

Continued

Four atoms		Four atoms	
C2H2 Acetylene [11] [12] Ethyne [10] Sol Liq Gas [12] Same charge position	H —C≡C— H [10]-[12] H—C—C—H 4 6 4 6	H2CN(+) or (−) Methyleneamino radical [10] Different charge position	[H₂C=N]⁺ [10] H₂C⁺—N 4 6 5 6
C2N2 Cyanogen [10]-[12] Sol Liq Gas [12] Same charge position	N≡C—C≡N [10]-[13] N—C—C—N 4 6 4 6 5 6		
C2F2 Difluoroacetylene [12] Difluorovinylene [10] Sol Liq [10]	F—C≡C—F [10] [12] F—C—C—F 7 8 4 6 4 6 7 8	C3N(−) Cyanoacetylene ion [12]	[C≡C—C≡N]⁻ C—C—C—N 4 6 4 6 4 6 5 6
N2H2 Diazene [10]-[12] Sol Liq [12] Different charge position	H—N=N—H [10]-[12] H₂—N⁻—N⁺—H 1 2 5 8 5 6		
COCl2 or CCl2O Phosgene [10]-[12] Sol Liq Gas [12] Different charge position	Cl₂C=O [10]-[12] Cl₂C⁺—O⁻ 4 6 6 8	CHO2(−) Formic ion Format [11]	[H—C(=O)(O⁻)]⁻ H—C⁺(O⁻)(O⁻) 1 2 4 6 6 8
HN3 Hydrazoic acid [10]-[12] Sol Liq Gas [10] [12] Different charge position Other structure	H—N=N⁺=N⁻ [10]-[12] H—N⁻—N⁺≡N [12] H—N⁻—N⁺—N 5 8 5 6 5 6 N—N—N 5 6 5 8 5 6	N2O2(2−) Hyponitrite ion [10]-[12] From hyponitrous acid Solution [12] No charge position [12] Different charge position	[O⁻—N=N—O⁻]²⁻ [10]-[12] O⁻—N⁻—N⁺—O⁻ 6 8 5 8 5 6 6 8
HCNO Fulminic acid [10]-[12] Same or different Charge position	H—C≡N⁺—O⁻ [10]-[12] H—C⁺=N—O⁻ [12] H₂—C—N⁺—O⁻ 1 2 4 6 5 6 6 8		

Continued

HNCO Isocyanic acid [10]-[12] Sol Liq Gas [12] Different charge position	H—N=C=O [10] [12] H—N⁺—C—O⁻ 1 2 5 6 4 6 6 8	H2N2(2+) [10] Diazynediium [10] Same charge position	[H—N≡N—H]²⁺ [10] H₂—N⁺—N⁺—H₂ 1 2 5 6 5 6 1 2
HNO2 Nitrous acid [10] [11] Sol Liq [12] Different charge position	H—O—N=O [10] [12] O⁻—N⁺—O—H₂ 6 8 5 6 6 8 1 2		

Five atoms | **Five atoms**

CH2N2 Cyanamide [10]-[12] Sol Liq Gas [12] Same charge position	N≡C—NH2 [10] [11] N—C—N(H)(H) 5 6 4 6 5 8	N2H3(+) Diazenium cation Different charge position	[H—N=N⁺(H)(H)]⁺ [10] [11] H—N⁺—N(H)(H) 1 2 5 6 5 8
Isomer CH2N2 HNCNH Carbodiimide [10] [12] Different charge position	H—N=C=N—H [10] H—N⁻—C—N⁺—H 1 2 5 8 4 6 5 6	H2PO2(−) [10] [12] Hypophosphorous acid [10] [12] Different charge position	[H₂P(=O)(O⁻)]⁻ H₂—P⁺(O⁻)— 1 2 5 8 6 8
Isomer CH2N2 Diazomethane [10]-[12] Sol Liq Gas [12]	H2C⁻—N⁺≡N [12] H2C=N⁺=N⁻ [10] [11] H2C⁺—N⁻—N₆ 4 6 5 8 5 6	PO4(3−) Phosphate ion [10] [12] Different charge position	[HP(=O)(H)(O⁻)]⁻ [10] [12] O⁻—P⁺—O⁻ 6 8 5 8 6 8
Isomer CH2N2 Diazirine [10]-[12] No charge position Different charge position	(H)(H)C(N=N) [10]-[12] H₂C(N⁺)(N⁻) 1 2 4 8 5 6 5 8	CN4(2−) Tetrazole ion [12]	(tetrazole ring structure)

Continued

F3PO Phosphorus oxyfluoride [12] Phosphoryl fluoride [10] [11] No charge Different charge position	[10]-[12]	P(CN)2(−) [13] Dicyanophosphine	

More	More

CH3CN Acetonitrile [10]-[12] Sol Liq Gas [12] No charge Same charge position	[10]-[12]	C3H5 Allyl ion [10] [12] Anion Cation	[10] [12]

C3H3N Acrylonitrile [10]-[12]	[10]-[12]	C5H5(−) or (+) Cyclopentadienyl [10] [11] Gas [12] No charge position Different charge position Changing all charges is for C5H5(+)	[10] [11] [12]

C6H6 Benzene [10]-[12] Sol Liq Gas [12] No charge position Different charge position	[10]-[12]	C5H5O(+) Pyrylium cation [10]-[12] In solution [12] Charge position on O Different charge position	

Continued

CH2N4

Tetrazole
[10]-[12]
Sol Liq Gas [12]

No charge position

[10]-[12]

Different charge position

C5H5NH(+)

Pyridinium cation
[10]-[12]
No charge position

Different charge position

C5H5N

Pyridine [10]-[12]
Sol Liq Gas [12]
Unknown charge position

[10]-[12]

Different charge position

C7H7(+)

Tropylium cation
[10] [12]
In solution [12]
No charge position

[12]

With charge position

[10] [11]

Different charge position

C3H7NO

Dimethyl formamide
[10] [11]

[10]-[12]

Different charge position

[12]

No resonance

C16H18N3S(+) Cl(−) Methylene blue   [12]

is equal to 4. This corresponds to the number of electrons in the outer-shell of the nitrogen molecule. The second structure in **Table 1**, uses the same structure but different elements. This molecule is resolved in the same operation.

The third molecule is di-boron. Two different classical structures are available. The first one uses a triple bond and the second one uses charges and a double bond. Both drawings end up as a unique even-odd structure after modification of multiple bonds. The left side number is 3 as in the periodic table for boron. The right side number is equal to 4. The difference is equal to 2 and gives the number of electrons in the outer shell of the boron atom.

Still in **Table 1**, situated right below Boron, the di-oxygen molecule, classically represented with a double bond, is also drawn using the proposed modification rule. The resulting structure is composed of a single covalent-bond and two oxygen atoms bearing opposite charges. Both opposite charges cancel on a molecular scale to give a neutral total charge. Both atoms have a valence number of 6, according to the periodic table. The negatively charged atom thus has an effective valence number of 7, which, by adding the number of covalent bonds, gives a right-side number of 8. The effective valence number of the positively charged atom is 5, which gives a right-side number equal to 6. Both values are even, in agreement with the even-odd rule. The difference between the effective valence number and the number of covalent bonds gives 6 electrons in the outer shell for the negatively charged oxygen and 4 electrons for the positively charged oxygen.

The same procedure is applied to all other neutral molecules in **Table 1**. It is of interest to note that in **Table 1**, the even-odd structure of PF3O is the only molecule following Lewis' octet rule.

Below in **Table 1**, one of the most complicated molecules is diazomethane (not diazirine): CH2N2. It is classically represented with two different structures, where charge locations depend on the position of multiple bonds. Applying the even-odd procedure on both classical structures result in only one possibility for charge positions and a unique design.

Far below in **Table 1**, cyclic molecules are addressed, like for instance the benzene C6H6 molecule. In classical structure, the charge is not precisely located. In the even-odd structure on the other hand, the charges are assigned to specific atoms and they cannot migrate.

On the whole, no difficulty was met in **Table 1** to draw neutrally-charged multiple-bonded molecules.

In **Table 2**, chemical structural formulas of ions are listed. According to the proposed modification procedure, multiple covalent-bonds are replaced by single covalent-bond and assigned charges. The first ion has a triple bond which is replaced underneath by a single bond. This ion has one charge located on the carbon atom. According to the even-odd rule, the Nitrogen atom must be uncharged and the Carbon atom must be charged. The charge position for this ion is the same in both structures: classical or even-odd.

A little below, in both ions IO(+) and IO(−), denoted as IO(+/−), the charge is classically located on the iodine atom. The change from a double bond to a single bond implies shifting the charge from the iodine to the oxygen. In the new structure, the oxygen atom bears a charge, positive or negative. For this ion, the charge position is different according to the model used, classical or even-odd.

Two lines below, the nitric oxide ion NO(+/−) is represented with three classical structures and different positions for the charge. With the even-odd rule, this charge can only be assigned to the oxygen element, as explained in the following: The ion contains only one covalent bond and the valence number of the oxygen is 6, which would give 7 on the right-side. To make it an even number, the effective valence number could be 6 or 8, meaning that a positive or negative charge should be respectively assigned to the atom.

The same difficulty occurs below with the diazenyl ion HN2(+) in which the charge is assigned without ambiguity to the central atom by the even-odd rule.

In **Table 2**, this procedure is used for all other ions. On the whole, no difficulties were met to keep the same total charge and to reduce each classical multiple-bonded arrangement to a single-bonded one. Also in both **Tables**, the charge is always assigned without ambiguity to a specific element, resulting in outer-shells filled with electron-pairs as defined in the even-odd rule.

5. Discussion

5.1. The Octet Rule in the Even-Odd Rule

Historically, Lewis' octet rule greatly relies on multiple bonds. In **Table 1** and **Table 2**, all organic compounds as Nitrogen, Oxygen, Hydrogen cyanide, Nitroxyl, Acetylene, Phosgene, Diazirine, Benzene, follow the octet rule in the classical representation. Once the even-odd rule and its criterion of single covalent bonds is applied, there are only very few of these molecules in which all atoms have an outer-shell filled with eight electrons: phosphorus oxyfluoride F3PO and iodosyl radical IO(−).

Non-organic compounds were rarely compatible with Lewis' octet rule. The even-odd rule seems capable of addressing both types of compounds and unifies their representations.

5.2. Kekulé and Non-Kekulé Compounds

Kekulé precisely described the classical benzene structure about fifty years before the octet rule. He proposed to represent it with alternatively multiple-bonded and single-bonded connections (see **Table 1**) [14]. This procedure was extended to other molecules [15]. Shortly after, molecules described as "non-Kekulé" were also represented

[16] [17]. One of them appears in **Table 2**: tropylium cation, C7H7(+). This molecule is also drawn using the even-odd rule, *i.e.* without multiple bonds. As a consequence, the even-odd rule seems to encompass both types of Kekulé and non-Kekulé compounds.

5.3. Two Successive Double-Bonded Connections for Neutral Atoms

In tri-carbon, the center atom is classically represented with respectively two double covalent-bonds. While reducing each to single covalent-bonds and keeping an even right-side number, two charges must be assigned to the central atom. Two alternatives here: Either the same charge is assigned twice, resulting in the central atom bearing two charges that is electrostatically not very stable. Or two opposite charges are assigned resulting in a neutrally charged central carbon, thus electrically more stable. It seems possible that this procedure could be generalized but it has to be checked against several other ions and molecules. In both **Tables**, this pattern appears seven times.

5.4. Comparison between N2 and O2

It is common knowledge that O2 and N2 are chemically very different [12]. Elemental nitrogen, N2, found in the atmosphere, cannot be processed directly by living organisms whereas molecular di-oxygen, O2, is essential for cellular respiration in all aerobic organisms. This is not clear why because their classical structures are very similar: see **Table 1**. On the other hand, structures resulting from even-odd rule show a clear difference: Nitrogen is a neutral molecule with neutral atoms and the Oxygen has two active electrical charges. Differences in chemical properties could originate from charge location.

5.5. Resonance/Mesomerism and Charge Delocalization in Electronic Structures

Molecules classically represented using "resonance structures" also appear in **Table 1** and **Table 2**.

In 1928, L. Pauling introduced the concept of "charge delocalization" in chemistry [18] [19], in order to explain why some structures drawn using Lewis' rule have a higher potential energy than observed in the real compound. Diazenyl ion HN2(+) is for instance represented with two classical resonance structures: the charge can be either on the central nitrogen or on the outer one. By superposing both classical structures, both nitrogen bear half a charge. As an alternative, the even-odd structure assigns a charge to the central nitrogen. As a consequence, the resonance structure is not necessary for this ion.

A nearly identical pattern is proposed for dimethyl formamide; the last molecule in **Table 1**. Charges appear in one of the two classical structures. When applying the even-odd rule on the other hand, charges are clearly assigned to C(+) and O(−). No resonance structures are needed.

Another example: benzene, C6H6 molecule, in **Table 1**. Each carbon classically bears a charge of one half. With the even-odd rule, charges are assigned to atoms, alternatively positive and negative. They cannot shift from one atom to the next. Here as well, no contributing structure is necessary to represent the molecule.

The same can be written with all molecules classically represented using canonical forms in this paper. It would be of interest to apply the rule to other compounds, not listed in the present paper, in order to check if the same conclusion could be drawn.

5.6. Symmetrical Structures

In would seem natural that the charge position in a symmetrical ion should naturally respect this symmetry. This symmetry is unfortunately not preserved in many classical structures. In Allyl ion C3H5(+/−), azide anion N3(−) and methylene blue C16H18N3S(+), central atoms do not bear a charge in their classical structures. The even-odd structures of these ions preserve their symmetrical characteristic in charge positions. This symmetry preservation property supports the coherence of the even-odd rule.

5.7. Extension to Covalent Planar Layers in Crystals

The structure of benzene shown in **Table 1**, is planar. By replacing the hydrogen atoms by carbon rings and keeping the planar form, an infinite carbon-ring layer is obtained: it is the structural formula of graphite [12]. The even-odd conservation rule can also be applied to this structure, resulting in only single bonds between carbons and alternating positive and negative charges.

This planar layer structure, compatible with the even-odd rule, can also be found in alpha-boron-nitride crystals.

Another family of rhombohedral crystals has individual parallel layers following the even-odd rule; for iodides crystals, MnI2, CaI2 and PbI2; for bromides, MgBr2, MnBr2, CdBr2, FeBr2 and CoBr2; for chalcogenides, ZrS2 and alpa-TaS2 [12].

It seems thus plausible that the conversion of multiple bonds into single bonds is also applicable to classical crystal structures.

6. Conclusion

The hypothesis of the present paper is that all multiple covalent bonds in 2D structural formulas of ions and molecules could be replaced by single covalent-bonds. This modification causes internally a redistribution of charges. The scope of the even-odd rule, described in previous papers for well-known ions and molecules, is therefore expanded to multiple-bonded compounds and offers many advantages. This rule simplifies the drawing of compounds in several cases, for instance HN2 and HN3(+) are represented here with only one structure. It preserves the symmetry of large compounds like methylene blue. Compounds like C6H6, classically represented with resonance structures, have now a fixed configuration. The distinctive chemical properties of O2 and N2 can be explained by the presence or the absence of local charges. This rule can also be applied to covalent planar solids like graphite. Finally, as the even-odd rule was applied to many known compounds in past and present papers, it confirms the ability of this rule to address covalent-bonded compounds such as molecules, ions or planar crystals.

References

[1]   Couper, A.S. (1858) Sur une nouvelle théorie chimique. *Annales de chimie et de physique*, **53**, 488-489.

[2]   Loschmidt, J. (1861) Chemische Studien. Carl Gerold's Sohn, Vienna.

[3]   Abegg, R. (1904) Die Valenz und das periodische System. *Zeitschrift für anorganische Chemie*, **39**, 330-380. http://dx.doi.org/10.1002/zaac.19040390125

[4]   Lewis, G.N. (1916) The Atom and the Molecule. *Journal of the American Chemical Society*, **38**, 762-785. http://dx.doi.org/10.1021/ja02261a002

[5]   Langmuir, I. (1919) The Arrangement of Electrons in Atoms and Molecules. *Journal of the American Chemical Society*, **41**, 868-934. http://dx.doi.org/10.1021/ja02227a002

[6]   Musher, J.I. (1969) The Chemistry of Hypervalent Molecules. *Angewandte Chemie Internationale Edition*, **8**, 54-68. http://dx.doi.org/10.1002/anie.196900541

[7]   Gillespie, R.J. and Popelier, P.L.A. (2001) Chemical Bonding and Molecular Geometry. Oxford University Press, Oxford.

[8]   Auvert, G. (2014) Improvement of the Lewis-Abegg-Octet Rule Using an "Even-Odd" Rule in Chemical Structural Formulas: Application to Hypo and Hypervalences of Stable Uncharged Gaseous Single-Bonded Molecules with Main Group Elements. *Open Journal of Physical Chemistry*, **4**, 60-66. http://dx.doi.org/10.4236/ojpc.2014.42009

[9]   Auvert, G. (2014) Chemical Structural Formulas of Single-Bonded Ions Using the "Even-Odd" Rule Encompassing Lewis's Octet Rule: Application to Position of Single-Charge and Electron-Pairs in Hypo- and Hyper-Valent Ions with Main Group Elements. *Open Journal of Physical Chemistry*, **4**, 67-72. http://dx.doi.org/10.4236/ojpc.2014.42010

[10]   http://www.chemspider.com/

[11]   http://www.ncbi.nlm.nih.gov

[12]   http://en.wikipedia.org/

[13]   Greenwood, N.N. and Earnshaw, A. (1998) Chemistry of the Elements. 2nd Edition, Butterworth-Heinemann.

[14]   Kekulé, A. (1865) Sur la constitution des substances aromatiques. *Bulletin de la Société Chimique de Paris*, **3-2**, 98-110.

[15]   Wurtz, C.A. (1872) Sur un aldéhyde-alcool. *Compte Rendu de l'Académie des sciences*, **74**, 1361.

[16]   Dowd, P. (1972) Trimethylenemethane. *Accounts of Chemical Research*, **5-7**, 242-248. http://dx.doi.org/10.1021/ar50055a003

[17]   Tschitschibabin, A.E. (1907) Über einige phenylierte Derivate des p, p-Ditolyls. *Berichte der Deutschen Chemischen Gesellschaft*, **40-2**, 1810-1819. http://dx.doi.org/10.1002/cber.19070400282

[18]   Pauling, L. (1946) Resonance. Oregon State University Libraries Special Collections.

[19]   Pauling, L. (1960) The Concept of Resonance. The Nature of the Chemical Bond—An Introduction to Modern Structural Chemistry. 3rd Edition, Cornell University Press, Ithaca, 10-13.

9

A Method for Calculating the Heats of Formation of Medium-Sized and Large-Sized Molecules

Bing He[1,2], Hongwei Zhou[1,3*], Fan Yang[1], Wai-Kee Li[4]

[1]Molecular Design Institute, Chengdu Normal University, Chengdu, China
[2]State Key Laboratory of Biotherapy and Cancer Center, West China Hospital, Sichuan University, and Collaborative Innovation Center for Biotherapy, Chengdu, China
[3]IPNL, UMR IN2P3-CNRS-UCBL 5822, Villeurbanne, France
[4]Department of Chemistry, The Chinese University of Hong Kong, Hong Kong, China
Email: *jcbzhou@sina.com

Abstract

A calculation method for heats of formation (HOF, referred to as ΔH_f) based on the density functional theory (DFT) is presented in this work. Similar to Gaussian-3 theory, the atomic scheme is applied to calculate the heats of formation of the molecules. In this method, we have modified the formula for calculation of Gaussian-3 theory in several ways, including the correction for diffuse functions and the correction for higher polarization functions. These corrections are found to be significant. The average absolute deviation from experiment for the 164 calculated heats of formation is about 1.9 kcal·mol^{-1}, while average absolute deviation from G3MP2 for the 149 (among the 164 molecules, 15 large-sized molecules can not be calculated at the G3MP2 level) calculated heats of formation is only about 1.9 kcal·mol^{-1}. It indicates that the present method can be applied to predict the heats of formation of medium-sized and large-sized molecules, while the heats of formation of these molecules using Gaussian-3 theory are much difficult, even impossible, to calculate. That is, this method provides a choice in the calculation of ΔH_f for medium-sized and large-sized molecules.

Keywords

Heats of Formation, Gaussian-3 Theory, Energy, Absolute Deviation

---

*Corresponding author.

1. Introduction

Quantum chemical methods for the calculation of thermochemical data have been developed beyond the level of just reproducing experimental data and can now make accurate predictions where the experimental data are unknown or uncertain. The more accurate one in these methods is the Gaussian-n theory [1]-[8], which has been widely used to estimate the heats of formation [7] [8] of small-sized molecules. For example, in an assessment [9] of Gaussian-3 (G3) theory on the 148 calculated heats of formation of neutral molecules, the average absolute deviation from experiment is less than 1 kcal·mol^{-1}. This means that G3 theory can be used to predict heats of formation of molecules accurately. However, there are some deficiencies in G3 theory and its variation (commonly referred to as G3MP2 theory and G3B3 theory), such as, i) they can only be used to calculate the heats of formation of small-sized molecules, but become computationally intensive with the increasing number of atoms in molecules, and ii) there are large deviations for some molecules, especially for polynitrogen compounds, which are the potential candidates of high energy density materials. Especially, Gaussian-4 (G4) theory [8] and various modifications that recently come out show good accuracy for the calculation of heats of formations, the aforementioned deficiencies still exist.

The correlation method for calculation of heats of formation has drawn tremendous interest to find better ways to match the computational requirements of medium-sized and large-sized molecules, including isodesmic reaction schemes [9]-[14], group additive method, molecular mechanics and semiempirical methods [13] [15]-[17], and linear regression correction approach [18], etc. For the isodesmic reactions method, it is important to construct an appropriate bond separation reaction in which ΔH_f for all components, except the target component, are known. A bond separation reaction is a reaction which breaks down any molecule composed of three or more heavy atoms, and which can be represented in classical valence structure, into its simplest set of two heavy atom molecules containing the same type of bond, *i.e.* the number and types of all bonds are retained. Sometimes this approach is very difficult. Of cause, it does not incorporate the energy stabilization effect caused by conjugate bonds in polyene or aromatic compounds. For group additive method, molecular mechanics, semiempirical methods [13] [15]-[17] and linear regression correction approach [18], the results are strongly dependent on the parameters used and thus are less reliable because they are all parameterized methods. For example, the thermochemical parameters can be obtained easily by the semiempirical methods, but the heats of formation based on these parameters are either underestimated or overestimated. The deviations are so large that a set of terms are introduced to correct the heats of formation in agreement with experimental values. So, semiempirical methods cannot be used to predict heats of formation of compounds if the experimental data are unknown.

Ab initio MO method and density functional theory (DFT), on the other hand, are independent on the experimental results and parameters, and have emerged as a very reliable method to calculate geometries, energies, and frequencies of molecules. Hence, they have been used to evaluate the ΔH_f of interested molecules [15] [16] [19] [20]. Dunning's correlation consistent basis sets [21]-[23] (cc-pV*Z, where * denotes double, triple, quadruple, quintuple-zeta and sextuple-zeta, respectively) have the redundant functions removed and have been rotated [24] in order to increase computational efficiency. By combining the DFT with cc-pVDZ, the calculation results will be reliable. However, DFT/cc-pVDZ calculations do not produce ΔH_f directly, so special model reactions have to be designed to derive the ΔH_f (referred to as DFT ΔH_f) from the calculated total energy and vibrational analysis results [25]-[27]. This is also the goal we will pursue.

Our objective is to develop a procedure applicable to any molecular system in an unambiguous manner, which can reproduce experimental data to an accuracy of about of 2 kcal·mol^{-1} even to species having larger experimental uncertainty. Recently, we have investigated the relative stabilities of N_{2n} (N_6 (D_{3h}), N_8 (O_h), N_{10} (D_{5h}), N_{12} (D_{6h}), N_{12} (D_{3d}), N_{16} (D_{4d}), N_{18} (D_{3h}), N_{20} (I_h), N_{24} (D_{3d}), N_{24} (D_{4h}), N_{24} (D_{6d}), N_{30} (D_{3h}), N_{30} (D_{5h}), N_{32} (D_{4d}), N_{36} (D_{3d}), N_{40} (D_{4h}), N_{42} (D_{3h}), N_{48} (D_{4d}), N_{48} (D_{3d}), N_{54} (D_{3h}), N_{56} (D_{4h}), N_{60} (D_{3d}) and $N_{72}(D_{3d})$) [28] [29] molecules at B3LYP/cc-pVDZ. As the potential candidates of high energy density materials, one important issue is to calculate the ΔH_f of the molecules. However, the calculations of ΔH_f of the molecules from N_{16} to N_{72} are very difficult, even impossible using Gaussian-n theory because these molecules are medium-sized or large-sized and the experimental energies have not been well established. Furthermore, we found that Gaussian-n theory performed poorly on the polynitrogen compounds (about 2 kcal·mol^{-1} for each nitrogen atom in the molecules). In such case, the computational method for heats of formation based on DFT (referred to as DFT method) was conceived as the first in a series of well defined methods that could be routinely applied to the calculation of molecular energies of these medium-sized and large-sized molecules in a systematic manner and indeed, the results agreed with experimental values and so were reliable.

2. Theoretical and Computational Method

For the reaction Reactants → Product:

The heats of formation at 298 K (ΔH_f) can be calculated by Equation (1).

$$\Delta H_f = H_{rxn} + \Delta H_{exp,0} + \Delta H_m - \Delta H_{atom} \tag{1}$$

where $H_{rxn} = E_{product,0} - \Sigma E_{atom,0}$;

$H_{exp,0} = \Sigma H_{atom,0}$, which can be obtained from Ref. [30];

$\Delta H_m = H_{product,0} - \Sigma E_{product,0}$;

$\Delta H_{atom} = \Sigma H_{atom}$, which can be obtained from Ref. [30].

Thereof, terms $H_{exp,0}$ and ΔH_{atom} in Equation (1) are constants for the specified product whatever calculation methods are used to obtain the thermodynamic data. While terms H_{rxn} and ΔH_m vary with different computational levels.

Equation (1) is applied to calculate the ΔH_f of a compound in G3 theory and G3MP2 theory (referred to as G3MP2 ΔH_f), where total energy of the product ($E_{product,0}$) and total energy of each atom of the reactants ($\Sigma E_{atom,0}$) are referred to as "G3 (0 K)" or and "G3MP2 (0 K)". G3 (0 K) or G3MP2 (0 K) are modified by a series of corrections (referred to as E_c) from additional calculations, including a correction for diffuse functions [9] [10]

$$\Delta E(+) = E\left[MP4/6\text{-}31+G(d)\right] - E\left[MP4/6\text{-}31G(d)\right] \tag{2}$$

and a correction for higher polarization functions on nonhygrogen atoms and p-functions on hydrogens, [9] [10] etc.

$$\Delta E(2df,p) = E\left[MP4/6\text{-}31G(2df,p)\right] - E\left[MP4/6\text{-}31G(d)\right]. \tag{3}$$

It can be found that the key issue is to obtain $E_{product,0}$ and $E_{atom,0}$. In our work, only the total energy at the level B3LYP/cc-pVDZ can be obtained. Similar to the G3 theory and G3MP2 theory, the total energy at the level B3LYP/cc-pVDZ is modified by a correction for diffuse functions

$$\Delta E(+) = E\left[B3LYP/6\text{-}31+G(d)\right] - E\left[B3LYP/6\text{-}31G(d)\right] \tag{4}$$

and a correction for higher polarization functions on nonhygrogen atoms and p-functions on hydrogens.

$$\Delta E(2df,p) = E\left[B3LYP/6\text{-}31G(2df,p)\right] - E\left[B3LYP/6-31G(d)\right]. \tag{5}$$

Comparing to the 6-31G (d) basis set, the cc-pVDZ basis set has the redundant functions removed. So, the corrected total energy is described as

$$E_0(DFT) = E\left[B3LYP/cc\text{-}pVDZ\right] - \Delta E(+) - \Delta E(2df,p) \tag{6}$$

where E_0 (DFT) is the energy of each atom of the reactants that Equation (1) requires. The correction energy is defined as E_c, which can be written as

$$E_c = \Delta E(+) + \Delta E(2df,p). \tag{7}$$

Note that for H (Hydrogen) to O (Oxygen) atoms, ΔE (+) will be removed from E_0 (DFT), for fluorine atom, and ΔE (2df, p) will be removed from E_0 (DFT). According to the above corrections, E_c for the first and second row atoms are listed in **Table 1**.

A number of deficiencies in the method should be noted and future developments to alleviate them are proposed. In particular, this method works poorly on dissociation energies of ionic molecules such as LiF, on inorganic molecules such as CO_2 (5.6 kcal·mol^{-1} too low), NH_3 (3.7 kcal·mol^{-1} too large). Also, it works poorly on the hypervalent species, such as -SO_2 group and -NO_2 group, where their energies are high by 19 - 21 kcal·mol^{-1} for the -SO_2 group and low by 9 - 10 kcal·mol^{-1} for the -NO_2 group. It was found that additional group corrections might reduce discrepancy so that experimental values could be fitted perfectly.

Now, the total energy and the enthalpy of the product can be obtained from quantum chemistry calculation

Table 1. The atomic energies of the first row and the second row.

Atom	E_0 (au)	H_0 (kcal·mol^{-1})	H_m (kcal·mol^{-1})	E_c (au)
H	−0.501258	51.63	1.01	−0.001393
He	−2.907054	0.00	0.00	−0.002036
Li	0.000000	37.69	1.10	−0.000211
Be	0.000000	76.48	0.46	−0.002312
B	−24.660873	136.20	0.29	−0.005346
C	−37.851975	169.98	0.25	−0.005383
N	−54.589136	112.53	1.04	−0.003555
O	−75.068499	58.99	1.04	−0.007213
F	−99.726602	18.47	1.05	−0.001885
Ne	−128.909463	0.00	0.00	−0.021749
Na	0.000000	25.69	1.54	−0.000032
Mg	0.000000	34.87	1.19	−0.001603
Al	−242.382859	78.23	1.08	−0.003215
Si	−289.388651	106.60	0.76	−0.013175
P	−341.276438	75.42	1.28	−0.003958
S	−398.125081	65.66	1.05	−0.006948
Cl	−460.158464	28.59	1.10	−0.006929
Ar	−527.542275	0.00	0.00	−0.006574

E_0: energy of each atom of the reactants (au); H_0: the experimental heats of each atom of the reactants (kcal·mol^{-1}); H_m: the correction value of the experimental heat of each atom of the reactants (kcal·mol^{-1}); E_c: the correction energy (au).

directly. The $\Delta H_{\exp,0}$ and ΔH_{atom} can be obtained from correlative books [30]. The ΔH_f of a molecule at the level B3LYP/cc-pVDZ can be calculated by Equation (1) via Equation (6).

3. Results and Discussion

In this work, 164 compounds are selected for testing. They are divided into four test sets: i) **G2/97** test set, ii) **CH** test set, iii) **NOS** test set, and iv) **LARGE** test set.

3.1. G2/97 Test Set

There are 70 neutral molecules in this test set. The structures are taken from Ref. [9]. All calculations are carried out using the GAUSSIAN 98 program package [31]. Density Function Theory has been applied to optimize the structures at basis set cc-pVDZ. The basis sets are the correlation-consistent basis sets of Dunning, specifically the polarized valence double-ζ (cc-pVDZ). The convergence criterion is 10^{-8}. The optimized structures of the 70 species at the level B3LYP/cc-pVDZ and G3MP2 are shown in **Table 2**. The harmonic vibrational frequencies have been predicted in these optimized structures. All the vibrational frequencies of the molecules both at the levels B3LYP/cc-pVDZ and G3MP2 are positive (not listed). This indicates that the molecules are at a local minimum at the levels B3LYP/cc-pVDZ and G3MP2.

In **Table 2**, the experimental ΔH_f (Exp. column) are taken from Ref. [9]. Some values have been updated by values from Ref. [30], such as the experimental value for **02** is changed from −118.4 kcal·mol^{-1} to −119.4 kcal·mol^{-1}, for **48** from 8.9 kcal·mol^{-1} to 5.0 kcal·mol^{-1}, etc.

It can be found that the DFT ΔH_f deviations from experiment in some molecules are comparatively high (It is noted that the absolute ΔH_f deviations which are greater than 2.5 kcal·mol^{-1} are in bold and italic in **Table 2** and the subsequent tables): **01** (−7.1 kcal·mol^{-1}), **12** (−2.5 kcal·mol^{-1}), **13** (−3.1 kcal·mol^{-1}), **31** (−6.0 kcal·mol^{-1}), **37** (−2.5 kcal·mol^{-1}), **41** (−2.6 kcal·mol^{-1}), **50** (−2.7 kcal·mol^{-1}), **51** (−3.1 kcal·mol^{-1}), **53** (−4.9 kcal·mol^{-1}), **56**

Table 2. The ΔH_f and the deviations from experiment of the 70 selected molecules of the **G2/97** test set. All are in kcal·mol^{-1}.

No.	Mol.	Exp.	G3MP2	DFT	G3Dev	DFTDev
01	C_2F_4	−158.0	−161.1	−165.1	*−3.1*	*−7.1*
02	C_2Cl_4	−3.0	−7.4	−2.5	*−4.4*	0.5
03	CF_3CN	−119.4	−118.8	−121.1	0.6	−1.7
04	CH_3CCH (propyne)	44.2	44.5	46.3	0.3	2.1
05	$CH_2=C=CH_2$ (allene)	45.5	44.6	43.2	−0.9	−2.3
06	C_3H_4 (cyclopropene)	66.2	68.4	68.1	2.2	1.9
07	$CH_3CH=CH_2$ (propylene)	4.8	4.9	4.3	0.1	−0.5
08	C_3H_6 (cyclopropane)	12.7	14.1	13.5	1.4	0.8
09	C_3H_8 (propane)	−25.0	−24.5	−25.4	0.5	−0.4
10	$CH_2CHCHCH_2$ (butadiene)	26.3	29.1	29.3	*2.8*	*3.0*
11	C_4H_6 (2-butyne)	34.8	35.5	35.9	0.7	1.1
12	C_4H_6 (methylene cyclopropane)	47.9	46.5	45.4	−1.4	−2.5
13	C_4H_6 (bicyclobutane)	51.9	54.8	55.0	*2.9*	*3.1*
14	C_4H_8 (cyclobutane)	6.8	7.4	7.0	0.6	0.2
15	C_4H_{10} (trans butane)	−30.0	−29.6	−29.8	0.4	0.2
16	C_5H_8 (spiropentane)	44.3	45.4	45.2	1.1	0.9
17	C_6H_6 (benzene)	19.7	19.2	17.8	−0.5	−1.9
18	CH_2F_2	−107.7	−107.3	−106.6	0.4	1.1
19	C_4H_6 (cyclobutene)	37.4	39.4	38.9	2.0	1.5
20	C_4H_8 (isobutene)	−4.0	−3.8	−3.5	0.2	0.5
21	CHF_3	−166.6	−165.9	−166.2	0.7	0.4
22	CH_2Cl_2	−22.8	−22.3	−23.4	0.5	−0.6
23	C_4H_{10} (isobutane)	−32.1	−31.5	−30.6	0.6	1.5
24	CH_3Cl	−19.6	−19.0	−20.8	0.6	−1.2
25	$CHCl_3$	−24.7	−24.7	−24.0	0.0	0.7
26	CH_3NH_2 (methylamine)	−5.5	−3.5	−4.5	2.0	1.0
27	CH_3CN (methyl cyanide)	18.0	18.5	17.0	0.5	−1.0
28	CH_3NO_2 (nitromethane)	−17.8	−16.1	−17.5	1.7	0.3
29	CH_3SiH_3 (methyl silane)	−7.0	−6.1	−8.9	0.9	−1.9
30	HCOOH (formic acid)	−90.5	−85.4	−89.4	*5.1*	1.1
31	$HCOOCH_3$ (methyl formate)	−85.0	−85.1	−91.0	−0.1	*−6.0*
32	CH_3CONH_2 (acetamide)	−57.0	−54.3	−59.0	*2.7*	−2.0
33	C_2H_4NH (aziridine)	30.2	32.2	30.3	2.0	0.1
34	NCCN (cyanogen)	73.3	74.1	71.1	0.8	−2.2
35	$(CH_3)_2NH$ (dimethylamine)	−4.4	−2.4	−5.3	2.0	−0.9
36	$CH_3CH_2NH_2$ (trans ethylamine)	−11.3	−10.4	−11.0	0.9	0.3

Continued

37	CH$_2$O	−26.2	−25.9	−28.7	0.3	−2.5
38	CH$_3$CHO (acetaldehyde)	−39.8	−38.9	−42.9	0.9	*−3.1*
39	HCOCOH (glyoxal)	−50.7	−46.0	−52.9	*4.7*	−2.2
40	CH$_3$CH$_2$OH (ethanol)	−56.2	−55.2	−55.1	1.0	1.1
41	CH$_3$OCH$_3$ (dimethylether)	−44.0	−43.1	−46.6	0.9	*−2.6*
42	C$_2$H$_4$S (thiirane)	19.6	18.7	20.5	−0.9	0.9
43	(CH$_3$)$_2$SO (dimethyl sulfoxide)	−36.2	−34.2	−35.7	2.0	0.5
44	C$_2$H$_5$SH (ethanethiol)	−11.1	−10.8	−9.9	0.3	1.2
45	CH$_3$SCH$_3$ (dimethyl sulfide)	−8.9	−8.9	−7.4	0.0	1.5
46	CH$_2$=CHF (vinyl fluoride)	−33.2	−34.0	−34.8	−0.8	−1.6
47	C$_2$H$_5$Cl (ethyl chloride)	−26.8	−26.3	−27.6	0.5	−0.8
48	CH$_2$=CHCl (vinyl chloride)	5.0	5.1	4.3	0.1	−0.7
49	CH$_2$=CHCN (acrylonitrile)	43.2	45.0	43.7	1.8	0.5
50	CH$_3$COCH$_3$ (acetone)	−51.9	−50.9	−54.6	1.0	*−2.7*
51	CH$_3$COOH (acetic acid)	−103.4	−101.9	−106.5	1.5	*−3.1*
52	CH$_3$COF (acetyl fluoride)	−106.7	−104.6	−108.5	2.1	−1.8
53	CH$_3$COCl (acetyl chloride)	−58.0	−57.4	−62.9	0.6	*−4.9*
54	CH$_3$CH$_2$CH$_2$Cl (propyl chloride)	−31.5	−31.4	−31.8	0.1	−0.3
55	(CH$_3$)$_2$CHOH (isopropanol)	−65.2	−64.5	−63.5	0.7	1.7
56	C$_2$H$_5$OCH$_3$ (methyl ethyl ether)	−51.7	−51.4	−54.4	0.3	*−2.7*
57	(CH$_3$)$_3$N (trimethylamine)	−5.7	−4.6	−7.4	1.1	−1.7
58	C$_4$H$_4$O (furan)	−8.3	−7.7	−11.6	0.6	*−3.3*
59	C$_4$H$_4$S (thiophene)	27.5	26.1	31.6	−1.4	*4.1*
60	C$_4$H$_5$N (pyrrole)	25.9	27.0	24.0	1.1	−1.9
61	H$_2$S	−4.9	−4.9	−5.6	0.0	−0.7
62	CH$_4$	−17.9	−17.3	−18.2	0.6	−0.3
63	HCN	31.5	31.8	31.7	0.3	0.2
64	CO	−26.4	−26.8	−26.6	−0.4	−0.2
65	HCO	−26.0	−25.9	−28.7	0.1	*−2.7*
66	ClH	−22.1	−21.8	−22.8	0.3	−0.7
67	H$_3$COH	−48.0	−47.1	−47.0	0.9	1.0
68	C$_2$H$_4$	12.5	12.5	12.4	0.0	−0.1
69	C$_2$H$_6$	−20.1	−19.5	−21.0	0.6	−0.9
70	H$_2$NNH$_2$	22.8	25.9	24.2	*3.1*	1.4

Exp.: experimental ΔH_f taken form Ref. [30]; G3MP2: ΔH_f obtained at the level G3MP2; DFT: ΔH_f obtained at the level B3LYP/cc-pVDZ; G3Dev: G3MP2 ΔH_f deviation from experiment; DFTDev: DFT ΔH_f deviation from experiment. Deviations which exceed 2.5 are in bold and italic.

(-2.7 kcal·mol^{-1}), **58** (-3.3 kcal·mol^{-1}) and **65** (-2.7 kcal·mol^{-1}). The **01** (C_2F_4) is a halide. As known, Gaussian-n theory and other method work poorly on this species. For example, the calculated enthalpy of formation of C_2F_4 at G3 [9] is too negative by 4.9 kcal·mol^{-1}, whereas at G3MP2 [10] is too negative by 3.1 kcal·mol^{-1}. Our method works poorly on the molecules **05**, **10**, which contain cumulated double-bond (-X=C=Y-) because the cumulated double-bond -X=C=Y- can also be written as >X-C≡Y. There should be different ΔH_f between -X=C=Y- and >X-C≡Y. It can be found that the present method works poorly on the species which contain functional group >C=O. The calculated enthalpies of formation are underestimated too negative by 2.5 to 5.6 kcal·mol^{-1}. The molecules **31**, **37**, **41**, **50**, **51**, **53**, **56**, **58** and **65** belong to this category. It can also be found that the present method works poorly on the inorganic species. The molecules **01**, **03** and **65** belong to this category. The sum of absolute deviation from experiment for the 70 calculated heats of formation is 110.1 kcal·mol^{-1}. The average absolute deviation from experiment is about 1.6 kcal·mol^{-1}.

The G3MP2 ΔH_f deviations of some molecules from experiment value are also comparatively high: **01** (-3.7 kcal·mol^{-1}), **02** (-4.4 kcal·mol^{-1}), **10** (2.8 kcal·mol^{-1}), **13** (2.9 kcal·mol^{-1}), **30** (5.1 kcal·mol^{-1}), **32** (2.7 kcal·mol^{-1}), **39** (4.7 kcal·mol^{-1}) and **70** (3.1 kcal·mol^{-1}). It can be found that G3MP2 does poorly on the halides, too. **01** and **02** belong to this category. G3MP2 also works poorly on the molecules which contain cumulated double-bond (-X=C=Y-), **01** (-0.9 kcal·mol^{-1}) and **10** belong to this category. Both DFT and G3MP2 work poorly on the bicyclobutane (**13** in **Table 2**). The sum of absolute deviation from experiment for the 70 calculated heats of formation is only 78.6 kcal·mol^{-1}. The average absolute deviation from experiment is about 1.1 kcal·mol^{-1}.

The G3MP2 ΔH_f deviations and the DFT ΔH_f deviations from experiment value are shown in **Figure 1**. It can be found that the trends of the two lines are identical for the same molecule if the deviation is neglected. Most of the G3MP2 ΔH_f deviations from experiment are positive, while most of the DFT ΔH_f deviations from experiment are negative.

It is noted that the molecule structures are taken from the original test set of G3 theory [9] (**G2/97** test set), where a "higher level correction" (HLC) [9] is added to take into account some deficiencies in the energy calculations.

$$Ee(\text{G3}) = E(\text{combined}) + E(\text{HLC})$$

The HLC is $-An_\beta - B(n_\alpha - n_\beta)$ for molecules and $-Cn_\beta - D(n_\alpha - n_\beta)$ for atoms (including atomic ions). The n_β and n_α are the number of β and α valence electrons, respectively, with $n_\alpha \geq n_\beta$. The number of valence electron pairs corresponds to n_β. Thus, A is the correction for pairs of valence electrons in molecules, B is the correction for unpaired electrons in molecules, C is the correction for pairs of valence electrons in atoms, and D is the correction for unpaired electrons in atoms. The use of different corrections for atoms and molecules can be justified,

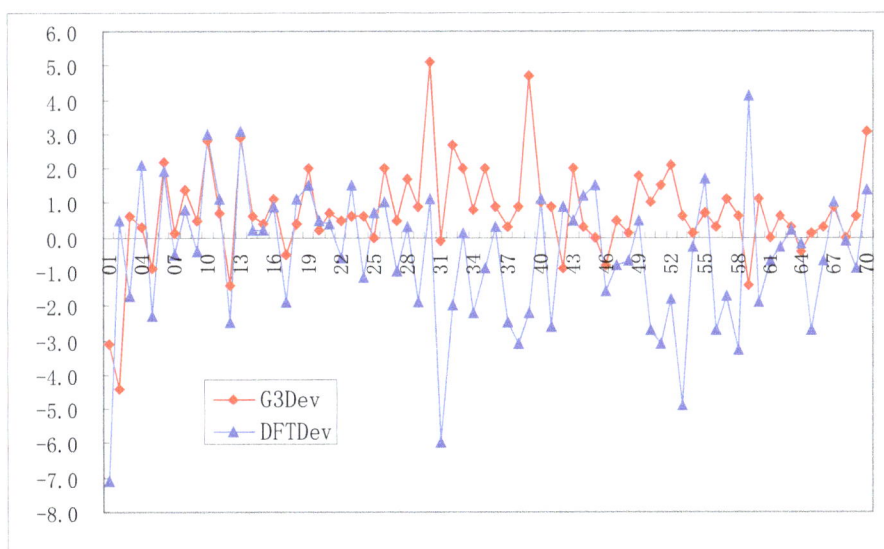

Figure 1. DFT ΔH_f and G3MP2 ΔH_f deviations from experiment of the **G2/97** test set.

in part, by noting that these extrapolations take some account of effects of basis functions with higher angular momentum, which are likely to be of more importance in molecules than in atoms. For G3 theory, A = 6.386 mhartrees, B = 2.977 mhartrees, C = 6.219 mhartrees, D = 1.185 mhartrees. The A, B, C, D values are chosen to give the smallest average absolute deviation from experiment for the **G2/97** test set. Obviously, A, B, C and D are the fit parameters which are taken into account the electron structures of molecules in **G2/97** test set, and in turn, they are used to calculate the energies of molecules in the same test set. That is, the precisions for calculation energies, especially for the molecules in the test set, are improved by introducing the fit parameters A, B, C, D. In this circumstances, it is not strange that the average absolute deviation of G3MP2 ΔH_f from experiment is less than that of DFT ΔH_f.

3.2. CH Test Set

There are 20 neutral molecules which are all typical hydrocarbons in this test set. All calculations are carried out using the GAUSSIAN 98 program package. Density Function Theory (DFT) has been applied to optimize the structures at basis set cc-pVDZ. The optimized structures of the 20 species at the levels B3LYP/cc-pVDZ and G3MP2 are shown in **Table 3**. The harmonic vibrational frequencies have been predicted in these optimized structures. All the vibrational frequencies of the molecules both at the levels B3LYP/cc-pVDZ and G3MP2 are positive (not listed). This indicates that the molecules are at local minimum at the levels B3LYP/cc-pVDZ and G3MP2.

In **Table 3**, the experimental ΔH_f (**Exp.** column) are taken from Ref [30]. It can be found that the DFT ΔH_f deviations of some molecules from experiment value are comparatively large: **01** (3.7 kcal·mol^{-1}), **15** (−4.5 kcal·mol^{-1}) and **18** (4.0 kcal·mol^{-1}). For **01** and **18**, both contain a functional group -C≡C-. It indicates that this method works poorly on the species. As is known, the isodesmic method for calculation does not incorporate the energy stabilization effect caused by conjugated bonds in polyene or aromatic compounds. It can also be found that the present method works poorly on polyene or aromatic species. The molecules **02, 03, 05, 15, 18** belong to conjugated category. The sum of absolute deviation from experiment for the 20 calculated heats of formation is 36.3 kcal·mol^{-1}. The average absolute deviation from experiment is about 1.8 kcal·mol^{-1}.

In this test set, the G3MP2 ΔH_f deviations of some molecules from experiment are also comparatively high: **02** (−4.2 kcal·mol^{-1}), **04** (−3.3 kcal·mol^{-1}), **13** (−3.3 kcal·mol^{-1}), **15** (−5.7 kcal·mol^{-1}), **16** (3.0 kcal·mol^{-1}), **19** (−5.6 kcal·mol^{-1}) and **20** (−3.8 kcal·mol^{-1}). These results show that G3MP2 theory, which is known as the isodesmic method, for calculation does not incorporate the energy stabilization effect caused by conjugated bonds in polyene or aromatic compounds. Whereas **02, 04, 013, 15, 16, 19, 20** belong to conjugated category. It can be found that the number of G3MP2 ΔH_f deviations is more than that of the DFT ΔH_f deviations. And comparing the G3MP2 ΔH_f deviations and the DFT ΔH_f deviations, one can find that the former is higher than that of the later. The sum of absolute deviation from experiment for the 20 calculated heats of formation is 45.4 kcal·mol^{-1}. The average absolute deviation is about 2.3 kcal·mol^{-1}.

The G3MP2 ΔH_f deviations and the DFT ΔH_f deviations from experiment are shown in **Figure 2**. It can be found that the trends of the two lines are identical for the same molecule if the deviation sign is neglected. Most of the G3MP2 ΔH_f deviations from experiment are negative, while most of the DFT ΔH_f deviations from experiment are possibly negative or positive. From the view of point of average absolute deviation from experiment, DFT ΔH_f method is more preferable than the G3MP2 ΔH_f method in this test set.

3.3. NOS Test Set

There are 60 neutral molecules in this test set. All calculations are carried out using the GAUSSIAN 98 program package. Density Function Theory (DFT) has been applied to optimize the structures at basis set cc-pVDZ. The optimized structures of the 60 species at the levels B3LYP/cc-pVDZ and G3MP2 are shown in **Table 4**. The harmonic vibrational frequencies have been predicted in these optimized structures. All the vibrational frequencies of the molecules both at the level B3LYP/cc-pVDZ and G3MP2 are positive (not listed). This indicates that the molecules are at local minimum at B3LYP/cc-pVDZ and G3MP2.

In **Table 4**, the experimental ΔH_f (**Exp.** column) are taken from Ref [30]. In this test set, we selected some typical molecules which contain special functional groups, such as -NO$_2$, -SO$_2$, -X=C=Y-, etc. The calculation results show that each -NO$_2$ group may be low by 9.6 kcal·mol^{-1}, and each -SO$_2$ group may be high by 20.0 kcal·mol^{-1}. In order to fit for the experimental values, 9.6 kcal·mol^{-1} is added for the DFT ΔH_f for each -NO$_2$

Table 3. The ΔH_f and the deviations from experiment of the 20 molecules of the **CH** test set. All are in kcal·mol^{-1}.

No.	Mol.	Exp.	G3MP2	DFT	G3Dev	DFTDev
01		54.5	54.0	58.2	−0.5	*3.7*
02		73.0	68.8	70.6	*−4.2*	−2.4
03		128.0	129.5	130.2	1.5	2.2
04		85.0	81.7	82.8	*−3.3*	−2.2
05		90.0	88.4	87.9	−1.6	−2.1
06		19.8	19.2	17.8	−0.6	−2.0
07		53.5	50.8	52.2	*−2.7*	−1.3
08		80.4	80.7	80.9	0.3	0.5
09		135.7	134.6	136.8	−1.1	1.1
10		53.0	51.4	53.5	−1.6	0.5
11		35.3	34.3	34.5	−1.0	−0.8
12		48.0	46.9	47.1	−1.1	−0.9
13		150.0	146.7	150.9	*−3.3*	0.9
14		49.0	47.0	47.0	−2.0	−2.0
15		66.0	60.3	61.5	*−5.7*	*−4.5*
16		60.0	57.0	58.6	*−3.0*	−1.4
17		64.0	62.0	65.7	−2.0	1.7
18		28.0	28.5	32.0	0.5	*4.0*
19		33.0	27.4	31.8	*−5.6*	−1.2
20		36.0	32.2	35.1	−3.8	−0.9

Exp.: experimental ΔH_f taken form Ref. [30]; G3MP2: ΔH_f obtained at the level G3MP2; DFT: ΔH_f obtained at the level B3LYP/cc-pVDZ; G3Dev: G3MP2 ΔH_f deviation from experiment; DFTDev: DFT ΔH_f deviation from experiment.

group a molecule contains, and 20.0 kcal·mol^{-1} is subtracted from the DFT ΔH_f for each -SO$_2$ group a molecule contains. The listed DFT ΔH_f values in **Table 4** are corrected by the two values, 9.6 kcal·mol^{-1} and 20.0 kcal·mol^{-1}.

In **Table 4**, it can be found that the DFT ΔH_f deviations of some molecules from experiment are comparatively large: **03** (3.7 kcal·mol^{-1}), **07** (−4.0 kcal·mol^{-1}), **08** (−6.3 kcal·mol^{-1}), **11** (−5.7 kcal·mol^{-1}), **12** (−3.1 kcal·mol^{-1}), **18** (−3.2 kcal·mol^{-1}), **23** (3.3 kcal·mol^{-1}), **25** (−4.0 kcal·mol^{-1}), **27** (−4.1 kcal·mol^{-1}), **28** (−3.2 kcal·mol^{-1}), **29** (−4.6 kcal·mol^{-1}), **33** (3.5 kcal·mol^{-1}), **34** (−2.7 kcal·mol^{-1}), **41** (−2.7 kcal·mol^{-1}), **42** (−2.7 kcal·mol^{-1}), **48** (−3.4 kcal·mol^{-1}), **50** (3.4 kcal·mol^{-1}), **51** (−6.0 kcal·mol^{-1}), **53** (−5.0 kcal·mol^{-1}), **54** (−4.9

Table 4. The ΔH_f and the deviations from experiment of the 60 molecules of the **NOS** test set. All are in kcal·mol^{-1}.

No.	Mol.	Exp.	G3MP2	DFT	G3Dev	DFTDev
01	CH$_3$NO$_2$	−17.9	−15.6	−17.5	2.3	0.4
02	CH$_3$ONO	−15.9	−13.3	−15.7	*2.6*	0.2
03	CH$_3$ONO$_2$	−29.5	−26.8	−25.8	*2.7*	*3.7*
04	(NH$_2$)$_2$CS	4.5	6.0	2.9	1.5	−1.6
05	CH$_3$SiH$_3$	−8.0	−6.1	−8.9	1.9	−0.9
06		111.0	117.5	111.5	*6.5*	0.5
07	HOOCCOOH	−175.7	−169.5	−179.7	*6.2*	*−4.0*
08	CH$_2$=C=S	39.0	45.3	45.3	*6.3*	*6.3*
09	CH$_3$NHNH$_2$	22.7	27.8	24.0	*5.1*	1.3
10	CH$_3$NCO	−31.0	−24.7	−32.6	*6.3*	−1.6
11	CH$_3$NCS	31.0	29.7	25.3	−1.3	*−5.7*
12	CH$_3$SCN	38.0	32.8	34.9	*−5.2*	*−3.1*
13		59.0	65.7	59.6	*6.7*	0.6
14		45.5	48.7	43.6	*3.2*	−1.9
15	NH$_2$COCONH$_2$	−95.0	−90.4	−96.5	*4.6*	−1.5
16		19.0	20.2	21.0	1.2	2.0
17		−7.0	−3.3	−7.2	*3.7*	−0.2
18	CH$_3$COOH	−103.3	−101.9	−106.5	1.4	*−3.2*
19	S=C=S	19.9	18.7	20.4	−1.2	0.5
20	C$_2$H$_5$NO$_2$	−24.4	−22.8	−25.1	1.6	−0.7
21	NH$_2$CH$_2$COOH	−92.0	−89.9	−94.2	2.1	−2.2
22	C$_2$H$_5$ONO	−25.0	−21.5	−23.7	*3.5*	1.3
23	C2H$_5$ONO$_2$	−36.8	−35.1	−33.5	1.7	*3.3*
24	(CH$_3$)$_2$NNO	15.0	16.4	13.1	1.4	−1.9
25	(CH$_3$S)$_2$	−5.9	−5.8	−1.9	0.1	*4.0*
26	(CH$_3$)$_2$SO	−36.2	−34.2	−35.7	2.0	0.5
27	(CH$_3$)$_2$SO$_2$	−89.0	−84.5	−84.9	*4.5*	*4.1*
28	(CH$_3$O)$_2$SO	−115.5	−106.3	−118.7	*9.2*	*−3.2*
29	(CH$_3$O)$_2$BH	−138.8	−134.3	−143.4	*4.5*	*−4.6*
30	(CH$_3$)$_2$SiH$_2$	−22.0	−20.7	−20.1	1.3	1.9
31	NCCN	73.3	74.5	71.1	1.2	−2.2
32	N≡C-S-S-C≡N	85.0	89.2	84.1	*4.2*	−0.9
33	CF$_2$=C=CF$_2$	−142.0	−131.8	−138.5	*10.2*	3.5
34	CH$_2$(CN)$_2$	63.5	62.1	60.8	−1.4	*−2.7*
35		−3.6	−8.3	−4.5	*−4.7*	−0.9

Continued

#	Structure	Exp	G3MP2	DFT	G3Dev	DFTDev
36		20.8	21.5	18.9	0.7	−1.9
37		40.0	37.5	41.3	−2.5	1.3
38		35.0	35.6	37.1	0.6	2.1
39		−29.0	−30.6	−27.1	−1.6	1.9
40		22.4	24.8	24.8	2.4	2.4
41	C_2H_5COOH	−107.5	−106.1	−110.2	1.4	*−2.7*
42	$HCOOC_2H_5$	−92.0	−88.3	−94.7	*3.7*	*−2.7*
43	$n\text{-}C_3H_7NO_2$	−29.7	−27.9	−29.1	1.8	0.6
44	$i\text{-}C_3H_7NO_2$	−33.2	−31.7	−31.9	1.5	1.3
45	$n\text{-}C_3H_7ONO$	−28.0	−26.6	−28.0	1.4	0.0
46	$i\text{-}C_3H_7ONO$	−32.0	−30.8	−31.4	1.2	0.6
47	$B(CH_3)_3$	−28.0	−23.1	−27.9	*4.9*	0.1
48		12.5	-	9.1	0.0	*−3.4*
49		33.0	31.6	34.7	−1.4	1.7
50		96.0	93.8	99.4	−2.2	*3.4*
51		15.0	15.5	9.0	0.5	*−6.0*
52		21.0	21.6	19.6	0.6	−1.4
53		14.0	14.0	9.0	0.0	*−5.0*
54		14.0	14.0	9.1	0.0	*−4.9*
55		16.1	15.0	11.8	−1.1	*−4.3*
56		48.0	47.6	47.9	−0.4	−0.1
57		48.0	49.4	46.3	1.4	−1.7
58		52.0	50.9	49.6	−1.1	−2.4
59		85.0	84.9	83.7	−0.1	−1.3
60		64.0	65.0	66.0	1.0	2.0

Exp.: experimental ΔH_f taken form Ref. [30]; G3MP2: ΔH_f obtained at the level G3MP2; DFT: ΔH_f obtained at the level B3LYP/cc-pVDZ; G3Dev: G3MP2 ΔH_f deviation from experiment; DFTDev: DFT ΔH_f deviation from experiment.

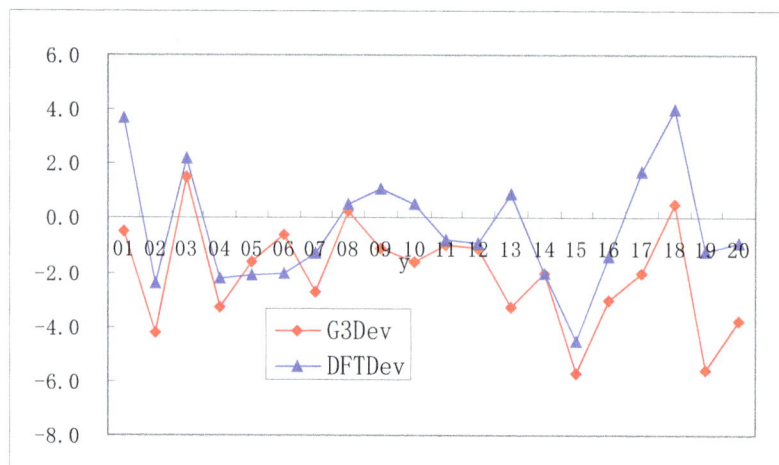

Figure 2. DFT ΔH_f and G3MP2 ΔH_f deviations from experiment of the **CH** test set.

kcal·mol^{-1}) and **55** (-4.3 kcal·mol^{-1}). Among these molecules, **03**, **23**, **48**, **51**, **53**, **54**, **55** contain the -NO$_2$ group, **27** and **28** contain the -SO$_2$ group, and **07**, **18**, **41**, **42** contain the -CO$_2$ group, while **08**, **11**, **12**, **33**, **34** contain the -X=C=Y- group. As mentioned above, the DFT ΔH_f method works poorly on these species. The sum of absolute deviation from experiment for the 60 calculated heats of formation is 132.3 kcal·mol^{-1}. The average absolute deviation is about 2.2 kcal·mol^{-1}.

In this test set, the G3MP2 ΔH_f deviations of some molecules from experiment value are also comparative high: For the molecules contain the -NO$_2$ group, **02** (2.6 kcal·mol^{-1}), **03** (2.7 kcal·mol^{-1}) and **22** (3.5 kcal·mol^{-1}); for the molecules contain the -SO$_2$ group, **17** (3.7 kcal·mol^{-1}), **27** (4.5 kcal·mol^{-1}), **28** (9.2 kcal·mol^{-1}) and **32** (4.2 kcal·mol^{-1}); for the molecules contain the -X=C=Y- group, **07** (6.2 kcal·mol^{-1}), **08** (6.3 kcal·mol^{-1}), **10** (6.3 kcal·mol^{-1}), **12** (-5.2 kcal·mol^{-1}), **33** (10.2 kcal·mol^{-1}) and **35** (-4.7 kcal·mol^{-1}); For the molecules contain the -CO$_2$ group, **07** (6.2 kcal·mol^{-1}), **15** (4.6 kcal·mol^{-1}) and **42** (3.7 kcal·mol^{-1}). Furthermore, the G3MP2 ΔH_f deviations of polynitrogen compounds, **06** (6.5 kcal·mol^{-1}), **09** (5.1 kcal·mol^{-1}), **10** (6.3 kcal·mol^{-1}), **13** (6.7 kcal·mol^{-1}), **14** (3.2 kcal·mol^{-1}), **15** (4.6 kcal·mol^{-1}) and **32** (4.2 kcal·mol^{-1}), and of boron compounds, **29** (4.5 kcal·mol^{-1}) and **47** (4.9 kcal·mol^{-1}), are high. These results show that G3MP2 theory works poorly on these species. The sum of absolute deviation from experiment for the 59 calculated heats of formation, wherein the molecule **48** cannot be calculated at G3MP2, is 157.0 kcal·mol^{-1}. The average absolute deviation from experiment for the 59 calculated G3MP2 ΔH_f is 2.7 kcal·mol^{-1}.

The G3MP2 ΔH_f deviations and the DFT ΔH_f deviations from experiment value are shown in **Figure 3**. It can be found that most of the G3MP2 ΔH_f deviations from experiment value are positive, while most of the DFT ΔH_f deviations from experiment value are possibly negative or positive. From the judgment of average absolute deviation from experiment value, the DFT ΔH_f method is more preferable than that of G3MP2 ΔH_f method in the test set because the average absolute deviation from experiment of the DFT ΔH_f is lower than that of the G3MP2 ΔH_f.

The sum of the absolute deviations from experiment is 278.7 for the above 150 calculated DFT ΔH_f. While the sum of the absolute deviations from experiment is 281.0 for the above 149 calculated G3MP2 ΔH_f. Both of the average absolute deviations are about 1.9 kcal·mol^{-1} (1.89 kcal·mol^{-1} for G3MP2 theory, 1.86 kcal·mol^{-1} for DFT method). The average absolute deviation of G3MP2 theory for the 70 molecules in **G2/97** test set is only 1.1 kcal·mol^{-1}, while the average absolute deviations of the remaining two test sets are very high (2.3 kcal·mol^{-1} for **CH** test set, and 2.7 kcal·mol^{-1} for **NOS** test set) because the former is the original test set while the later are not. Whereas, the average absolute deviations of DFT method the results are from 1.6 kcal·mol^{-1} to 2.2 kcal·mol^{-1} for all the three test sets. By taking this into account, we can conclude that the DFT method is the same effective as the G3MP2 theory in predication of ΔH_f of compounds.

3.4. LARGE Test Set

There are 14 neutral molecules in this test set. All calculations are carried out using the GAUSSIAN 98 program package. DFT has been applied to optimize the structures at basis set cc-pVDZ. The optimized structures of the

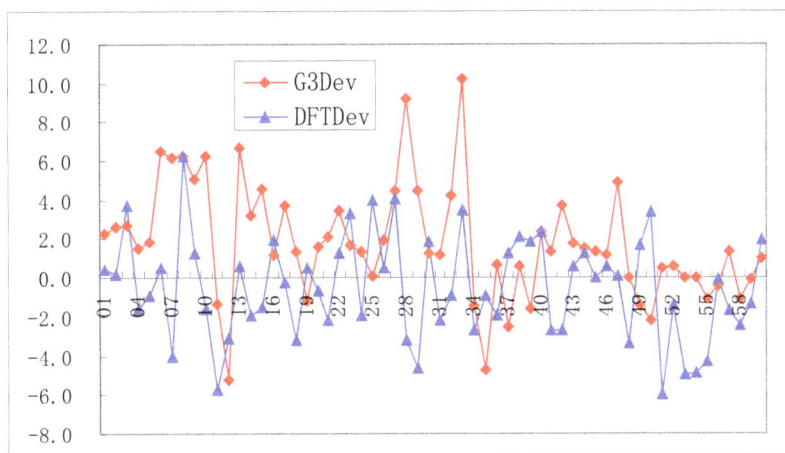

Figure 3. DFT ΔH_f and G3MP2 ΔH_f deviations from experiment of the **NOS** test set.

Table 5. The ΔH_f and the deviations from experiment of the 14 molecules of the LARGE test set. All are in kcal·mol^{-1}.

No.	Mol.	Mol.	Exp.	DFT	DFTDev
01	$C_{10}H_9NO_3$		−10.0	−9.3	0.7
02	$C_{10}H_{10}O_3$		−94.0	−98.9	*−4.9*
03	$C_{11}H_8O_2$		−55.9	−57.6	−1.7
04	$C_{11}H_9F_3O$		−168.0	−171.1	*−3.1*
05	$C_{12}H_8F_2$		−48.0	−53.8	*−5.8*
06	$C_{12}H_8O_2$		−15.0	−20.6	*−5.6*
07	$C_{12}H_{10}OS$		26.7	30.1	*3.4*
08	$C_{13}H_9N$		59.0	61.7	*2.7*
09	$C_{14}H_8O_2$		−24.4	−26.1	−1.7
10	$C_{14}H_{10}$		49.2	49.2	0.0
11	$C_{18}H_{30}$		−32.6	−31.5	1.1
12	$C_{20}H_{12}$		75.0	78.3	*3.3*
13	$C_{22}H_{14}$		81.0	82.0	1.0
14	$C_{24}H_{12}$		77.0	75.5	−1.5

Exp.: experimental ΔH_f taken form Ref. [30]; DFT: ΔH_f obtained at the level B3LYP/cc−pVDZ; DFTDev: DFT ΔH_f deviation from experiment.

14 species at the level B3LYP/cc-pVDZ are shown in **Table 5**. The harmonic vibrational frequencies have been predicted in these optimized structures. All the vibrational frequencies of the molecules at the level B3LYP/cc-pVDZ are positive (not listed). This indicates that the molecules are at local minimum at the level B3LYP/cc-pVDZ.

In **Table 5**, the experimental ΔH_f (**Exp.** column) are taken from Ref. [30]. In this test set, we selected some medium-sized and large-sized molecules, of which the calculation of heats of formation of these molecules using G3 or G3MP2 theory is much difficult, even impossible.

From **Table 5**, it can be found that the DFT ΔH_f deviations of some molecules from experiment value are comparative large: **02** (-4.9 kcal·mol^{-1}), **04** (-3.1 kcal·mol^{-1}), **05** (-5.8 kcal·mol^{-1}), **06** (-5.6 kcal·mol^{-1}), **07** (3.4 kcal·mol^{-1}), **08** (2.7 kcal·mol^{-1}) and **12** (3.3 kcal·mol^{-1}). Among them, the deviations of the molecules **04** and **05** are mainly caused by the halogen atoms in the molecules. While the deviations of the molecules **02** and **06** are mainly caused by the -CO$_2$ group. The sum of absolute deviation from experiment for the 14 calculated heats of formation is 36.5 kcal·mol^{-1}. The average absolute deviation from experiment for the 14 calculated heats of formation is about 2.6 kcal·mol^{-1}. It seems that the average absolute deviation is comparatively high in this test set. However, the high absolute deviation 5.8 kcal·mol^{-1}, for example in 05, is acceptable because the molecules are the medium-sized and large-sized.

4. Conclusion

In this work, we have developed a method for calculating the heats of formation of medium-sized and large-sized molecules. This method has the following characteristics: i) The calculation formula for the heats of formation is derived from the famous G3 and G3MP2 theory. The atomic energies are obtained from the calculated results. There are no empirical parameters or fit parameters to be introduced to eliminate the deficiencies in the calculation of the heats of formation except the corrections of the chemical functional groups -NO$_2$ and -SO$_2$. ii) The average absolute deviation from experiment for the 150 calculated DFT ΔH_f is 1.5 kcal·mol^{-1}. While the average absolute deviation from experiment for the 149 calculated G3MP2 ΔH_f is 1.7 kcal·mol^{-1}. The average absolute deviation from experiment for the whole 164 calculated DFT ΔH_f is also 1.9 kcal·mol^{-1}. The G3MP2 ΔH_f and DFT ΔH_f can be used to predict the heats of formation when the experimental data are unknown or uncertain. iii) The present method can be applied to predict the heats of formation of medium-sized and large-sized molecules. The heats of formation of a molecule containing 100 up to 200 heavy atoms can be calculated by this method. Under economical consideration, this method is expected to impact the applications in the calculations of heats of formation of large-sized molecules.

References

[1] Curtiss, L.A., Raghavachari, K., Trucks, G.W. and Pople, J.A. (1991) Gaussian-2 Theory for Molecular Energies of First- and Second-Row Compounds. *Journal of Chemical Physics*, **94**, 7221-7230. http://dx.doi.org/10.1063/1.460205

[2] Curtiss, L.A. and Raghavachari, K. (1995) In: Langhoff, S.R., Ed., *Quantum Mechanical Electronic Structure Calculations with Chemical Accuracy*, Kluwer Academic, Netherlands, 139.

[3] Pople, J.A., Head-Gordon, M., Fox, D.J., Raghavachari, K. and Curtiss, L.A. (1989) Gaussian-1 Theory: A General Procedure for Prediction of Molecular Energies. *Journal of Chemical Physics*, **90**, 5622-5629. http://dx.doi.org/10.1063/1.456415

[4] Curtiss, L.A., Jones, C., Trucks, G.W., Raghavachari, K. and Pople, J.A. (1990) Gaussian-1 Theory of Molecular Energies for Second-Row Compounds. *Journal of Chemical Physics*, **93**, 2537-2545. http://dx.doi.org/10.1063/1.458892

[5] Curtiss, L.A., Raghavachari, K., Redfern, P.C. and Pople, J.A. (1997) Assessment of Gaussian-2 and Density Functional Theories for the Computation of Enthalpies of Formation. *Journal of Chemical Physics*, **105**, 1063-1079. http://dx.doi.org/10.1063/1.473182

[6] Curtiss, L.A., Redfern, P.C., Raghavachari, K. and Pople, J.A. (1998) Assessment of Gaussian-2 and Density Functional Theories for the Computation of Ionization Potentials and Electron Affinities. *Journal of Chemical Physics*, **109**, 42-55. http://dx.doi.org/10.1063/1.476538

[7] Lau, C.-K., Li, W.-K., Wang, X., Tian, A.M. and Wong, N.B. (2002) A Gaussian-3 Study of N_7^+ and N_7^- Isomers. *Journal of Molecular Structure (THEOCHEM)*, **617**, 121-131. http://dx.doi.org/10.1016/S0166-1280(02)00411-6

[8] Curtiss, L.A., Redfern, P.C. and Raghavachari, K. (2011) Gn Theory. *Wireless Communications & Mobile Computing*,

1, 810-825.

[9]   Curtiss, L.A., Raghavachari, K., Redfern, P.C., Rassolov, V. and Pople, J.A. (1998) Gaussian-3 (G3) Theory for Molecules Containing First and Second-Row Atoms. *Journal of Physical Chemistry*, **109**, 7764-7775. http://dx.doi.org/10.1063/1.477422

[10] Curtiss, L.A., Redfern, P.C., Raghavachari, K., Rassolov, V. and Pople, J.A. (1999) Gaussian-3 Theory Using Reduced Moller-Plesset Order. *Journal of Chemical Physics*, **110**, 4703-4709. http://dx.doi.org/10.1063/1.478385

[11] Haworth, N.L. and Bacskay, G.B. (2002) Heats of Formation of Phosphorus Compounds Determined by Current Methods of Computational Quantum chemistry. *The Journal of Chemical Physics*, **117**, 11175-11187. http://dx.doi.org/10.1063/1.1521760

[12] Gong, X.D., Zhang, J. and Xiao, H.M. (1999) Studies on the Synthesis of (2S,3R)-3-Hydroxy-3-Methylproline via C-2-N Bond Formation. *Proceedings of the 26th International Pyrotechnics Seminar*, 136.

[13] Chen, Z.X., Xiao, J.M., Xiao, H.M. and Chiu, Y.N. (1999) Studies on Heats of Formation for Tetrazole Derivatives with Density Functional Theory B3LYP Method. *The Journal of Chemical Physics*, **103**, 8062-8066. http://dx.doi.org/10.1021/jp9903209

[14] Hehre, W.J. (1995) Practical Strategies for Electronic Structure Calculation. Wavefunction, Inc., Irvine, 102-134.

[15] Xu, X.J., Xiao, H.M., Ma, X.F. and Ju, X.H. (2006) Looking for High-Energy Density Compounds among Hexaazaadamantane Derivatives with Bond CN, Bond NC, and Bond ONO_2 Groups. *International Journal of Quantum Chemistry*, **106**, 1561-1568. http://dx.doi.org/10.1002/qua.20909

[16] Wang, G.X., Gong, X.D. and Xiao, H.M. (2009) Theoretical Investigation on Density, Detonation Properties, and Pyrolysis Mechanism of Nitro Derivatives of Benzene and Aminobenzenes. *International Journal of Quantum Chemistry*, **109**, 1522-1530. http://dx.doi.org/10.1002/qua.21967

[17] Ruzsinszky, A., van Alsenoy, C. and Csonka, G.I. (2002) Optimal Selection of Partial Charge Calculation Method for Rapid Estimation of Enthalpies of Formation from Hartree-Fock Total Energy. *The Journal of Physical Chemistry*, **106**, 12139-12150. http://dx.doi.org/10.1021/jp026913s

[18] Duan, X.M., Song, G.L., Li, Z.H., Wang, X.J., Chen, G.H. and Fan, K.N. (2004) Accurate Prediction of Heat of Formation by Combining Hartree-Fock/Density Functional Theory Calculation with Linear Regression Correction Approach. *The Journal of Chemical Physics*, **121**, 7086-7095. http://dx.doi.org/10.1063/1.1786582

[19] Jursic, B.S. (2003) Density Functional Calculation of the Heats of Formation for Various Aromatic Nitro Compounds. *Journal of Molecular Structure* (*THEOCHEM*), **634**, 215-224. http://dx.doi.org/10.1016/S0166-1280(03)00345-2

[20] Chen, P.C., Chieh, Y.C. and Tzeng, S.C. (2000) Computing Heats of Formation for Cubane and Tetrahrane with Density Functional Theory and Complete Basis Set *ab Initio* Methods. *Journal of Molecular Structure* (*THEOCHEM*), **499**, 137-140. http://dx.doi.org/10.1016/S0166-1280(99)00293-6

[21] Dunning, T.H. (1989) Gaussian Basis Sets for Use in Correlated Molecular Calculations. I. The Atoms Boron through Neon and Hydrogen. *The Journal of Chemical Physics*, **90**, 1007-1023. http://dx.doi.org/10.1063/1.456153

[22] Peterson, K.A., Woon, D.E. and Dunning Jr., T.H. (1994) Benchmark Calculations with Correlated Molecular Wave Functions. IV. The Classical Barrier Height of the $H + H_2 \rightarrow H_2 + H$ Reaction. *The Journal of Chemical Physics*, **100**, 7410-7415. http://dx.doi.org/10.1063/1.466884

[23] Wilson, A., van Mourik, T. and Dunning Jr., T.H. (1997) Gaussian Basis Sets for Use in Correlated Molecular Calculations. VI Sextuple Zeta Correlation Consistent Basis Sets for Boron through Neon. *Journal of Molecular Structure* (*THEOCHEM*), **388**, 339-349. http://dx.doi.org/10.1016/S0166-1280(96)80048-0

[24] Davidson, E.R. (1996) Comment on "Comment on Dunning's Correlation-Consistent Basis Sets". *Chemical Physics Letters*, **220**, 514-518. http://dx.doi.org/10.1016/0009-2614(96)00917-7

[25] Berry, R.J., Burgess Jr., D.R.F., Nyden, M.R., Zacharian, M.R., Melius, C.F. and Schwarz, M. (1996) Halon Thermochemistry: Calculated Enthalpies of Formation of Chlorofluoromethanes. *The Journal of Physical Chemistry*, **100**, 7405-7410.

[26] Raghavachari, K., Stefanov, B.B. and Curtiss, L.A. (1997) Accurate Thermochemistry for Larger Molecules: Gaussian-2 Theory with Bond Separation Energies. *The Journal of Chemical Physics*, **106**, 6764-6767. http://dx.doi.org/10.1063/1.473659

[27] Baboul, A.G., Curtiss, L.A., Redfern, P.C. and Raghavachari, K. (1999) Gaussian-3 Theory Using Density Functional Geometries and Zero-Point Energies. *The Journal of Chemical Physics*, **110**, 7650-7657. http://dx.doi.org/10.1063/1.478676

[28] Zhou, H.W., Wong, N.B., Zhou, G. and Tian, A.M. (2006) Theoretical Study on "Multilayer" Nitrogen Cages. *The Journal of Physical Chemistry A*, **110**, 3845-3852. http://dx.doi.org/10.1021/jp056435w

[29] Zhou, H.W., Wong, N.B., Zhou, G. and Tian, A.M. (2006) What Makes the Cylinder-Shaped N_{72} Cage Stable? *The*

Journal of Physical Chemistry A, **110**, 7441-7446. http://dx.doi.org/10.1021/jp062214u

[30] Lias, S.G., Bartmess, J.E., Liebman, J.F., Holmes, J.L., Levin, R.D. and Mallard, W.G. (1988) Gas-Phase Ion and Neutral Thermochemistry. *Journal of Physical and Chemical Reference Data*, **17**.

[31] Frisch, M.J., Trucks, G.W., Schlegel, H.B., Scuseria, G.E., Robb, M.A., Cheeseman, J.R., Scalmani, G., Barone, V., Mennucci, B., Petersson, G.A., Nakatsuji, H., Caricato, M., Li, X., Hratchian, H.P., Izmaylov, A.F., Bloino, J., Zheng, G., Sonnenberg, J. L., Hada, M., Ehara, M., Toyota, K., Fukuda, R., Hasegawa, J., Ishida, M., Nakajima, T., Honda, Y., Kitao, O., Nakai, H., Vreven, T., Montgomery Jr., J.A., Peralta, J.E., Ogliaro, F., Bearpark, M., Heyd, J.J., Brothers, E., Kudin, K.N., Staroverov, V.N., Kobayashi, R., Normand, J., Raghavachari, K., Rendell, A., Burant, J.C., Iyengar, S.S., Tomasi, J., Cossi, M., Rega, N., Millam, J.M., Klene, M., Knox, J.E., Cross, J.B., Bakken, V., Adamo, C., Jaramillo, J., Gomperts, R., Stratmann, R.E., Yazyev, O., Austin, A.J., Cammi, R., Pomelli, C., Ochterski, J.W., Martin, R.L., Morokuma, K., Zakrzewski, V.G., Voth, G.A., Salvador, P., Dannenberg, J.J., Dapprich, S., Daniels, A.D., Farkas, Ö., Foresman, J.B., Ortiz, J.V., Cioslowski, J. and Fox, D.J. (2009) Gaussian 09. Revision C.01. Gaussian, Inc., Wallingford.

10

Simulation of 5-Fluorouracil Intercalated into Montmorillonite Using Spartan '14: Molecular Mechanics, PM3, and Hartree-Fock

John H. Summerfield

Department of Chemistry and Physical Sciences, Missouri Southern State University, Joplin, USA
Email: summerfield-j@mssu.edu

Abstract

Molecular mechanics calculations, based on equations such as the one below, are used to investigate a colorectal cancer drug, 5-fluorouracil, intercalated into a clay, montmorillonite. This combination is currently being considered as a drug delivery system. The swelling of clays has been studied since the 1930s and is still not fully understood. Spartan '14 is used for the calculations. Semi-empirical and *ab initio* basis set scaling is also examined since there are roughly 300 atoms involved in the full model.

$$EB_{ij} = 143.9325 \frac{k_{ij}}{2} \Delta r_{ij}^2 \left(1 + c\Delta r_{ij} + \frac{7}{12} c^2 \Delta r_{ij}^2 \right)$$

Keywords

Quantum Chemistry, Clays, Spartan '14

1. Introduction

Colorectal cancer is expected to cause 50,000 deaths in the US in 2015 [1]. Surgery is typically the first option. If the disease is more advanced chemotherapy is relied on. The major alternative to chemotherapy is the drug 5-fluorouracil [2]. This drug is shown in **Figure 1**.

The objective of this work is to correctly model 5-fluoruracil as a pillared molecule between montmorillonite layers. The novelty of this study is to evaluate Spartan '14's ability to carry out this plan. In this regard molecular

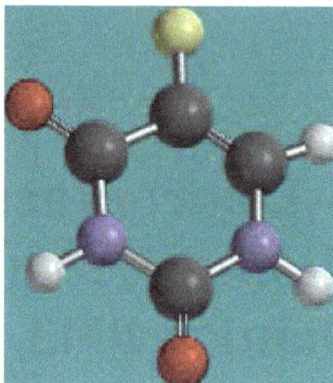

Figure 1. 5-fluorouracil. The carbons are gray. The oxygens are red. The nitrogens are blue. The fluorine is yellow. The hydrogens are white.

mechanics, semi-empirical, and Hartree-Fock computational approaches have been used to calculate the optimized structure of 5-fluorouracil intercalated between the layers of montmorillonite, a possible drug carrier.

Soils are typically 50% minerals by volume. Oxygen and silicon make up about 90% of this volume. For example, two common minerals are olivine (Fe_2SiO_4) and beryl ($Be_3Al_2Si_6O_{18}$).

Many soil minerals exist as sheets of repeating units. One type of these minerals is the phyllosilicates [3]. Phyllo is from the Greek word for "leaf". Not surprisingly, mica is a phyllosilicate. Mica is shown in **Figure 2**.

Phyllosilicates are divided into two types, 1:1 and 2:1 minerals. These ratios refer to how the sheets are arranged. Phyllosilicates consist of sheets of polymerized SO_4 tetrahedra. They are bound at three oxygen sites. The sheets are weakly bound by van der Waals forces. In addition to the tetrahedra, phyllosilicates have a sheet of octahedra (elements in six-fold coordination by oxygen) that balance out the basic tetrahedra, which have a negative charge. These tetrahedra and octahedra sheets are stacked in a variety of combinations to create the phyllosilicates. Clay minerals are a group of hydrous phyllosilicates characterized by sheets of corner sharing SiO_4 tetrahedra and/or AlO_4 octahedra [2].

The 1:1 layer structure consists of one tetrahedral sheet and one octahedral sheet as the repeating unit. An example of a 1:1 clay is kaolinite [4]. Kaolinite is shown in **Figure 3**.

Montmorillonite is an example of a 2:1 clay. It consists of two tetrahedral sheets sandwiching an octahedral sheet. Montmorillonite's structure is shown in **Figure 4** along with five intercalated 5-fluorouracil molecules.

Only oxides are present in these model molecules, no hydroxides. The distance from the top of one layer to the top of the other—the basal spacing—is 4.8×10^{-10} m in oven dried montmorillonite and $9.6 - 10.0 \times 10^{-10}$ m in the hydrated structure [5]. There are five 5-fluorouracil molecules intercalated between the layers so the interlayer distance is 12.5×10^{-10} m.

On closer examination of **Figure 4**, montmorillonite's tetrahedral sheets consist of $[SiO_4]^{4-}$ anions and the octahedral sheets of $[AlO_3(OH)_3]^{6-}$ anions. The isomorphous substitution of Al^{3+} with Mg^{2+} or Fe^{2+} in the octahedral sheets or replacement of Si^{4+} with Al^{3+} in the tetrahedral sheets generates a surface with negative charge. In nature, sodium, calcium, or magnesium intercalates the sheets to balance the charge. For 5-fluorouracil to displace the metal cations, it must form stronger intermolecular attractions than the cations [6]. Mica differs from montmorillonite. In mica, some of the silicons are replaced with aluminum. Ideally, every fourth silicon is replaced.

Currently montmorillonite is considered a good host material for controlled drug delivery. Controlled drug delivery has been of interest as a method of effective and targeted drug delivery as well as a method to reduce side effects. Particular attention has been paid to find a way to regulate the rate of drug release by a carrier where the drug is dispersed or incorporated in an inert matrix. To develop such carriers the interaction between the host (clay) and the guest (drug) compound has been of interest. It is important that the host material is harmless to the human body, will disintegrate, and be eliminated from the body once its drug delivery job is over. Due to flexible interlayer space, montmorillonite is a very good host for many guest molecules [7].

Clays were first ingested for their healing powers, medicinal effects, and as ceremonial offerings. The consumption of clays for antidiarrheal purposes has been known since the 1800s. The Chimayo clays of New Mexico

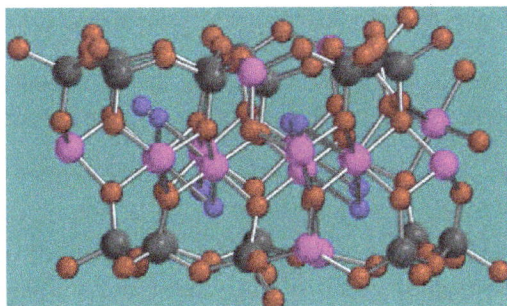

Figure 2. Mica. Silicon atoms are gray. Oxygen atoms are red. Aluminum atoms are pink. Magnesium atoms are violet.

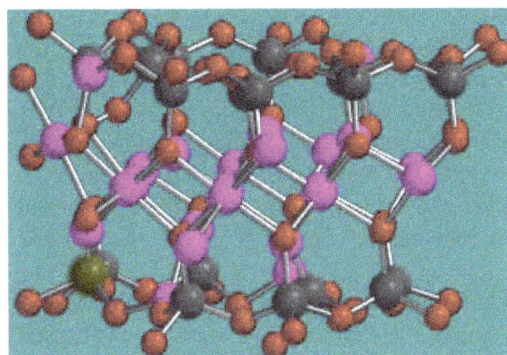

Figure 3. Kaolinite. Silicon atoms are gray. Oxygen atoms are red. Aluminum atoms are pink.

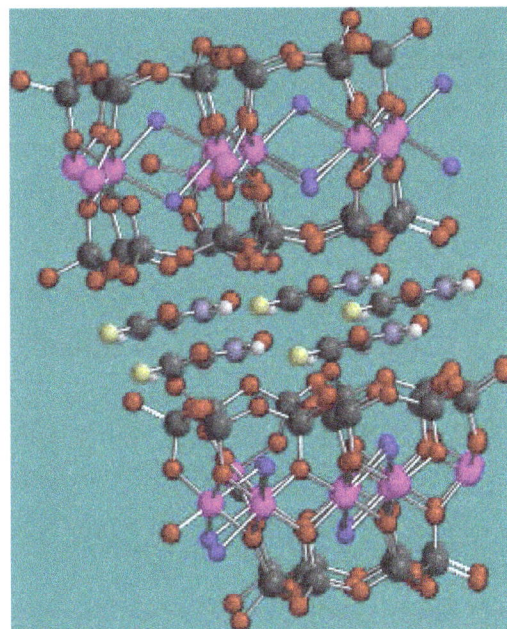

Figure 4. Montmorillonite's molecular structure along with five 5-fluorouracils. It consists of one octahedral sheet and two tetrahedral sheets. The hydroxides form at higher pH values while oxides are present at lower values. Oxygens are red. Silicons are gray. Aluminums are violet. Magnesiums are pink. Fluorines are yellow.

has been used for healing for so many years that the town's church is lined with abandoned crutches in testament to the medicinal value of this clay [8].

Montmorillonite has been shown to absorb a neutral base and then catalyze the base's conversion to its conjugate acid [9]. This reaction can be seen as another indication of the special quality of montmorillonite's surface.

2. Molecular Mechanics

Because of the combination of compositional disorder and the variety of stacking arrangements of the layers, complete information on the structural details of intercalated montmorillonite cannot usually be found using X-ray diffraction. Also, although a great deal of knowledge has been generated by various spectroscopic methods, such as infrared, these results typically require a degree of interpretation so are seldom unambiguous.

Force fields based on classical mechanics are a cornerstone in modern drug design. Molecular mechanic force fields have been used to calculate molecular geometries, energies, and thermodynamic properties [10] [11].

To begin, this work relies on the MMFF94 parameters [12] [13] as they are incorporated into *Spartan* '14 [14]. MMFF94 relies on seven energy terms. The energy expression includes a term for bond stretching, angle bending, stretch-bend interactions, out-of-plane bending, torsion interactions, van der Waals interactions, and electrostatic interactions.

The bond stretching energy expression is in the Abstract. In this expression k_{ij} is the force constant between the associated MMFF94 atom types. An example of an atom type is a C-N amide. This atom type has a calculated force constant of 5410 kJ/mol·Å^2. It is based on methylformamide and the Morse potential. The Δr_{ij} term is the difference between the actual and the associated MMFF94 atom type bond lengths, and $c = -2$ Å^{-1} is the cubic-stretch constant.

Molecular mechanics calculations are much faster than *ab initio* methods. Molecular mechanics calculations are so fast because instead of matrix diagonalization, a sum of terms is used to describe each atom. That is, individual electrons are not described but rather each atom and its electrons are units that are interconnected by the MMFF94 functions.

3. Molecular Mechanics Model Results

Figure 5 shows the swelling of the clay due to increased number of 5-fluorouracil molecules. Similar behavior is seen with a zinc intercalation simulation [15]. The interlayer spacing is on par with hydration simulations [16] [17].

Molecular mechanics calculations treats the rigid 5-fluorouracil as though it was a water molecule. That is, the guest molecules slide into the clay and form stacks.

4. Basis Set Methods

It has been shown that rigid molecules form pillars rather than stacks as they intercalate into the clay [18] [19].

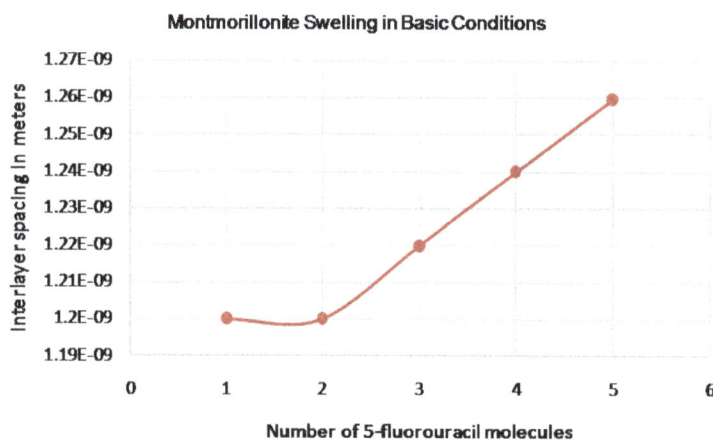

Figure 5. Intercalation of 5-fluorouracil at different interlayer spacings.

A more sophisticated calculation method or fewer atoms is required. More complex calculation methods treat the atoms and their electrons individually. The simplest of these are the semi-empirical methods. For example, parameterized model 3, PM3, uses Slater-type orbitals (STOs) to model the electrons. Each STO has the form

$$\psi = Nr^{n-1}\mathrm{e}^{-\zeta r}Y(\theta,\phi). \tag{1}$$

where N is the normalization factor, r is the electron-nucleus distance, n is the principle quantum number of the electron, ζ is the effective nuclear charge, and $Y(\theta, \phi)$ describes the angular motion of the electron [20]. Typically a linear combination of STOs is used for each electron used in the calculation. The effective nuclear charge is initially guessed from known molecules so energy minimization occurs quickly.

The problem with STOs is they increase steeply near the nucleus, where the electron-nucleus attraction is greatest. To speed calculations STOs are modelled using better behaved Gaussian functions. Two to twelve Gaussian-type orbitals are combined to model one STO. These two aspects are combined in naming basis sets. The STO-3G basis set relies on three Gaussian function to model each STO. The VSTO basis set only models the valence electrons. The 321-G uses three Gaussian functions for nonvalence electrons, two gaussians for small valence orbitals, and one for large valence orbitals. For example, the valence shell 2 s and 2 p orbitals are split into two pieces: the inner part is the sum of two gaussians while the outer part is modeled by one gaussian [21].

The next step up in calculation complexity is a Hartree-Fock calculation. In this case no experimental data is relied upon. Instead the effective nuclear charge is varied sequentially until the lowest energy is reached [22].

5. Basis Set Model Results

At the level of PM3, the current model calls for 1099 basis functions. CPU time scales as n^3 or n^4 where n is the basis set size [23]. Splitting the difference, this would call for about $1099^{3.5}$ s of CPU time which is over a million hours. For this reason, the model was reduced from 238 atoms to 58 atoms, about one-fourth the size. The modelled atoms are silicate rings and a 5-fluorouracil molecules. The simplified molecule is shown in **Figure 6**.

One change caused by simplifying the model is that now the 5-fluorouracil forms a pillar between the clay layers, as supported by experiment. The ground state energies found by different calculation methods are collected in **Table 1**. The interlayer distance 12×10^{-12} m.

6. Discussion

Spartan '14 converges for 58 small atoms. It only converges at the Hartree-Fock level using one of the smallest, diffuse basis sets, 3-21G*. This evaluation will hopefully provide guidance for future applications.

This evaluation of *Spartan* '14 is a reminder that experiment and computation go hand in hand. When too many molecules are used 5-fluorouracil is modeled as a hydrating molecule rather than a pillared molecule. Apparently, weak, short-range interactions are most important.

One possible extension is to investigate the protonated clay. It is shown in **Figure 7** along with one 5-fluorouracil molecule. This structure would offer the chance to investigate hydrogen bonding between the host and the guest.

7. Discussion

Spartan '14 easily models the intercalation of 5-fluorouracil into montmorillonite layers at the molecular

Table 1. Ground state energy of 58 atom model relying on a variety of calculations.

Method	Ground state energy in kJ/mol	Basis set	Basis function increase	Energy change in kJ/mol
MMFF	1.38×10^3			
PM3	1.08×10^3	VSTO-3G 223 functions		
Hartree-Fock	1.56×10^3	STO-3G 318 functions	95	480
Hartree-Fock	1.57×10^3	3-21G* 601 functions	283	10

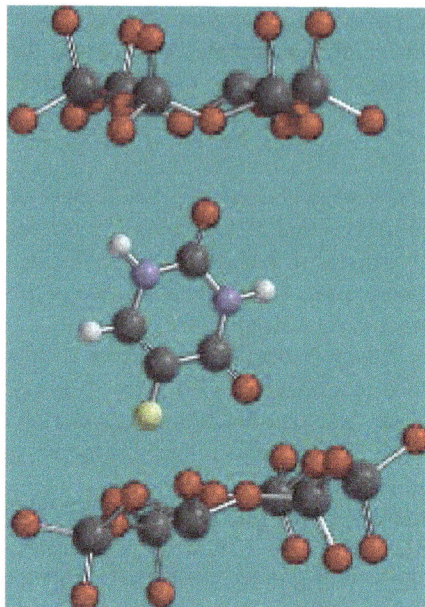

Figure 6. Silicate rings and a 5-fluorouracil molecule. Oxygens are red. Silicons are gray. Aluminums are violet. Magnesiums are pink. Fluorines are yellow. The interlayer distance 12×10^{-12} m.

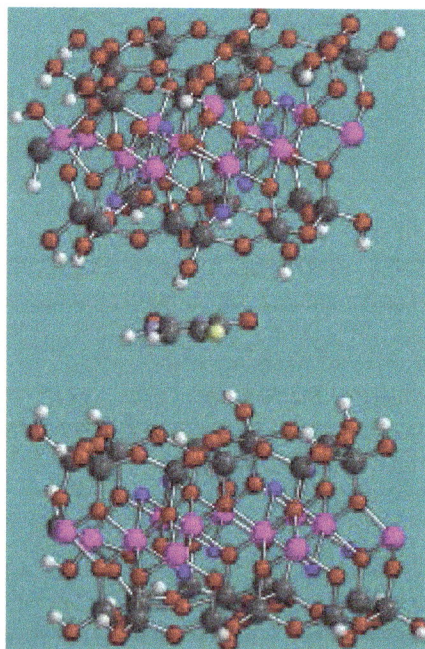

Figure 7. Protonated montmorillonite layers and a 5-fluorouracil molecule. Oxygens are red. Silicons are gray. Aluminums are violet. Magnesiums are pink. Fluorines are yellow.

mechanics level of computation. The 5-fluorouracil molecules slip between the layers and the clay swells.

Describing the correct behavior of the rigid guest molecule is more difficult. Using 58 atoms, rather than 300, the 5-fluorouracil stands vertically between the clay layers. At the Hartree-Fock level of computation, 318 basis functions yield near convergence.

References

[1]   Pyra, C. (2015) Colon Cancer. Colon Cancer Alliance, Washington DC, 3.

[2]   Lin, F.H., Lee, Y.H., Jian, C.H., Wong, J., Shieh, M. and Wang, C. (2002) A Study of Purified Montmorillonite Intercalated with 5-Fluorouracil as Drug Carrier. *Biomaterials*, **23**, 1981-1987.
       http://dx.doi.org/10.1016/S0142-9612(01)00325-8

[3]   Dixon, J.B. and Weed, S.B. (1989) Minerals in Soil Environments. 2nd Edition, Soil Science Society of America, Madison, 1.

[4]   Dixon, J.B. and Weed, S.B. (1989) Minerals in Soil Environments. 2nd Edition, Soil Science Society of America, Madison, 6-16.

[5]   Dixon, J.B. and Weed, S.B. (1989) Minerals in Soil Environments. 2nd Edition, Soil Science Society of America, Madison, 18-19.

[6]   Dixon, J.B. and Weed, S.B. (1989) Minerals in Soil Environments. 2nd Edition, Soil Science Society of America, Madison, 706.

[7]   Kevadiya, B.D., Patel, H.A., Joshi, G.V., Abdi, S.H.R. and Bajaj, H.C. (2010) Montmorillonite-Alginate Composites as a Drug Delivery System: Intercalation and *in Vitro* Release of Diclofenac Sodium *Indian Journal of Pharmaceutical Sciences*, **72**, 732-737.

[8]   Ferrell Jr., R.E. (2008) Medicinal Clay and Spiritual Healing. *Clays and Clay Minerals*, **56**, 751-760.
       http://dx.doi.org/10.1346/CCMN.2008.0560613

[9]   Helsen, J. (1982) Clay Minerals as Solid Acids and Their Catalytic Properties. *Journal of Chemical Education*, **59**, 1063-1065. http://dx.doi.org/10.1021/ed059p1063

[10]   Boyd, D.B. and Lipkowitz, K.B. (1982) Molecular Mechanics: The Method and Underlying Philosophy. *Journal of Chemical Education*, **59**, 269-274. http://dx.doi.org/10.1021/ed059p269

[11]   Cox, P.J. (1982) Molecular Mechanics: Illustrations of Its Application. *Journal of Chemical Education*, **59**, 275-277.
       http://dx.doi.org/10.1021/ed059p275

[12]   Halgren, T.A. (1996) Merck Molecular Force Field I. *Journal of Computational Chemistry*, **17**, 490-519.
       http://dx.doi.org/10.1002/(SICI)1096-987X(199604)17:5/6<490::AID-JCC1>3.0.CO;2-P

[13]   Halgren, T.A. (1996) Merck Molecular Force Field II. *Journal of Computational Chemistry*, **17**, 520-552.
       http://dx.doi.org/10.1002/(SICI)1096-987X(199604)17:5/6<520::AID-JCC2>3.0.CO;2-W

[14]   Hehre, W.J. and Ohlinger, W.A. (2014) Spartan '14. Wavefunction, Inc., Irvine.

[15]   Janeba, D., Čapková, P. and Schenk, H. (1998) Molecular Simulations of Zn-Montmorillonite. *Clay Minerals*, **33**, 197-204.

[16]   Skipper, N.T., Sposito, G. and Chang, F.C. (1998) Monte Carlo Simulation of Interlayer Molecular Structure in Swelling Clay Minerals. 2. Monolayer Hydrates. *Clays and Clay Minerals*, **43**, 294-303.
       http://dx.doi.org/10.1346/CCMN.1995.0430304

[17]   Sutton, R. and Sposito, G. (2002) Animated Molecular Dynamics Simulations of Hydrated Cesium-Smectite Interlayers. *Geochemical Transactions*, **3**, 73-80. http://dx.doi.org/10.1186/1467-4866-3-73

[18]   Lin, F., Chen, C., Cheg, W. and Kuo T. (2006) Modified Montmorillonite as Vector for Gene Delivery. *Biomaterials*, **17**, 3333-3338. http://dx.doi.org/10.1016/j.biomaterials.2005.12.029

[19]   Kameyama, H., Narumi, F., Hattori, T. and Kameyama, H. (2006) Oxidation of Cyclohexene with Molecular Oxygen Catalyzed by Calcium Porphyrin Complexes Immobilized on Montmorillonite. *Journal of Molecular Catalysis A: Chemical*, **258**, 172-177. http://dx.doi.org/10.1016/j.molcata.2006.05.022

[20]   Slater, J.C. (1930) Atomic Shielding Constants. *Physical Review*, **36**, 57-67. http://dx.doi.org/10.1103/PhysRev.36.57

[21]   Levine, I.N. (1991) Quantum Chemistry. Prentice Hall, Englewood Cliffs, 461-466.

[22]   Froese Fischer, C. (1987) General Hartree-Fock Program. *Computer Physics Communication*, **43**, 355-365.
       http://dx.doi.org/10.1016/0010-4655(87)90053-1

[23]   Strout, D. and Scuseria, G.A (1995) Quantitative Study of the Hartree-Fock Method. *Journal of Chemical Physics*, **102**, 8448-8452. http://dx.doi.org/10.1063/1.468836

11

Effect of Chloride Concentration on the Corrosion Rate of Maraging Steel

Hussam El Desouky[1], Hisham A. Aboeldahab[1,2*]

[1]Chemistry Department, University of Umm Al-Qura, Makka, KSA
[2]Chemistry Department, Faculty of Science, University of Alexandria, Alexandria, Egypt
Email: eldesouky4@gmail.com, *hdahab-41@hotmail.com

Abstract

The corrosion behavior of Maraging steel has been studied by using different techniques, including open circuit potential and polarization measurements in addition to microstructure examination such as optical microscopy and XRD (X-Ray Diffraction) investigation. The corrosion behavior of Maraging steel has been examined in sodium chloride solutions with different concentrations from 0.1 M to 2 M. It was found that the corrosion resistance of Maraging steel is inversely proportional with the concentration of sodium chloride solution. The corrosion resistance is directly proportional to the Mo and Ti content in the Maraging steel. Heat treatment of the Maraging steel improved its mechanical properties with no effect on the corrosion behavior as the precipitation of inter-metallic compounds leading to some galvanic action. However, sample IV having lower Mo content than sample V showed after heat treatment an improvement in the corrosion resistance.

Keywords

Maraging Steel, Polarization, Microstructure Examination, Optical Microscopy, XRD, Heat Treatment and Corrosion Resistance

1. Introduction

Studying the corrosion behavior of different kinds of steels is important being the most widely-used materials for constructions in our life, primarily due to the fact that they can be manufactured relatively cheaply in large quantities and to very precise specifications. They also provide an extensive range of mechanical properties from moderate strength levels with excellent ductility and toughness, to very high strength with adequate ductility. Iron and steels comprise well over 80% by weight of the alloys in general industrial uses. Maraging steels

---

*Corresponding author.

are a class of high strength steels characterized by very low carbon contents and the use of substitutional elements to produce age hardening in iron-nickel martensites. The very low carbon content and the use of intermetallic precipitation to achieve hardening produce served unique characteristics that set maraging steel apart from conventional steels.

Maraging steel work well in electromechanical components where ultra-high strength is required, along with good dimensional stability during heat treatment. Several desirable properties of maraging steel are [1]:

- Ultra-high strength at room temperature.
- Simple heat treatment, which results in minimum distortion.
- Superior fracture toughness compared to quenched and tempered similar strength level.
- Low carbon content, which precludes decarburization problems.
- Section size is an important factor in the hardening process.
- Easily fabricated.
- Good weld ability.

Maraging steel is characterized by a combination of good ductility, very high strength, good mechanical properties and simple heat treatment, so that Maraging steel considers the most important steel in the heavy industry.

The useful characteristics of maraging steel are:

1) Excellent mechanical properties comprising high-to-weight ratios and high strength combined with good toughness.

2) Good processing characteristics including good hot and cold workability.

3) Simple heat treatment with no decarburization, no liquid quenching and high dimensional stability.

4) Good fabrication characteristics including a low rate of work hardening, good machinability and excellent weldability in annealed or aged conditions.

They are suitable for engine components, such as crank shafts and gears, and the firing pins of automatic weapons. It also used in surgical components and hypodermic syringes.

The present work is aimed to study the following: a) effect of adding different alloying elements on the corrosion behavior of maraging steel; b) the corrosion behavior in different concentrations of Sodium Chloride; c) study the corrosion behavior of heat treated Maraging steel in Sodium Chloride solution.

The techniques used in this works included open circuit potential, linear polarization, microstructure investigation (Optical Microscope) and X-ray analysis.

2. Materials and Experimental Techniques

The corrosion behavior of maraging steel electrodes was studied in details using open circuit potential and potentiodynamic polarization measurements as well as optical microscopy and X-ray analysis.

2.1. Materials

Table 1 illustrates that there are five solution testament samples of maraging steel had been prepared and used in the corrosion testing of maraging steel. The solution treatment samples are I, II, III, IV and V. Only three heat treated samples of maraging steel (aged) were used in the experiments, these three samples are (I, IV and V).

2.2. For Electrochemical Measurements

The eight maraging steel electrodes (five solution treatment samples and three aged samples) used in the electrochemical experiments were cut as cylindrical shape. Copper wire welded on the top of the sample for the electrical contact. All samples (solution treatment samples and aged samples) were mounted in glass tubes by two component araldite leaving a surface area of ~0.8 cm^2 to contact the test solution.

2.3. Polarization Measurements

Polarization measurements were carried out by the same electrodes in 0.6 M Sodium Chloride solution. Polarization studies were done using the same specimens in the same previous solutions in the polarization cell. E vs. log I curves were recorded at temperature 25°C. Spiral platinum was used as the counter electrode (C); saturated calomel electrode (SCE) was used as reference electrode (RE). A constant quantity of the test solutions (100 ml) was taken in the polarization cell in order to make the conditions identical in all experiments. The corrosion kinetic parameters such as corrosion current (I_{corr}), corrosion potential ($E_{corr.}$), cathodic Tafel slope (βc), and anodic Tafel slope (βa) were derived from the curves. The polarization data were recorded from potentiostate model **IM6e (Zahner Elektrik)**.

2.4. Electrolytes

Sodium chloride solutions were prepared from chemically pure grade ADWIC and five concentrations were used from Sodium Chloride 0.1 M, 0.3 M, 0.6 M, 1 M and 2 M. Second distilled water was used throughout the experiments for the preparation of the solutions. All experiments were approximately made at the same temperature of 25°C ± 0.5°C.

3. Results and Discussion

3.1. Open Circuit Polarization Measurements

The open circuit (OC) potential is the potential of the working electrode relative to the reference electrode when no potential or current is being applied to the cell. Although Potentiometric experiments are very simple, they have many important applications. Potentiometric measurements are based on Nernst equation, which relates the concentration of electroactive species at the electrode surface (Cs) to the potential (E) at that electrode; that is, for the reaction: $O + e^- = R$

$$E = Eo + \frac{0.059}{n} \log \frac{Co}{Cr}$$

Where Eo is the formal redox potential of the electron transfer reaction. The potential E is measured between two electrodes: the working electrode and the reference electrode.

The way in which a metal changes its potential upon immersion in solutions indicates the nature of reaction taking place at its surface. Whilst a shift in potential towards more positive values denotes film formation and thickening, a shift in the negative direction signifies film destruction and the exposure of more of the bare metal to the aggressive solution. The results obtained were made to be used to discuss the mechanism of oxide film growth or corrosion of the metal in solutions.

The open circuit potentials of steel electrodes with different compositions are followed, as a function of time in different solutions till steady state value (E_{ss}). The concentrations of the test solutions were varied from 0.1 M to 2 M of Sodium Chloride.

3.1.1. Effect of Sodium Chloride Concentration on the Solution Treatment Maraging Steel

Figures 1-5 illustrate the effect of Sodium Chloride concentrations varied from 0.1 M to 2 M on solution treatment maraging steel, the open circuit potentials of the following electrodes I, II, III and IV have a tendency to shift towards more negative values with increasing Sodium Chloride concentrations. On the other hand, the open circuit potentials of V sample changing towards more noble values immediately on immersion in Sodium Chloride medium at any concentration varied from 0.1 M to 1 M. In sodium chloride solutions with concentration higher than 1 M the open circuit potential of this sample will tend to metal dissolution like the other samples.

As shown in **Table 1**, V electrode is the sample that contains the highest molybdenum content (4.48%) than the other samples (I = 0.0074%, III = 0.112 %, II = 0.0484% and IV 2.95%).

A theory for film thickening on the surface of metals and alloys based on open circuit potential-time measurements has been developed by *A. M. Shams El-Din and Paul* [2]. The essence of the theory is based on the idea that the potential is determined by a simultaneous anodic (film formation) and cathodic (oxygen reduction) couple, in which the anodic reaction is rate limiting. By presenting the data in the form of potential-log (time) curves, straight lines were obtained satisfying the relation

$$E = cons \tan t + 2.303 \frac{\delta}{\beta} \log t$$

where t was the time from the moment of immersion in solution, δ^- was the rate of oxide film thickening per-

Table 1. Chemical composition and designation of the tested samples.

	I	II	III	IV	V
C	0.031	0.0483	0.036	0.0386	0.0497
Si	0.0008	0.485	0.0935	0.318	0.253
Mn	0.421	0.308	0.246	0.345	0.398
P	0.0191	0.0214	0.0289	0.013	0.0274
S	0.0119	0.0145	0.0115	0.0145	0.0203
Cr	0.004	5.22	5.4	4.97	5.43
Mo	0.0074	0.0484	0.112	2.95	4.48
Ni	11.75	11.333	12.48	12.54	12.64
Al	0.0013	0.136	0.0001	0.113	0.125
Ti	0.003	0.995	0.028	0.677	0.629

The balance is the wt% of Fe in each sample.

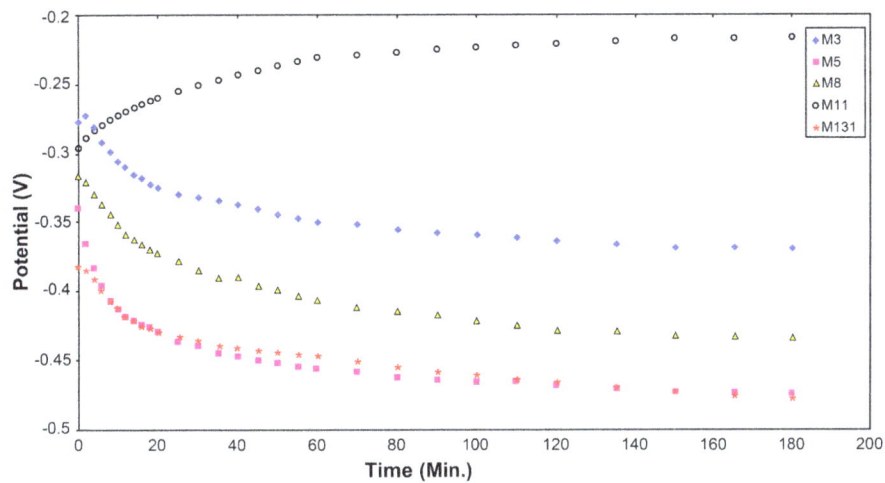

Figure 1. Potential time curve for solution treatment Maraging steel samples in 0.1 M Sodium Chloride solution at room temperature.

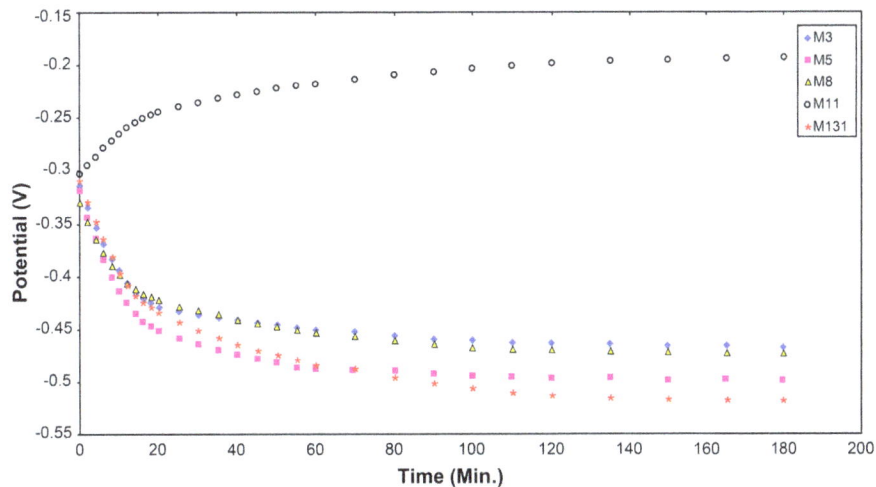

Figure 2. Potential time curve for solution treatment Maraging steel samples in 0.3 M Sodium Chloride solution at room temperature.

decade of time, and β was given by:

$$\beta = \frac{nF}{RT}\alpha\delta t$$

α is a transference coefficient similar to that found in electrochemical kinetic rate expressions ($0 < \alpha < 1$) and δ is the width of the activation energy barrier to be traversed by the ion during oxide formation, R is the gas constant and T is the absolute temperature (at 25°C). Assuming α to have the value of 0.5 and δ the value of 1.0 nm, values of δ for the various steel examined can be calculated. By analogy with the case of Fe-Cr and molybdenum containing steels, it is assumed that the trivalent cations diffuse through the film to the oxide-solution interphase. The constant n in the equation is set equal to 3, and β acquires the value of 58.6 nm V^{-1}.

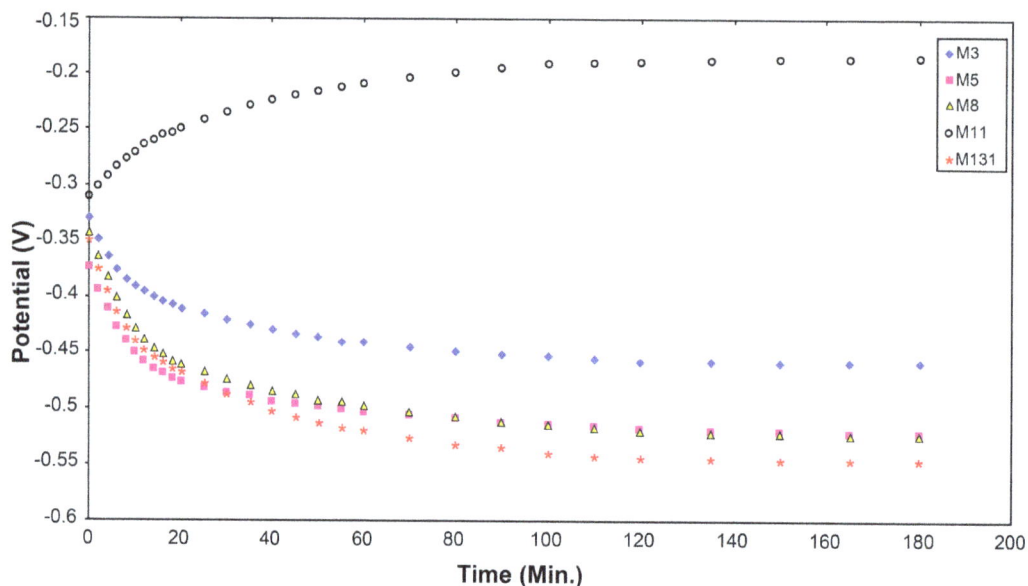

Figure 3. Potential time curve for solution treatment Maraging steel samples in 0.6 M Sodium Chloride solution at room temperature.

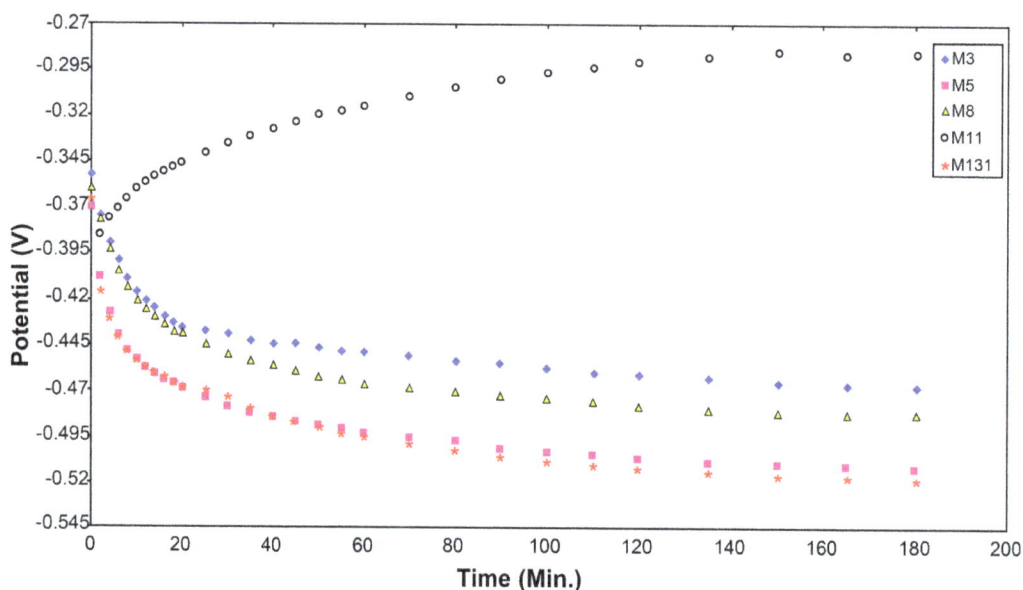

Figure 4. Potential time curve for solution treatment Maraging steel samples in 1 M Sodium Chloride solution at room temperature.

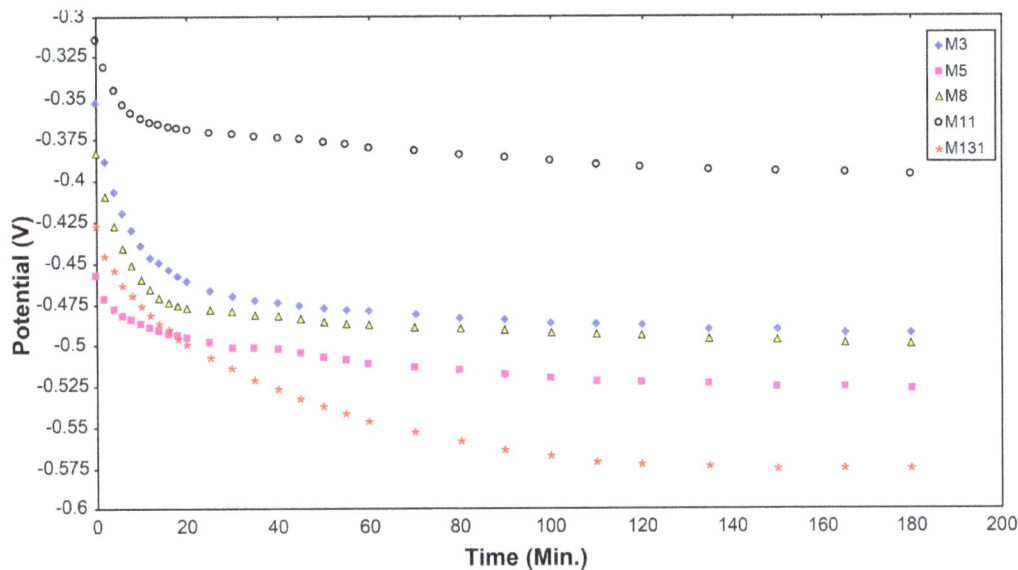

Figure 5. Potential time curve for solution treatment Maraging steel samples in 2 M Sodium Chloride solution at room temperature.

The presence of Chloride ions in solution initiates breakdown of passivity and the reaction is diffusion controlled. With increasing chloride concentrations, the diffusion components disappear and the interfacial reactions become charge transfer controlled, so that the tendency for steel to corrode in solutions increases with increasing Sodium Chloride concentration [3].

The film thickening of V increases at Sodium Chloride solutions from 0.1 M to 1 M but after this range the steel tends to breakdown the layer, the rate of oxide thickening representative curve in the figure, amounted along the fires segment $\delta 1$, was always higher than that along the second $\delta 2$. This may be correlated with the creation of ionic current which would decrease with increased film thickness with subsequent decrease of the growth rate [2].

In Sodium Chloride solutions at concentrations varied from 0.1 M to 1 M, the film thickening rate δ increased gradually with increasing the concentration of Sodium Chloride. This may be explained by assuming that the adsorption of these ions initiates a larger field strength which promotes the development of thicker oxide film but at 1 M Sodium Chloride the film starts to breakdown [4].

3.1.2. Effect of Aging on Maraging Steel
Effect of Sodium Chloride on Aged Maraging Steel

The open circuit potential of aged maraging steel electrodes will be inspected in **Figure 6** in 0.1 M Sodium Chloride solution, which represent that all samples even V, will be corroded in 0.1M Sodium Chloride after heat treatment.

The steady state potential of the solution treatment maraging steel in 0.1 M Sodium Chloride for I, IV and V are –460 mV, –370 mV and –220 mV respectively, but in case of 0.1 M Sodium Chloride aged maraging steel the steady state potential for I, IV and V are –475 mV, –400 mV and –370 mV respectively, so that by comparison the open circuit potential of the aged steel samples will shift towards more negative values than the solution treatment samples.

From the previous results it can be concluded that the corrosion behavior of solution treatment maraging steel is better than the aged maraging steel. On the other hand aged steel has a good mechanical properties than the solution treatment steel, because intermetallic compounds which formed inside the alloy have different electro negativity between each ether, so that galvanic corrosion may be appeared, which can be lead to increase the corrosion of the aged maraging steel more than the corrosion of solution treatment maraging steel.

Table 2 represents the average results of the room temperature mechanical testing of investigated steels after solution treatment at 820°C for 1 hour and air cooling. The data representing the 0.2% offset yield (Rp0.2) and

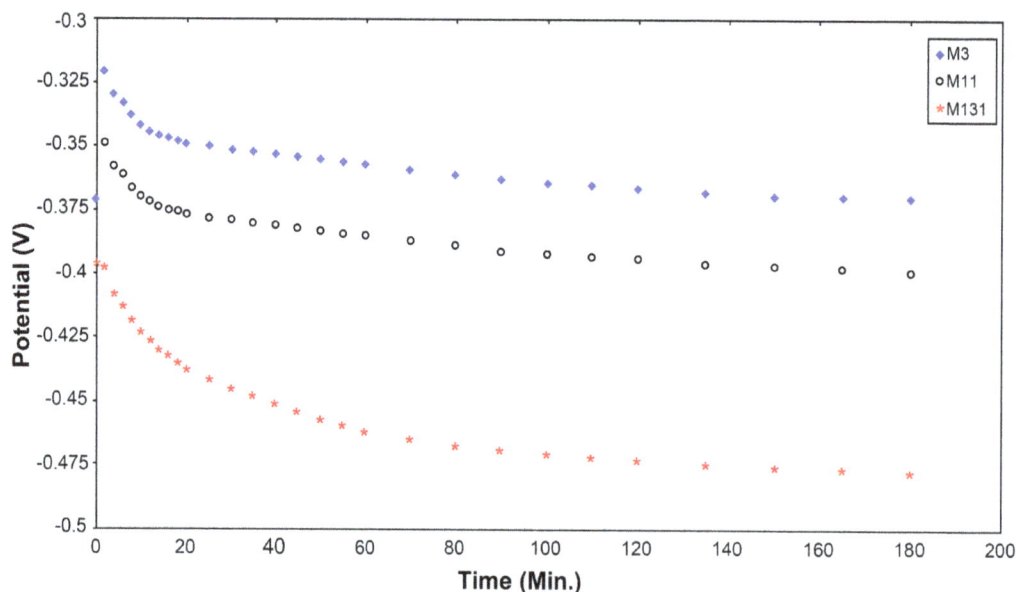

Figure 6. Potential time curve for aged Maraging steel samples in 0.1 M Sodium Chloride solution at room temperature.

Table 2. Mechanical properties of Maraging steel before and after aging.

	Aging	V	IV	III	II	I
Yield Strength	Before	890	897	858	918	645
	After	1427	1211	1046	1725	701
Ultimate Tensile Strength	Before	1092	1008	1142	1052	757
	After	1429	1225	1096	1746	733
Elongation (A)%	Before	14	13	15	10	19
	After	12	15	18	8	24

ultimate tensile strength (Rm) as well as tensile elongation (A). The ultimate tensile and yield strength increased from 757 and 645 to 1092 and 890 N/mm^2, respectively, with increasing Ti up to 1.6%. **Figures 7-9** illustrate the mechanical properties of the solution treatment and aged Maraging steels.

3.2. Polarization measurements

3.2.1. Effect Sodium Chloride Concentrations on Solution Treatment Maraging Steel

The resistance of metals and alloys to corrosion is dependent upon multitude of factors. It is therefore difficult to predict the behavior of a metal or alloy in a special environment or to establish the optimal choice of alloy in a give process. The comparison of the polarization curves for different metals in the same solution provides information, which can also be applied to other solutions. Alloys can be ranked with regard to corrosion resistance and the influence of different alloying element can be determined with respect to different corrosion parameters. However, the appearance of the polarization curves depends on the method used.

In different concentrations of Sodium Chloride solutions varied from (0.1 M to 2 M), the polarization measurements of the five maraging steel electrodes (I, II, III, IV and V) were started at −1000 mV. At this potential, the electrodes are activated, since surface oxide is reduced or pealed off as a result of hydrogen evolution. Dissolution of metal may take place, since the potential is well above the equilibrium potential for iron but dissolution is very slow. Green and Leonard and others [5] reported that the time for activation and the potential at which this occurs are of considerable importance for the shape of the polarization curve. The current density (I) (mA/cm^2) was plotted against polarization potential (E) (Vs. SCE) (I/E curves) from the figures. In all concen

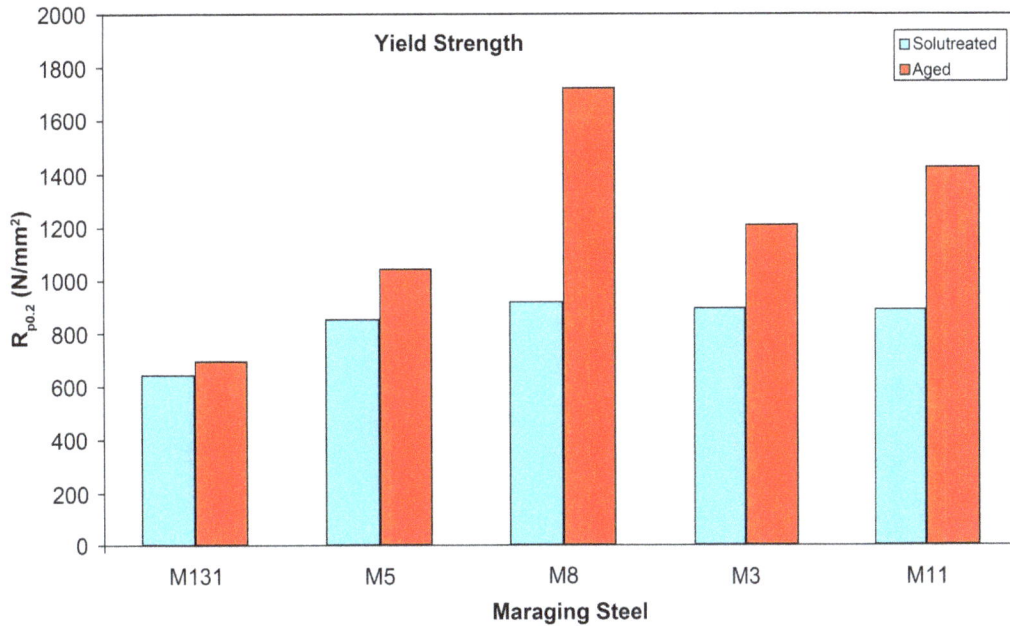

Figure 7. Comparison between mechanical properties (yield strength) of solution treatment and aged Maraging steel.

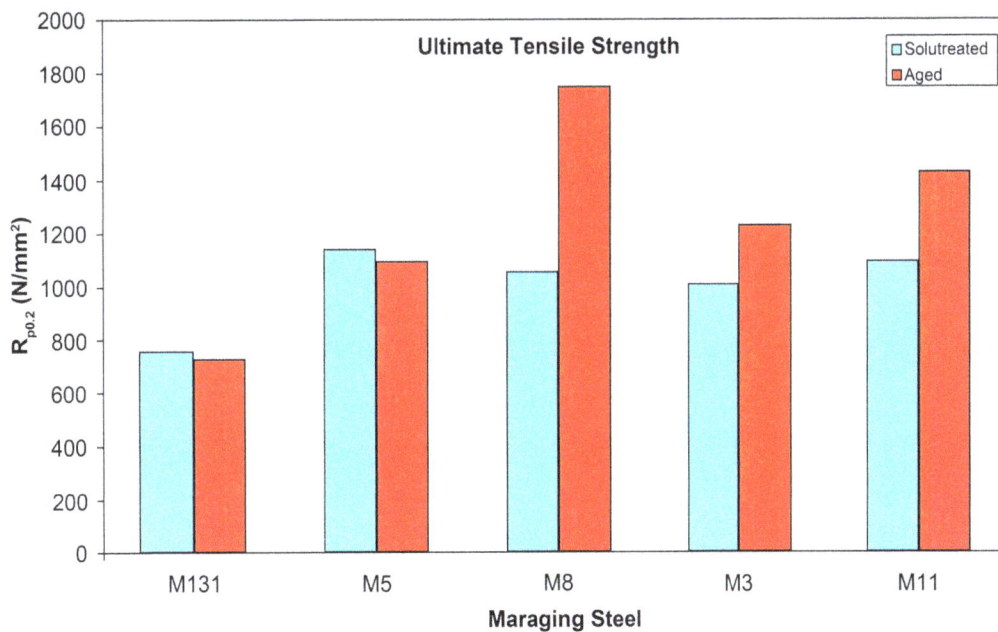

Figure 8. Comparison between mechanical properties (ultimate tensile strength) of solution treatment and aged Maraging steel.

trations of Sodium Chloride solution (0.1 M to 2 M). V and IV samples have less negative potential values than II, III and I samples.

Figures 10-14 display the potential-current density curves from each sample individually in different concentrations of Sodium Chloride (from 0.1 M to 2 M).

Log concentration of Sodium Chloride solutions against Log anodic corrosion current (Log I_{corr}) was plotted **Figure 15** in order to confirm the polarization results and I_{corr} values are tabulated in **Table 3**, electrode V is the lowest I_{corr} value than the other samples.

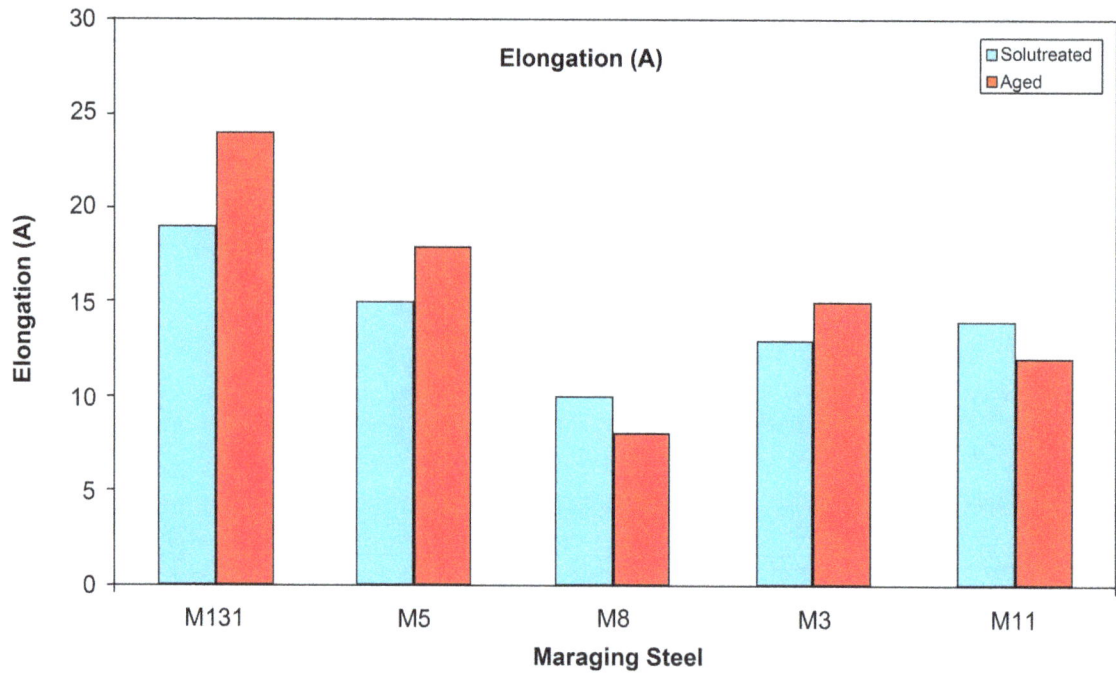

Figure 9. Comparison between mechanical properties (Elongation) of solution treatment and aged Maraging steel.

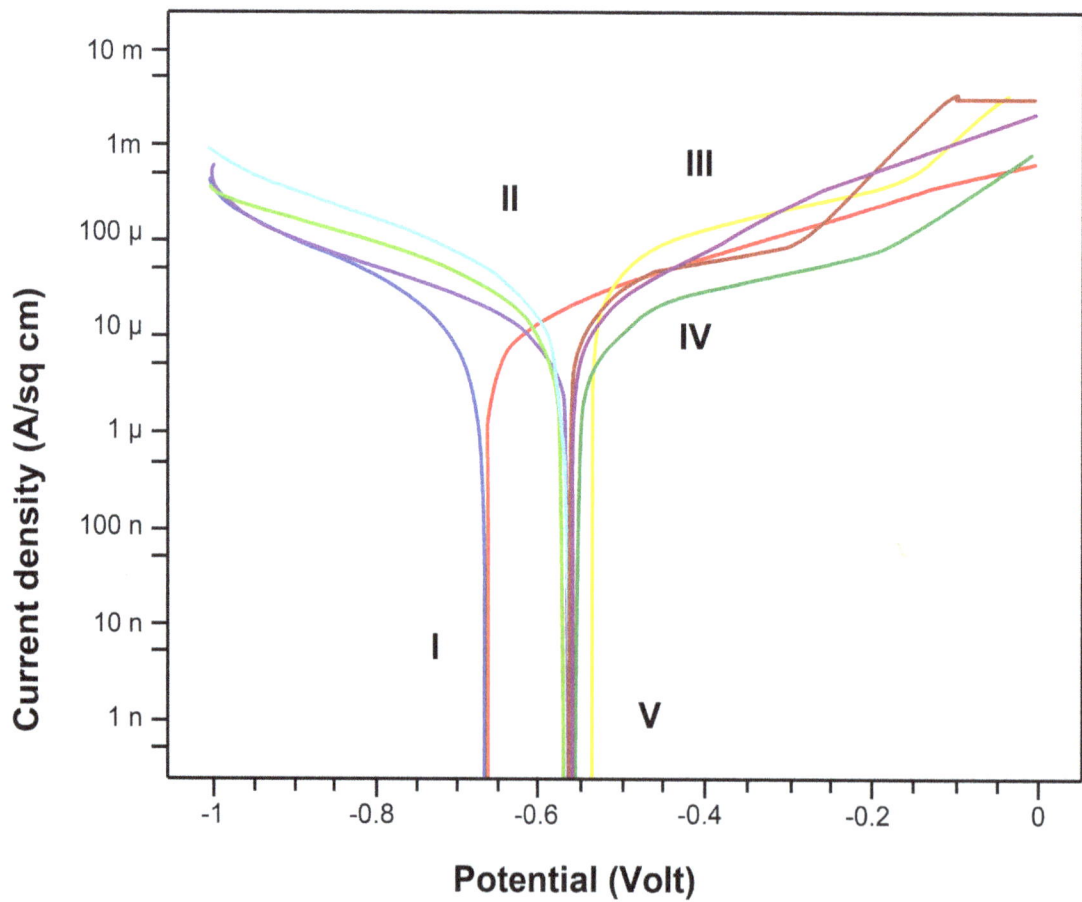

Figure 10. Polarization curve for Maraging steel samples in 0.1 M Sodium Chloride solution at room temperature.

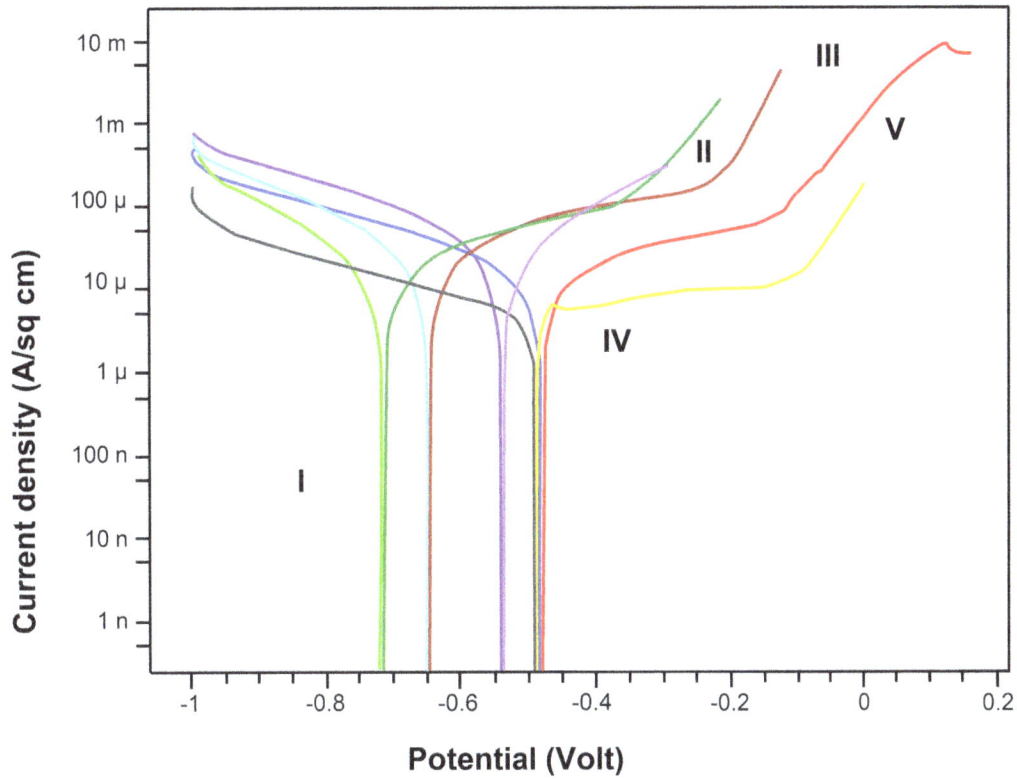

Figure 11. Polarization curve for Maraging steel samples in 0.3 M Sodium Chloride solution at room temperature.

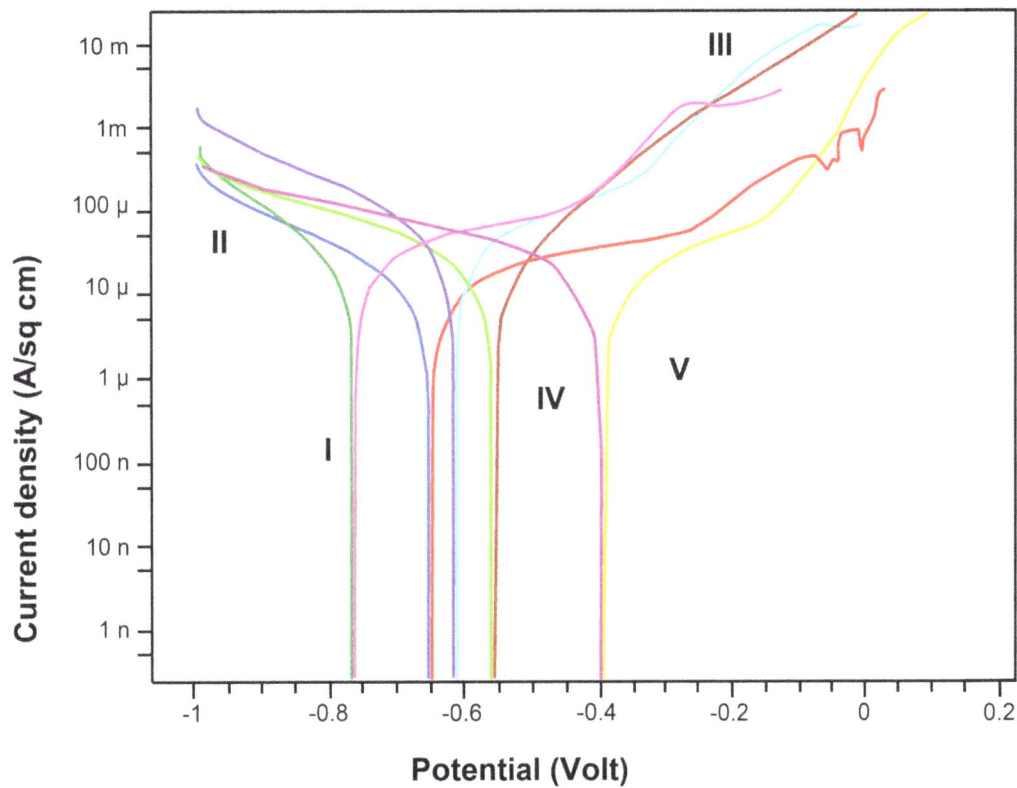

Figure 12. Polarization curve for Maraging steel samples in 0.6 M Sodium Chloride solution at room temperature.

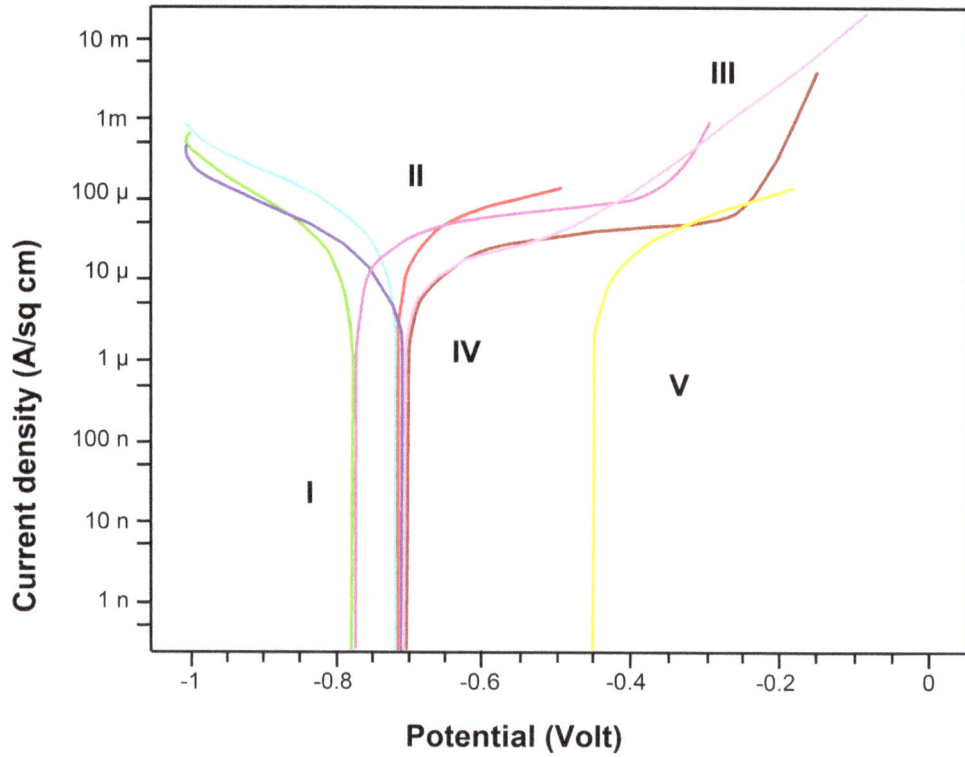

Figure 13. Polarization curve for Maraging steel samples in 1 M Sodium Chloride solution at room temperature.

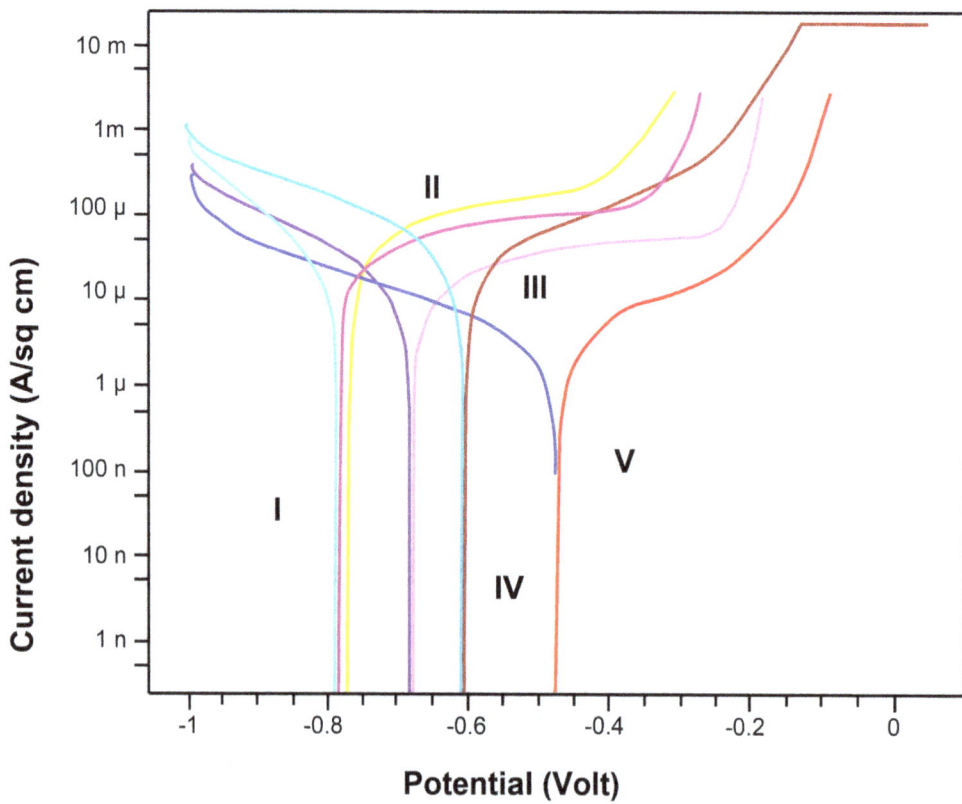

Figure 14. Polarization curve for Maraging steel samples in 2 M Sodium Chloride solution at room temperature.

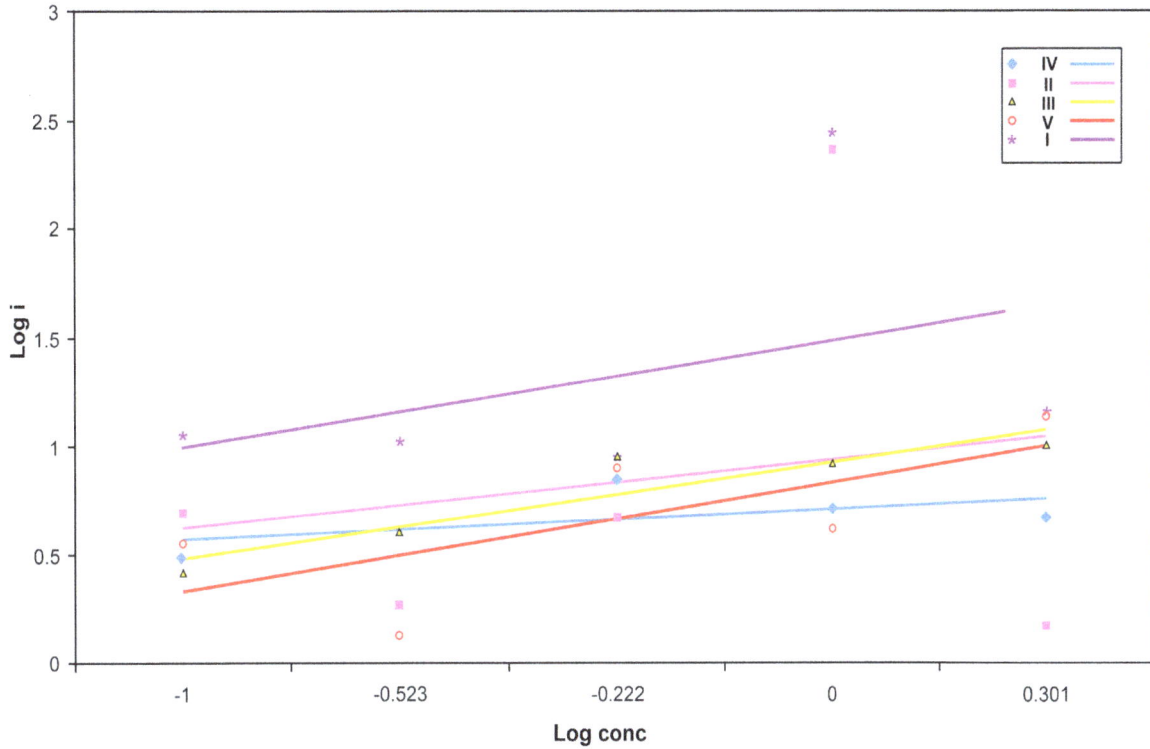

Figure 15. Log i-Log concentration of Sodium Chloride solution curve for solution treatment Maraging steel samples at room temperature.

Table 3. Polarization of solution treatment maraging steel in different concentrations of NaCl.

I_{corr}	0.1 M	0.3 M	0.6 M	1 M	2 M
V	3.73	1.41	3.50	175	1.12
IV	2.40	3.02	5.45	3.91	3.63
III	2.08	3.19	7.09	6.55	7.96
II	2.80	1.06	6.16	3.28	10.90
I	9.35	8.69	7.36	230	11.90

Log concentration of Sodium Chloride solutions against corrosion potential (E_{corr}) was plotted in the **Figure 16** in order to illustrate the corrosion behavior of the steel in different concentration of Sodium Chloride, the results is summarized in **Table 4**. Electrode V is the most corrosion resistive specimen then IV, II, III and I because it is the highest E_{corr} values than the other samples.

At low concentrations of Chloride, the curves exhibit a passive region that disappears as the potential increases in the noble direction. The passive film breakdown potential moves in the active direction as the concentration of Chloride is increased to such extent that, at higher concentrations the passive region completely disappears. The free corrosion potential, E_{corr} decreases with an increase in Sodium Chloride concentration [3].

The passive current density obtained from polarization curves is not a stationary current density. The stationary passive current density is reached after prolonged polarization [6]-[10].

3.2.2. Effect of Aging on Maraging Steel
Effect of Sodium Chloride concentrations on Aged Maraging Steel
The polarization potential of aged maraging steel electrodes can be inspected in the figure in 0.1 M Sodium

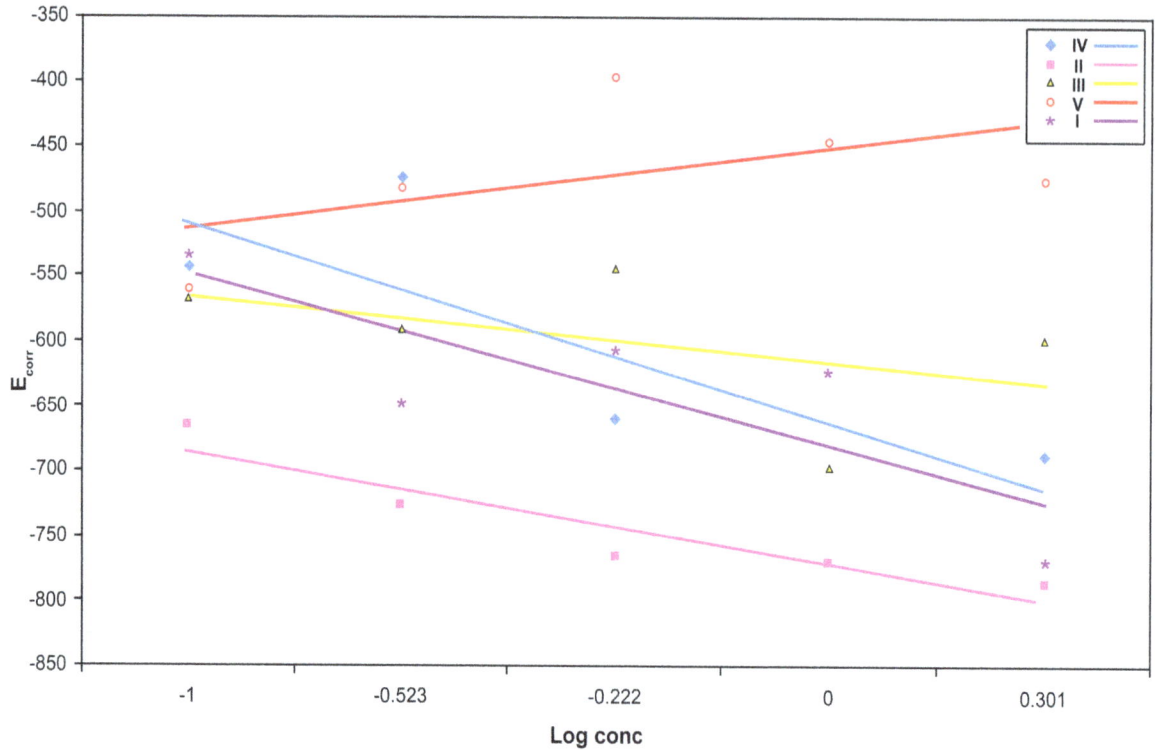

Figure 16. E_{corr}-Log concentration of Sodium Chloride solution curve for solution treatment Maraging steel samples at room temperature.

Table 4. Polarization of solution treatment maraging steel in different concentration of NaCl.

E_{corr}	0.1 M	0.3 M	0.6 M	1 M	2 M
V	−559.8	−482.7	−396.5	−446.8	−475.6
IV	−541.5	−475.2	−659.2	−697.3	−687.1
III	−666.2	−724.9	−765.6	−769.5	−786.3
II	−566.9	−590.2	−544.1	−696.6	−636.3
I	−534.4	−647.8	−606.3	−622.3	−769.7

Chloride solution, which represent that IV sample has a more positive values than V and I respectively.

Figures 17-20 will illustrate comparison between the polarization curves of solution treatment and aged maraging steel in 0.1 M Sodium Chloride solution.

3.3. Metallographic Examination of Corroded Alloys

3.3.1. Optical Microscopy

Figure 21 shows the results of optical microscopic examination after different treatment for different samples. Microscopic examination was carried out to investigate the effect of alloying elements additions and heat treatments on the microstructure changes after different treatments, investigated that, heat treatment of the maraging steel has a good influence on its mechanical properties but has no effect on the corrosion behavior of it. Because of the precipitation of intermetallic compounds which will tend to some kind of the galvanic corrosion.

3.3.2. X-Ray Diffraction Analysis

Figures 22-25 show the results of X-ray diffraction analysis for the two samples IV and V (after solution treatment and after aging). For IV XRD figures, in solution treatment sample martensite phase only is appeared be-

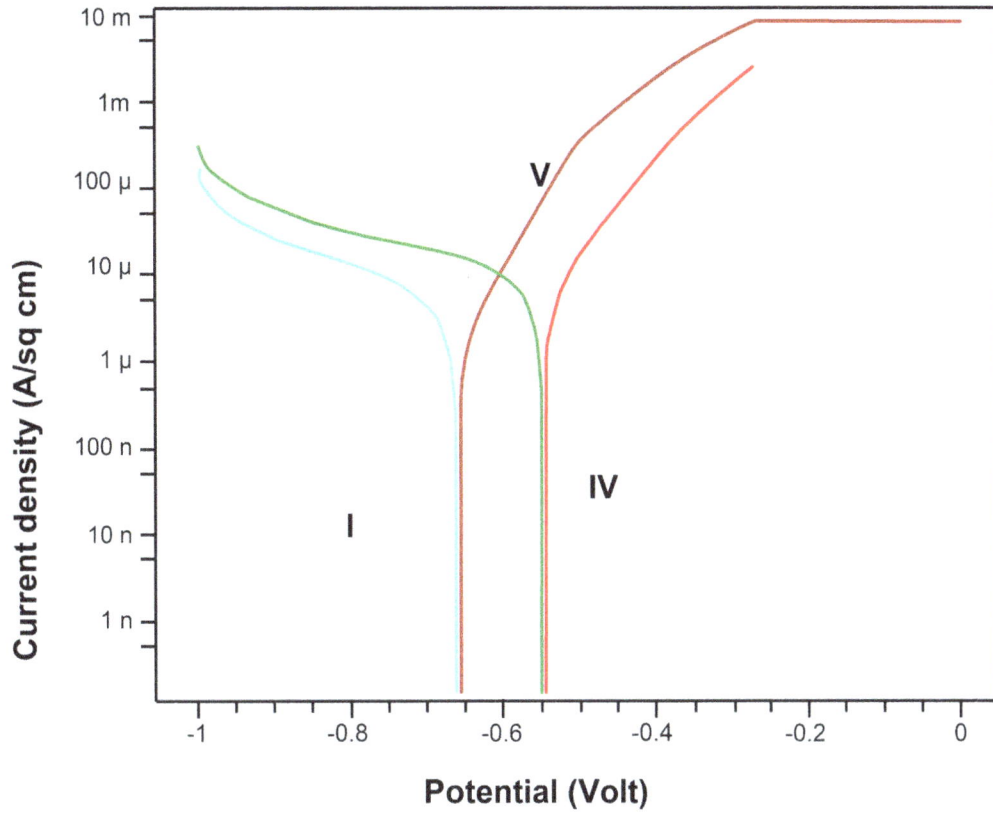

Figure 17. Polarization curve for aged Maraging steel in 0.6 M Sodium Chloride solution at room temperature.

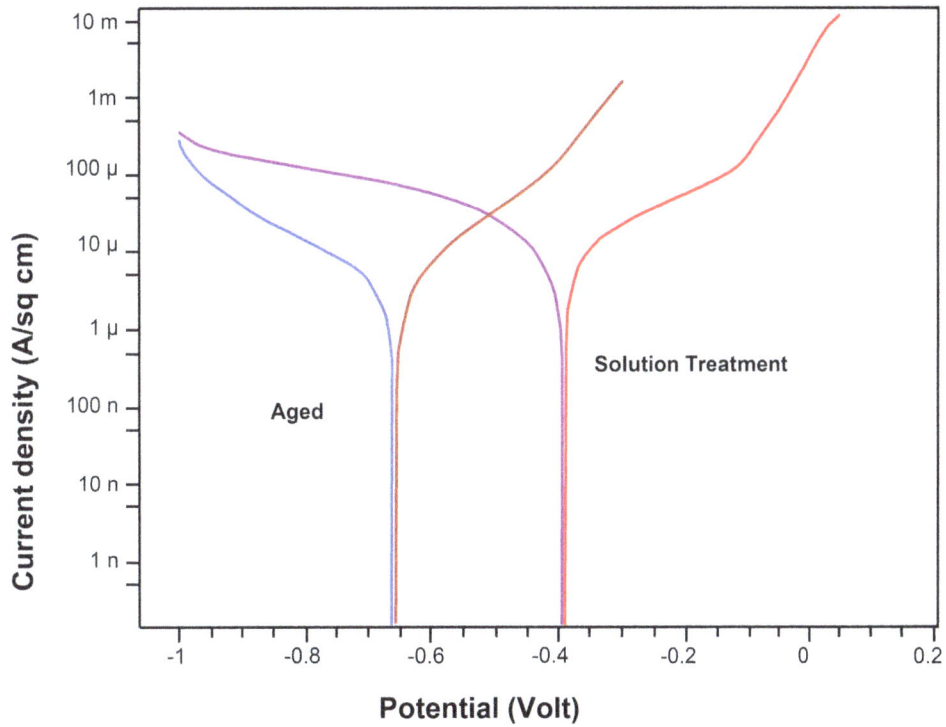

Figure 18. Polarization curve for V sample before and after heat treatment in 0.6 M Sodium Chloride solution at room temperature.

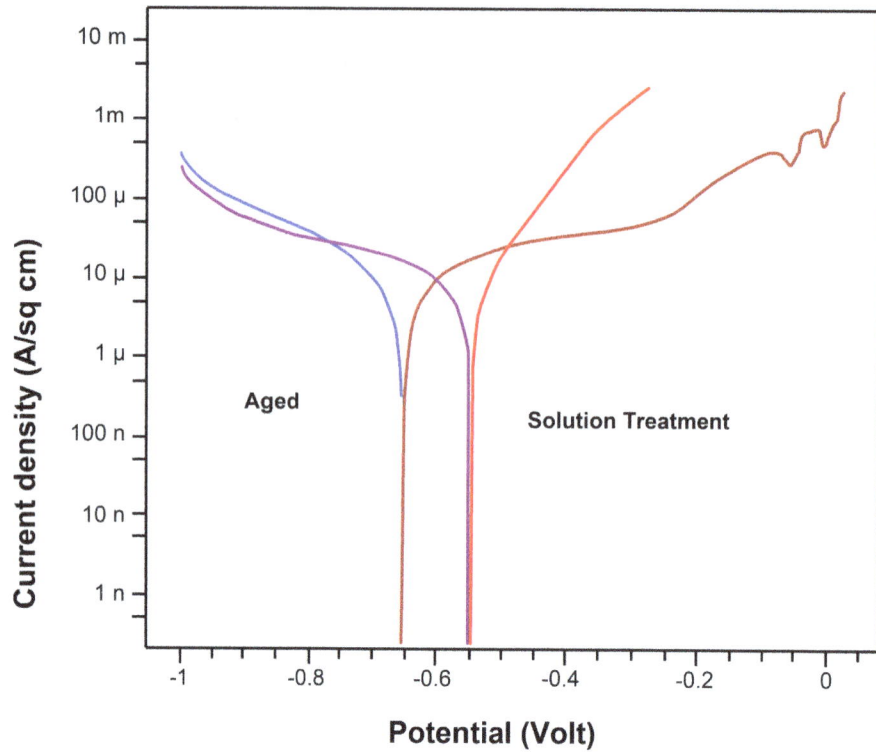

Figure 19. Polarization curve for IV sample before and after heat treatment in 0.6 M Sodium Chloride solution at room temperature.

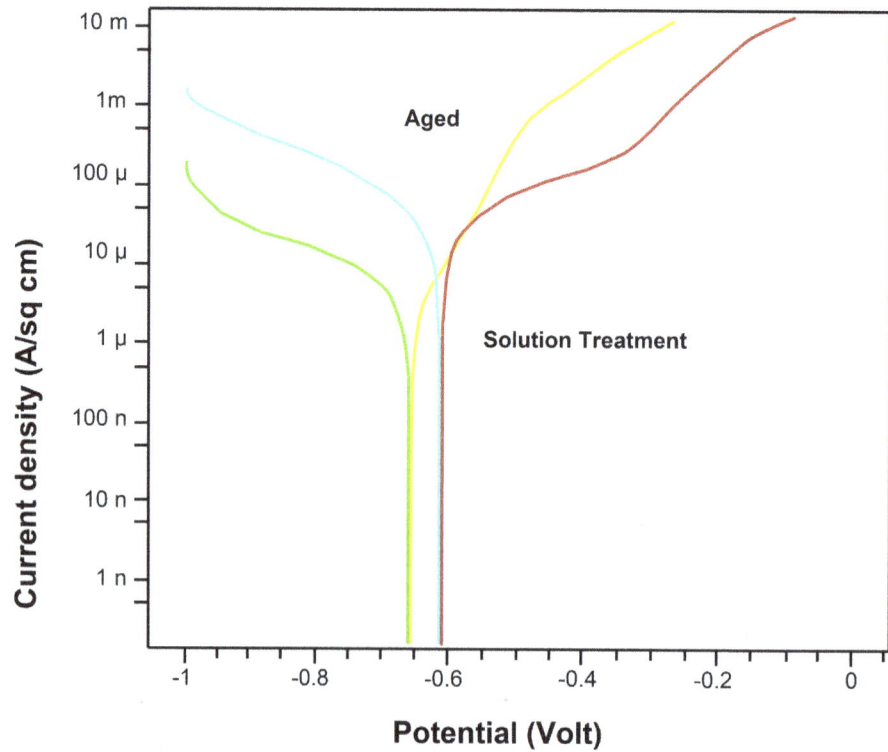

Figure 20. Polarization curve for I sample before and after heat treatment in 0.6 M Sodium Chloride solution at room temperature.

Sample I

Sample II

Sample III

Sample IV

Sample V

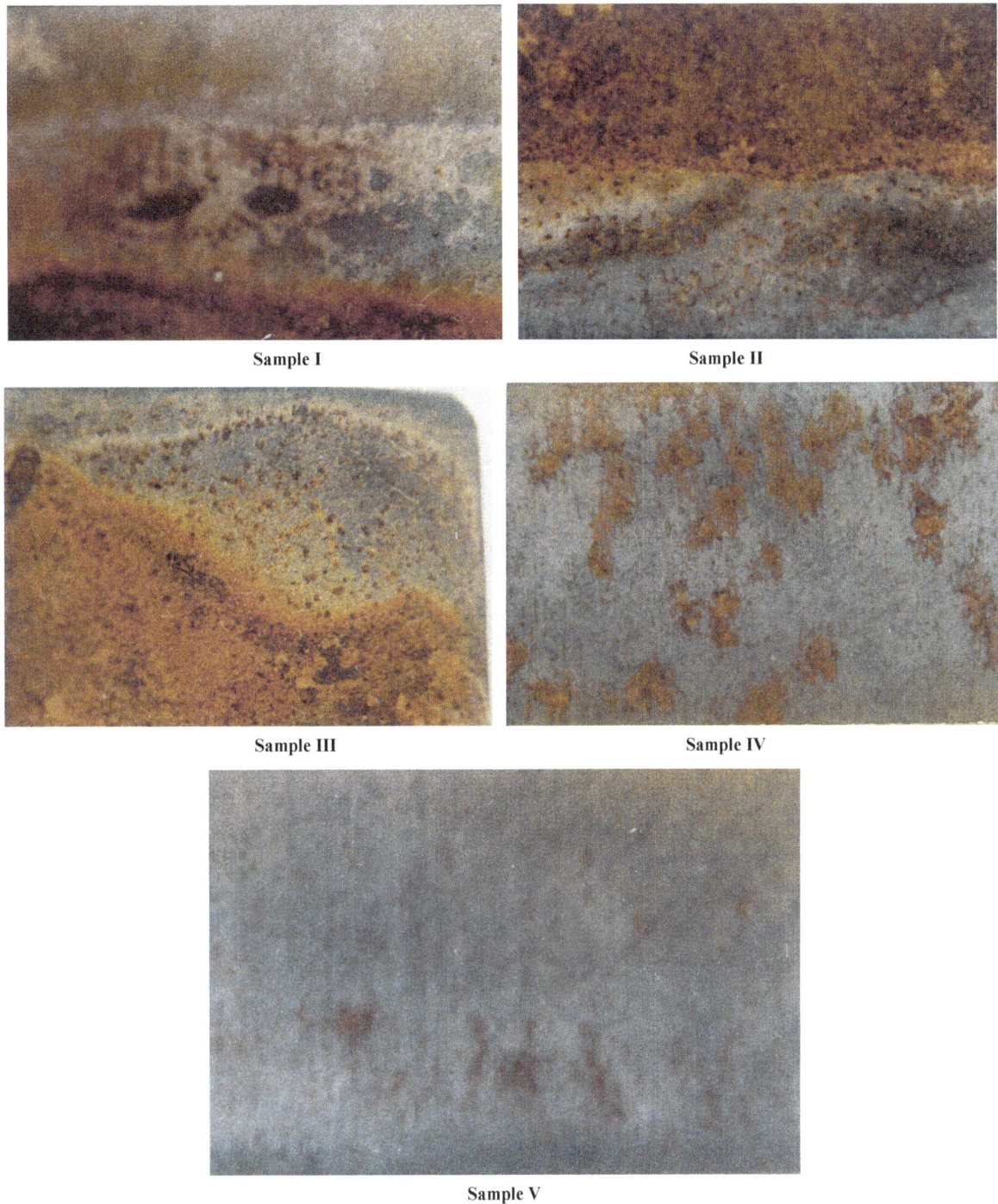

Figure 21. Optical microscopic examination for different samples.

fore and after aging. So that, solution treatment samples will be more corrosion resistance than aged sample as aresult of the precipitation of the intermetallic compounds which leads to form more galvanic cells. For V sample XRD figures, in solution treatment sample martensite phase is appeared in addition to small quantity from the iron austenite (Fe, C) phase but after aging martensite phase is only appeared. So that, solution treatment samples will be more corrosion resistance than aged sample as a result of the precipitation phenomena.

Microscopic examination and X-ray investigation were carried out to investigate the effect of alloying elements additions and heat treatments on the microstructure changes after different treatments, investigated that,

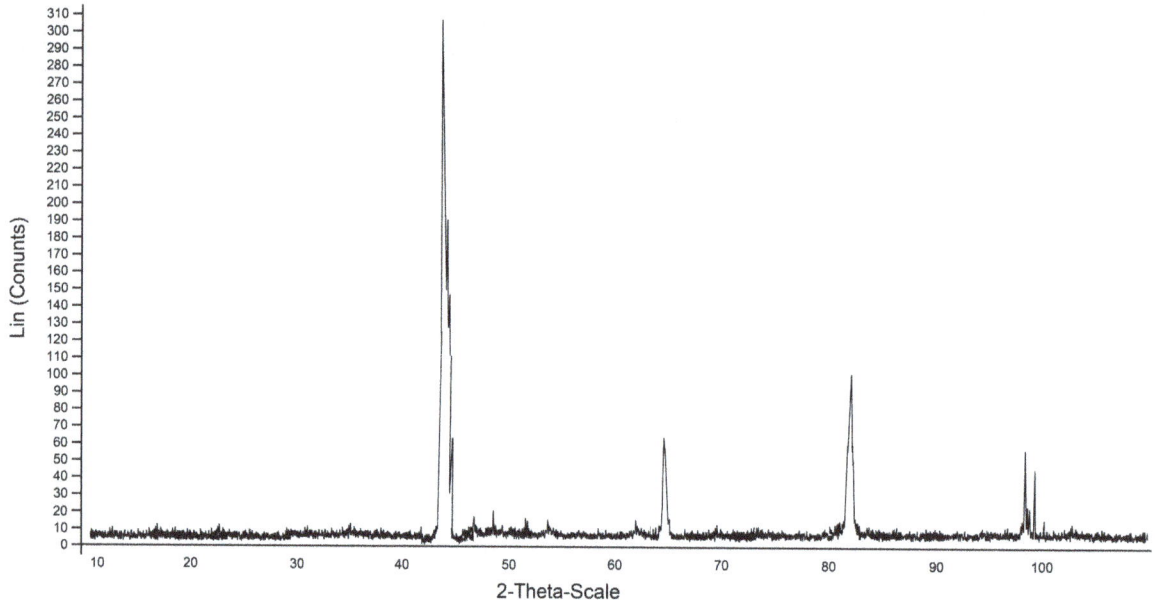

Figure 22. X-ray diffraction analysis for solution treatment IV.

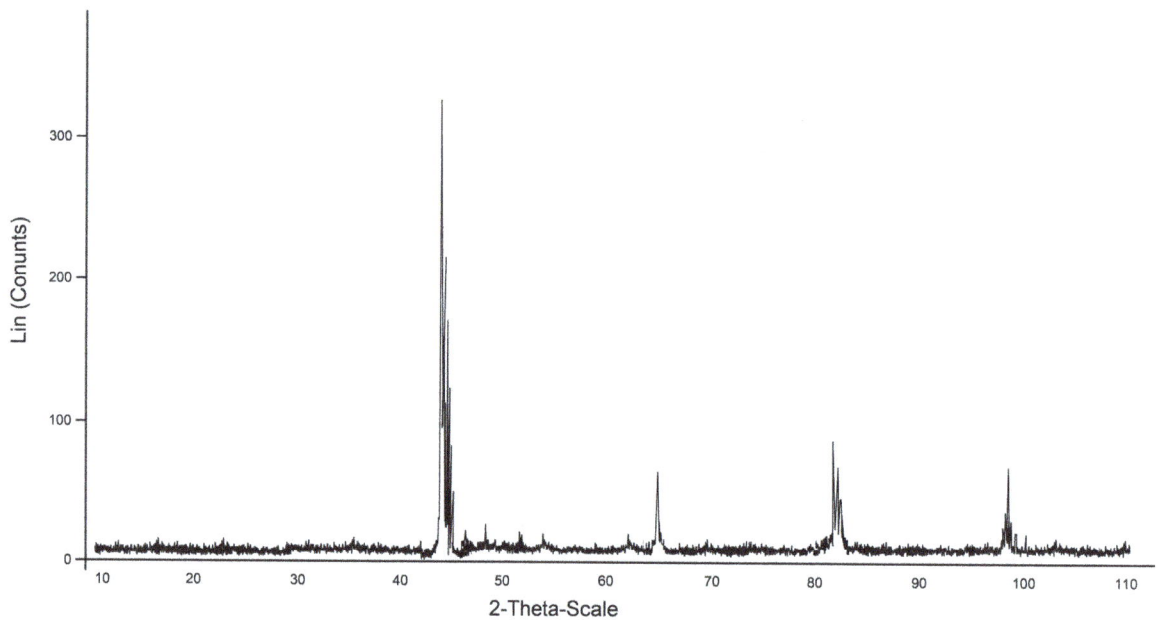

Figure 23. X-ray diffraction analysis for aged IV.

heat treatment of the maraging steel has a good influence on its mechanical properties but has no effect on the corrosion behavior of it. Because of the precipitation of intermetallic compounds which will tend to some kind of the galvanic corrosion. Finally, although IV sample is lower than V in Molybdenum content but IV after heat treatment, it has an observed improvement in mechanical properties and corrosion resistance [11]-[16].

4. Conclusions

Based on results of investigation, the following conclusions may be drawn:
1) The corrosion rate of solution treatment and aged maraging steel specimen in Sodium Chloride solutions with different concentrations from 0.1 M to 2 M is substantial.

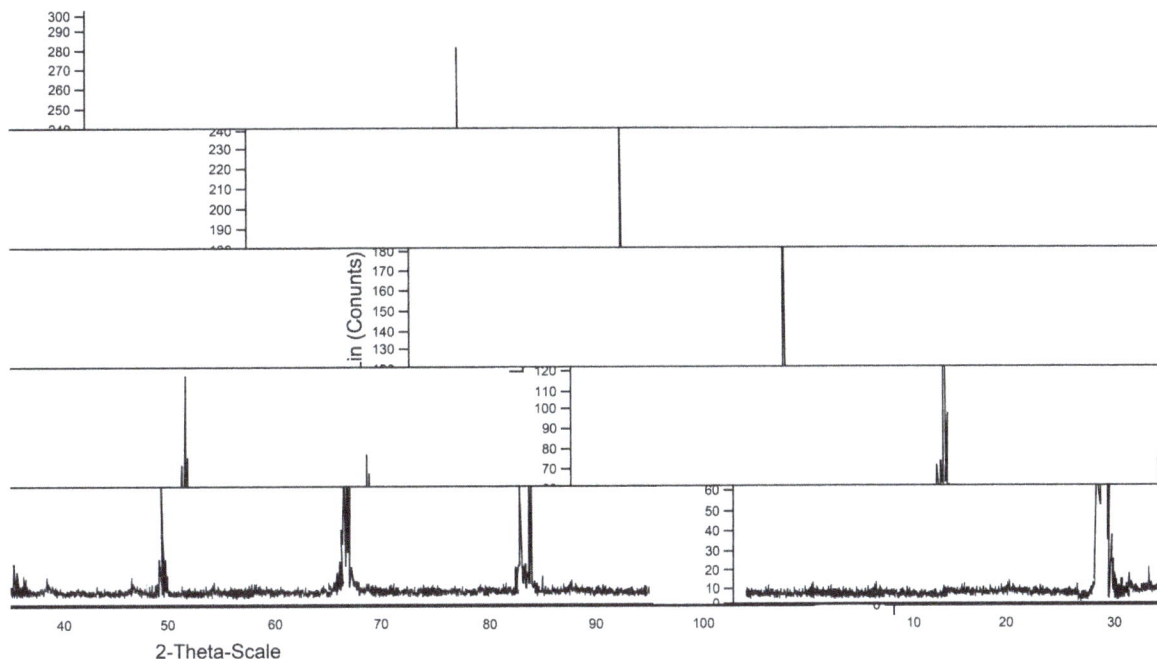

Figure 24. X-ray diffraction analysis for solution treatment V.

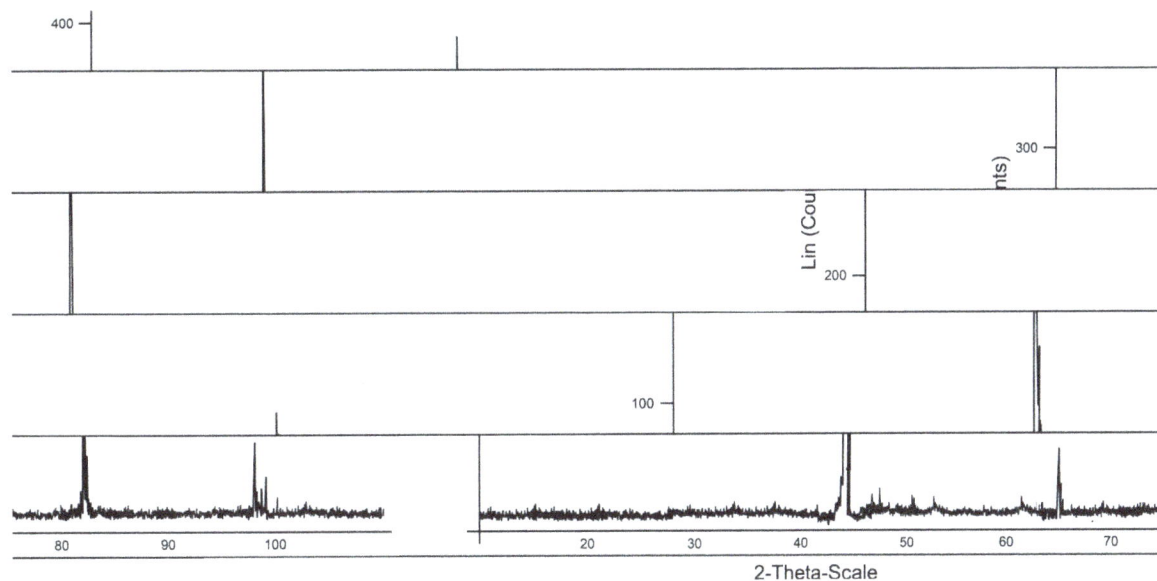

Figure 25. X-ray diffraction analysis for aged V.

2) The corrosion rates of the examined specimens are influenced by the concentration of the Sodium Chloride medium. The corrosion rate of the tested specimens under investigation increases with increase in concentration of chloride.

3) The corrosion rate of solution treatment samples is less than that of aged samples in Sodium Chloride solutions.

References

[1]   Fathy, A., El-Faramawy, H.S., Mattar, T., Eissa, M. and Bleck, W. (2004) *Metals*, 1.

[2]   Shams El-Din, A.M. and Paul, N.J. (1990) Oxide Film Thickening on the Surface of Metals in Aqueous Solutions: A Critique of the Theory of Open-Circuit Potential Transients. *Thin Solid Films*, **189**, 205-216. http://dx.doi.org/10.1016/0040-6090(90)90449-N

[3]   Hegazy, M.M. and Eissa, M.M. (1992) *J. of the Faculty of Education*, **17**, 377.

[4]   Shams El-Din, A.M., Wang, L. and Saber, T.M.H. (1994) Behaviour of High Strength Molybdenum Containing Stainless Steels in Arabian Gulf Water. Part 1: Oxide Film Thickening. *British Corrosion Journal*, **29**, 58-64. http://dx.doi.org/10.1179/000705994798267935

[5]   Greene, N.D. and Leonard, R.B. (1964) Comparison of Potentiostatic Anodic Polarization Methods. *Electrochimica Acta*, **9**, 45-54. http://dx.doi.org/10.1016/0013-4686(64)80004-9

[6]   Fathy, A., Mattar, T., El-Faramawy, H.S. and Bleck, W. (2002) *Materials Technology*, **12**, 549.

[7]   Shams El-Din, A.M., Wang, L. and Saber, T.M.H. (1994) *Br.Corros. J.*, **29**, 58.

[8]   El-Meligi, A.A. and Ismail, N. (2009) Hydrogen Evolution Reaction of Low Carbon Steel Electrode in Hydrochloric Acid as a Source for Hydrogen Production. *International Journal of Hydrogen Energy*, **34**, 91-97. http://dx.doi.org/10.1016/j.ijhydene.2008.10.026

[9]   Fouda, A.S., El-TaibHeakal, F. and Radwan, M.S. (2009) Role of Some Thiadiazole Derivatives as Inhibitors for the Corrosion of C-Steel in 1 M H_2SO_4. *Journal of Applied Electrochemistry*, **39**, 391-402. http://dx.doi.org/10.1007/s10800-008-9684-2

[10]   Heo, N.H. and Lee, H.C. (1995) *Metals Mater*, **58**, 77.

[11]   Hondros, E.D. and Seah, M.P. (1977) *Int. Met. Rev.*, **22**, 262. http://dx.doi.org/10.1179/imr.1977.22.1.262

[12]   Seah, M.P. (1980) Adsorption-Induced Interface Decohesion. *Acta Metallurgica*, **28**, 955-962. http://dx.doi.org/10.1016/0001-6160(80)90112-1

[13]   Hossein Nedjad, S., Nili Ahmadabadi, M. and Furuhara, T. (2008) *Metall. Mater. Trans. A*, **39A**, 19.

[14]   Squires, D.R. and Wilson, E.A. (1972) Aging and Brittleness in an Fe-Ni-Mn Alloy. *Metallurgical Transactions*, **3**, 575-585. http://dx.doi.org/10.1007/BF02642065

[15]   Heo, N.H. (1996) Ductile-Brittle-Ductile Transition and Grain Boundary Segregation of Mn and Ni in an Fe-6Mn-12Ni Alloy. *Scripta Materialia*, **34**, 1517-1522. http://dx.doi.org/10.1016/1359-6462(96)00032-2

[16]   Hossein Nedjad, S., Nili Ahmadabadi, M. and Furuhara, T. (2008) Correlation between the Intergranular Brittleness and Precipitation Reactions during Isothermal Aging of an Fe-Ni-Mn Maraging Steel. *Materials Science and Engineering*: A, **490**, 105-112. http://dx.doi.org/10.1016/j.msea.2008.01.070

12

Effect of Surface Site on the Spin State for the Interaction of NO with Pd₂, Rh₂ and PdRh Nanoparticles Supported at Regular and Defective MgO(001) Surfaces

S. Abdel Aal

Department of Chemistry, Faculty of Science, Benha University, Benha, Egypt
Email: safabdelaal11@yahoo.com

Abstract

An attempt has been made to analyze the effect of surface site on the spin state for the interaction of NO with Pd₂, Rh₂ and PdRh nanoparticles that supported at regular and defective MgO(001) surfaces. The adsorption properties of NO on homonuclear, Pd₂, Rh₂, and heteronuclear transition metal dimers, PdRh, that deposited on MgO(001) surface have been studied by means of hybrid density functional theory calculations and embedded cluster model. The most stable NO chemisorption geometry is in a bridge position on Pd₂ and a top configuration of Rh₂ and PdRh with N-down oriented. NO prefers binding to Rh site when both Rh and Pd atoms co-exist in the PdRh. The natural bond orbital analysis (NBO) reveals that the electronic structure of the adsorbed metal represents a qualitative change with respect to that of the free metal. The adsorption properties of NO have been analyzed with reference to the NBO, charge transfer, band gaps, pairwise and non-pairwise additivity. The binding of NO precursor is dominated by the $E_{(i)}^{M_x-NO}$ pairwise additive components and the role of the support was not restricted to supporting the metal. The adsorbed dimers on the MgO surface lose most of the metal-metal interaction due to the relatively strong bond with the substrate. Spin polarized calculations were performed and the results concern the systems in their more stable spin states. Spin quenching occurs for Rh atom, Pd₂, Rh₂ and PdRh complexes at the terrace and defective surfaces. The adsorption energies of the low spin states of spin quenched complexes are always greater than those of the high spin states. The metal-support and dimer-support interactions stabilize the low spin states of the adsorbed metals with respect to the isolated metals and dimers. Although the interaction of Pd, Rh, Pd₂, Rh₂ and PdRh particles with Fs sites is much stronger than the regular sites O²⁻, the adsorption of NO is stronger when the particular dimers are supported on an anionic site than on an Fs site of the MgO(001). The encountered variations in magnetic properties of the adsorbed species at MgO(001) surface are

correlated with the energy gaps of the frontier orbitals. The results show that the spin state of adsorbed metal atoms on oxide supports and the role of precursor molecules on the magnetic and binding properties of complexes need to be explicitly taken into account.

Keywords

Surface Reactions, NO, Bimetallic Nanoparticles, Spin State and Charge Transfer

1. Introduction

Fundamental understanding of the electronic structure and activity of transition metal atoms and nanoclusters supported on metal-oxide surfaces is of great interest due to their broad applications in catalysis, coating for thermal applications, corrosion protection, and other technologically important fields [1]-[4]. Theoretical calculations have proved very helpful to gain insight into the mechanisms of growth of nanoclusters on oxide surfaces [5] [6]. It has been found that under typical conditions, formation of dimers constitutes the first step in the process of the growth of metal clusters on the oxide surface [7]. Even though in the gas phase there are dimers, trimers, etc., the cluster growth on the surface of the support is dominated by diffusion of adsorbed atoms and not by direct deposition of already existing gas phase clusters. It is observed that, diffusion is stopped at point defects, where the atoms are more strongly bound and nucleation takes place [8]. In addition, the properties of the deposited nanoclusters depend on the oxide substrate and in particular on the presence of point defects where the cluster can be stabilized. In general, there has been a consensus that defects not only can act as catalytic centers for chemisorption of small species but also as nucleation centers for growing metal clusters and can modify the catalytic activity of these adsorbed metal particles via the metal-support interaction at the interface [9].

The strength of interaction between metal and substrate is due to metal-substrate covalent bonding that implies a polarization of the metal orbitals or redistribution of the atomic orbital population. The metal s-orbital combines with the oxygen p-orbital perpendicular to the surface of an oxide material resulting in a bonding (occupied) and antibonding (unoccupied) combinations. This leads to a decrease in the atomic population of the metal atom [9]-[11]. When the free metal atom electronic configuration is $d^n s^2$, the resulting electronic configuration of the metal and atom may be expressed as $d^{n+1} s^1$ or even d^{n+2}. The strength of the metal-oxide interaction varies with the resulting d-population. This change in the electronic configuration of the adsorbed metal may result in a concomitant spin quenching with respect to the ground state multiplicity of the isolated metal atom.

On the basis of the performance of different density functionals, Markovits *et al.* [12] reported that the electronic state of Ni with the oxygen regular and defective sites of MgO is the result of a balance between the tendency of Hund's rule to preserve the atomic state and chemical covalent terms tending to form chemical bonds and hence to quench the atomic magnetic moment. Indeed, the stronger the interaction, the smaller the difference between the high and low spin states; in other words, the larger the interaction, the stronger the spin quenching.

Sousa *et al.* [13] calculated the low to high spin transition energy of Ni adsorbed on regular and defective sites of MgO and magnetic properties of first row transition metal oxides. The previous investigations suggest that the final spin state of an adsorbed metal can be different, when it interacts with an oxide support. However, the combined effect of oxide support and adsorbed species, such as NO on the final spin state is overlooked.

Bimetallic nanoparticles may create a synergistic catalytic effect that involves the change in local electronic properties of pure metal nanoparticles to modify the strength of the surface adsorption for oxygen reduction reactions [14] [15]. Although pure Pd and Rh clusters on MgO(001) and TiO_2 have been widely studied [16] [17], no reports are available on the geometrical and electronic structure of PdRh bimetallic that deposited on MgO surface. Efforts have been focused on the possibility of associating Pd with another noble metal, rhodium, to prepared bimetallic Pd-Rh/alumina catalysts and compared to reference Pd/alumina and Rh/alumina solids [18]. A. M. Ferrari [19], Shinkarenko *et al.* [20], Neyman and Illas [21], Nasluzov *et al.* [22], and Matveev *et al.* [23] have experimentally and theoretically studied the adsorption properties of different metal atoms and metal clusters deposited on the MgO (001) surface. Palladium and rhodium clusters of small size have been extensively studied at various semiempirical and *ab initio* levels of the theory by G. Berthier [24]. As the smallest cluster, homonuclear and heteronuclear transition metal dimers have been studied both experimentally and theoretically [25] [26]. For a systematic theoretical study, the homonuclear dimers of 4d transition metals were examined by use of diverse density functional methods [27]. The structures of AgPd clusters supported on MgO(001) are investigated via a combination of global optimization searches within an atom-atom potential model and density-functional calculations [28]. The reactions of H_2 with the heteronuclear dimers PdCu, PdAg, PdAu have been studied by the hybrid density functional method B3LYP [29]. CO adsorption on monometallic and bimetallic Au-Pd nanoparticles deposited onto well-ordered thin films of $Fe_3O_4(111)$, MgO(001), and $CeO_2(111)$ were studied by [30].

It is frequently observed that a transition metal atom doped in a small cluster of other metal can strongly change the properties of the host cluster [31] [32]. Previous theoretical calculations have been devoted to the study of heteroatomic or impurity-doped as well as homoatomic metal clusters, which indicate that the impurity atoms can strongly influence geometric, electronic, and bonding properties of mixed clusters [33]. The first objective of this work is to generalize the possibility that electron-rich MgO surface can be used to determine how the substrate could affect the structural, energetic and electronic properties of small bimetallic Rh-Pd dimers that belonging to a completely different valence structure, *i.e.* $Rh(4d^8\ 5s^1)$ with unfilled d and $Pd(4d^{10}\ 5s^0)$ with complete d shell. For this purpose, the simplest bimetallic particle, PdRh, is considered and the results are compared with monometallic Pd_2 and Rh_2 dimers. Second, to clarify the roles of defects as nucleation centers for the formation of dimers and represent how these defects can induce modifications in the electronic, geometric and chemical properties of the supported dimers. Third, to identify the bonding mechanism of NO with Pd_2, Rh_2 and PdRh nanoparticles that supported on regular and defective sites of MgO(001). Finally, to induce qualitatively different changes in the electronic states of the supported particles and on the transition energy required to switch from low spin to high spin state.

The intriguing heterogeneous processes associated with nitric oxide, NO, observed at transition metal and metal-oxide surfaces, are a continuous topic for research. The molecule, which is one of the simplest and most stable radicals, is spontaneously formed in combustion processes at elevated temperatures. Being a major environmental hazard, it is of vital importance to remove NO from the exhaust gases. The reduction of NO by CO on palladium is of practical interest and experimental investigations show that nanosized palladium clusters have significant capacity to catalyze the CO + NO reaction at low temperatures [34] [35]. As one of the key factors to understand the catalytic mechanism, the adsorption behaviors of NO on Pd clusters have been extensively studied [36] [37]. Viñes *et al.* performed a combined experimental and theoretical study on the adsorption of NO on Pd nanoparticels, using infrared reflection adsorption spectroscopy (IRAS) and calculations based on density functional theory (DFT) [36].

2. Computational Details and Surface Models

Hybrid density functional theory and embedded cluster models have been extensively employed in the description of the electronic and geometrical structures of Pd_2, Rh_2 and PdRh particles nucleated on regular and defect sites on the MgO(001) surface [5] [38] [39]. These models have demonstrated to be powerful in the description of the defective and non defective non polar oxide surface [40]. Sousa, *et al.* [13] used a cluster/periodic comparison within the same computational model (either DFT or HF) for the ionic systems (MnO, FeO, CoO, NiO, and CuO) to establish that embedded cluster models provide an adequate representation. They used a lattice parameter (421 pm) the same as was determined for the bulk, with no surface relaxation or rumpling in the defect-free system. The embedded cluster model considers a finite cluster embedded in the rest of the host crystal, by assuming that the electronic structure in this external region has remained the same as in the defect free system. This approach is adequate in principle, but is computationally demanding and requires an accurate analysis of the energy terms. Its flexibility is moderate and can describe the charged defects [41].

To represent the substrate, the ionic clusters Mg_9O_{14} and Mg_9O_{13} Fs have been embedded in arrays of point charges. This was done by following an embedding procedure previously reported for alkaline earth oxides [42]. A finite ionic crystal of 292 point charges was first constructed. The Coulomb potentials along the X and Y axes of this crystal are zero by symmetry as in the host crystal. The ±2 charges on the outer shells were then modified, by using a fitting procedure; to make the Coulomb potential at the four central sites closely approximates the Madelung potential of the host crystal, and to make the Coulomb potential at the eight points with coordinates (0,

±R, ±R) and (±R, 0, ±R) where R is half the lattice distance, which for MgO is 2.105, equal to zero as it should be in the host crystal. With these charges, 0.818566 and 1.601818, the Coulomb potential in the region occupied by the central ions is very close to that in the unit cell of the host crystal. The Coulomb potential was calculated to be (1.748) at the four central sites (compared with 1.746 for a simple cubic ionic crystal) and (0.0) at the previously defined eight points (compared with 0.0 for the same crystal). All charged centers with cartesian coordinates (±X), (±Y) and (Z > 0) were then eliminated to generate the (001) surface of MgO with 176 charged centers occupying the three dimensional space (±X), (±Y) and (Z ≤ 0). The clusters were then embedded within the central region of the crystal surface, and the electrons of the embedded clusters were included in the Hamiltonians of the *ab initio* calculations. Other crystal sites entered the Hamiltonian either as full or partial point charges as demonstrated in [42].

The density functional theory calculations were performed by using Becke's three parameter exchange functional B3 with LYP correlation functional [43] [44]. The B3LYP hybrid functional has been used since it provides a rather accurate description of the metal/oxide interaction [45]. Moreover, for the magnetic systems it provides a reasonable albeit not perfect picture which lies midway between the HF and pure GCA descriptions [46]. Even if the DFT has well known problems with the description of magnetic properties, hybrid functionals such as B3LYP provides a fair indication of the relative energies. B3LYP correctly reproduce the thermochemistry of many compounds including transition metal atoms [47] and seems to be able to properly describe the band structure of insulators [12]. B3LYP ensures a correct description of the electronic ground state of first row transition metal atoms and a reasonable description of the energy difference between low lying electronic states with different spin multiplicity. Finally, B3LYP is able to describe magnetic coupling in systems with localized spins although the magnetic coupling constant is too large [48].

The Stevens, Basch and Krauss Compact Effective Potential (CEP) basis sets [49] [50] were employed in the calculations. In the CEP basis sets, the double zeta calculations are referred to as CEP-31G, and similarly triple zeta calculations to as CEP-121G. It may be noted that there is only one CEP basis set defined beyond the second row, and the two basis sets are equivalent for these atoms. These basis sets have been used to calculate the equilibrium structure and spectroscopic properties of several molecules and the results are compared favorably with the corresponding all-electron calculations [51]. In the present calculations, the effective core potential of the cep-121g basis set was used for all atoms in the clusters.

The defect free surfaces exhibit very small relaxations only and therefore they have been kept fixed when studying deposition of metal atoms. A minimal energy search on a defect free surface does not usually include surface relaxation since this is experimentally very small, less than 5% [52]. Surface relaxation effects can be significant if discontinuities, like steps or point defects, are present [53]. Sousa *et al.* [13] focused in the problems when applying DFT methods to open-shell systems with particular emphasis on the consequences on the description of magnetic properties. They found for ionic systems with unfilled d- shells, such as the present Rh atom, that the resulting open-shell electrons are localized and hence it is possible to model these systems by means of embedded cluster models. All calculations are of spin unrestricted type and carried out by using Gaussian 98 system [54]. The figures were generated by using the corresponding Gauss View software.

The binding energy, E_a, of the Pd_2, Rh_2 and PdRh dimers at various sites of the metal oxide surface can be calculated as follows:

$$E_a(M_2) = -\left[E(M_2/MgO\_site) - E(M_2) - E(MgO\_site)\right] \quad (1)$$

Positive values of the binding energies mean that the formed dimers are stable.

The high to low spin transition energies were calculated from the relation

$$\Delta E_{complex}^{H-L} = E_{complex}^H - E_{complex}^L \quad (2)$$

where $E_{complex}$ is the total electronic energy of the complex.

The nucleation energy (E_{nucl}), dimer formation energy, is an important parameter to study the atom-by-atom growth of a particle from atoms in the gas phase. It is defined as the energy associated with the formation of homonuclear dimers Rh_2,Pd_2, and heteronuclear dimers, PdRh, when an atom of the gaseous phase, Rh or Pd bonds with a pre-adsorbed metallic particle, Rh/MgO_site or Pd/MgO_site [54], respectively:

$$E_{nucl} = -\left[E(M_2/MgO\_site) - E(M) - E(M/MgO\_site)\right] \quad (3)$$

where MgO_site indicates the nucleation site. These two quantities, Ea (M_2) and $E_{nucl.}$, measure the binding energy of gas phase Pd_2, Rh_2 and PdRh to a given MgO site [55].

The dimer binding energy, E_b, measures the stability of the adsorbed dimer with respect to Pd and Rh adatoms, where one of which is bound on a five coordinated terrace anion, O_{5c}. E_b is simply the difference between the adsorption energy of transition metal, TM, atom to the supported TM/MgO and the binding energy of the atom to the metal oxide terrace.

$$E_b = -E\left(M_2/MgO\_site\right) - E\left(MgO\_O_{5c}\right) + E\left(M/MgO\_site\right) + E\left(M/\left(MgO\_O_{5c}\right)\right) \quad (4)$$

The trapping energy, E_t, measures the energy gain when Pd and Rh atoms move from a terrace site to a strongly binding site, anion vacancy. The trapping energy is the difference in E_a between a regular and a defect site. E_t can be quite large on some specific defects, indicating their strong tendency to capture metal atoms. Thus, metal atoms have a high probability to find a defect in the diffusion process and to stick to this defect [56].

3. Results and Discussion

3.1. Adsorption of Single Pd and Rh Atoms

It was well established that small metal particles adsorb preferentially on sites where negative charge accumulates [6] [39] [57] [58]; more specifically the O^{2-} ionic sites for regular and the F_s centers for defective metal oxide. Experimentally, metal clusters are often formed on a surface by exposing it to a beam of gas-phase atoms. These atoms adsorb onto the surface and diffuse to the sites at O^{2-} or Fs centers. From these single adsorbed atoms, clusters are formed by nucleation. For this reason, the adsorption of a single Pd and Rh atoms on both O^{2-} and Fs of MgO is investigated as a first step in this study. The results provide a clear indication that the atomic Rh adsorbs more strongly on both sites than that of Pd. The interaction of Rh and Pd atom on F site is characterized by stronger binding energy (E_{ads} = 3.263 eV, 3.186 eV) with shorter equilibrium adsorption distance (1.62 Å, 1.54 Å) than on the surface O^{2-} site (E_{ads} = 1.301 eV, 1.251 eV) with equilibrium distance (2.06Å, 2.19 Å), respectively. The presence of trapped electrons at the defect site results in a more efficient activation of the supported Pd and Rh atoms. These results are in agreement with [5] [59] [60]. In general, a good agreement is established between the geometrical parameters obtained in this work and the reported theoretical values for Rh/MgO(001) surface (2.09 Å [61]) and for Pd/MgO(001) (2.15 Å [62]) at the low coordinated surface.

In addition, it is not a trivial task to conclude a priory which one of the ^4F ($4d^8s^1$) and ^2D (d^9) states of Rh determines the ground state energy of the unit cell of MgO(001) surface with the adsorbed Rh atom. Therefore, the effect of the substrate on the electronic states of the adsorbate and the energy required to switch from high-spin to low-spin state are analyzed. By using the B3LYP calculation, high- to low-spin transition energies of Rh atoms free, ΔE^{H-L}, and supported on O^{2-} and Fs sites of MgO (001), $\Delta E_{complex}^{H-L}$, are summarized in **Table 1**. Since $\Delta E_{complex}^{H-L}$ is negative value, the spin-polarized structure with one unpaired electron is the most stable state, in agreement with [62]. Consequently, the interaction of Rh at MgO (001) surface induces a quenching of the magnetic moment, which results in a doublet ground state, separated by 0.267 eV from the lowest quartet. Whereas, upon interaction with O^{2-} and Fs surface sites, the high to low spin transition energies $\Delta E_{complex}^{H-L}$ of Pd atom is positive indicating that the spin states are preserved and the low spin states are favored. Hence, the number of unpaired electrons in the Pd adatom tends to be the same as in the gas phase and the ground state of Pd–MgO is spin singlet. The interaction of Rh and Pd atoms with the F_S center merits a separate discussion since results show that in almost all cases $\Delta E_{complex}^{H-L}$ exhibits the largest increase. Hence, it is clear that, the low-spin state is more favored because of the formation of a direct bond between the adsorbed species and the electronic levels corresponding to the oxygen vacancy electrons.

Nevertheless, the important point here is the trend of the adsorbed atoms from one site to another. The analysis of these results clearly show that, there is a change in the transition energy required to switch from high spin to low spin, $\Delta E_{complex}^{H-L}$, which is induced by the MgO (001) surface. It is observed that the transition energy from the high-spin to the low-spin state increases when the oxide support is present. This is a clear indication that the metal-support interaction tends to stabilize the low-spin state with respect to the isolated atom. The difference in the equilibrium distance perpendicular to the surface was calculated. Notice that, as expected, there is an inverse correlation between adsorption energy and equilibrium distance, the larger the former the shorter the later, **Table 1**.

Table 1. Transition energy, $\Delta E^{H-L}_{complex}$, required to excite adsorbed monomer, Pd and Rh from the high- to low-spin state. A positive sign indicates that the ground state is provided by the low-spin coupling. Δd^{H-L} are the change in the equilibrium distances of Pd and Rh atoms that supported to the regular (O^{2-}) and oxygen vacancy (Fs) site at MgO (001) going from high- to low-spin state. A positive value indicates that d is larger in the high-spin state. Energies are expressed in eV, the corresponding equilibrium distances (d) in (Å), and charges in electron units H: High spin. L: Low spin. Ne: number of unpaired electrons.

	Pd/MgO (O^{2-}) site	Pd/MgO (Fs) site	Rh/MgO (O^{2-}) site	Rh/MgO (Fs) site
ΔE^{H-L}	0.904		−0.034	
E^H_{ads} (M)	0.919	1.933	1.034	1.812
d^H(M-S)	2.39	1.82	2.54	2.06
$\Delta E^{H-L}_{complex}$	1.234	2.156	0.232	1.417
Δd^{H-L}	0.2	0.28	0.48	0.44
Electronic configuration	$5s^{0.28}4d^{9.74}5p^{0.01}6p^{0.02}$	$5s^{0.79}4d^{9.86}5p^{0.14}5d^{0.01}6p^{0.02}$	$5s^{0.43}4d^{8.6}5p^{0.03}5d^{0.01}6p^{0.01}$	$5s^{0.83}4d^{8.81}5p^{0.13}5d^{0.02}6p^{0.02}$
Ne	10.05	10.82	9.05	9.81

There are some differences in the adsorption heights between the lowest spin state and the highest spin state of the adsorbed Rh and Pd atoms on both the O_{5c} and oxygen vacancy sites. In particular, for the perfect surface, the adsorption height of 2.54 Å and 2.39 Å for the quartet Rh and triplet Pd that is larger by 0.48 Å and 0.2 Å than for the doublet Rh and singlet Pd, respectively. The similar phenomena for the vacancy surface are also observed. This observation may result from the larger overlaps between the highest occupied molecular orbital, HOMO, of MgO cluster and the lowest unoccupied molecular orbital, LUMO, of the adsorbed metal atoms in the doublet and singlet state than those in the quartet and triplet states of the adsorbed Rh and Pd atoms that have anti-bonding character as shown from **Figure 1**. Therefore, a short bond distance and a strong interaction between the O^{2-} and Fs sites and the Rh and Pd atoms at the lowest spin state have been observed (**Table 1**).

The increase in the adsorption heights of supported Pd and Rh can contribute to the Pauli repulsion of the valence s orbital of the Pd and Rh that is almost empty with those in the p orbitals of the surface oxygen atoms. As well as, the HOMO for the oxygen vacancy has a large s-like character, which would also lead to repulsive interaction with the metal s orbital; this orbital is occupied by ~−0.8e due to the charge transfer. The adsorption heights of supported Pd and Rh increased at oxygen anion and oxygen vacancy sites also the energy gain of 1.90 eV and 2.0 eV due to the electron occupying this bonding orbital of the Pd and Rh atom, where the binding energy is calculated to be 3.192 eV and 3.263 eV for the adsorption complexes Pd/Mg$_9$O$_{13}$Fs and Rh/Mg$_9$O$_{13}$ Fs, respectively.

It is interesting to explore the electronic configuration for the interaction of Pd and Rh atoms with the regular site at MgO(001) surface. The only appreciable change with respect to the free atom is the hybridization between 5s and 4d orbitals with negligible contribution of the 5p subshell, $5s^{0.28}4d^{9.74}5p^{0.01}6p^{0.02}$ and $5s^{0.43}4d^{8.6}5p^{0.03}5d^{0.01}6p^{0.01}$ for the supported Pd and Rh atoms and no appreciable charge transfer, −0.054e and −0.07e, respectively, These results are consistent with [63]. The increased adsorption energy of Pd and Rh atoms with F_s centers is accompanied by notable changes in the electronic configuration of the adsorbed metal, which is progressively changed by a charge transfer, −0.799e and −0.807e, with an increasing participation of the 5p orbitals, for the supported Pd and Rh atoms, $5s^{0.79}4d^{9.86}5p^{0.14}$ and $5s^{0.83}4d^{8.81}5p^{0.13}$, respectively, **Table 1**.

As it has been shown later, the interaction of NO with supported Pd and Rh depends strongly on the metal-oxide interaction and it is essential to dispose of an adequate substrate model for the subsequent NO adsorption [64] [65]. Hence, as a first step in our approach, Rh$_2$/MgO(001), Pd$_2$/MgO(001) and PdRh/MgO(001) systems were studied using the above detailed cluster models.

3.2. Rh$_2$, Pd$_2$ and PdRh Particles Deposited on MgO (001)

To underscore the most stable configuration of Rh$_2$, Pd$_2$ and Pd-Rh dimer on the MgO(001) surface, two confi-

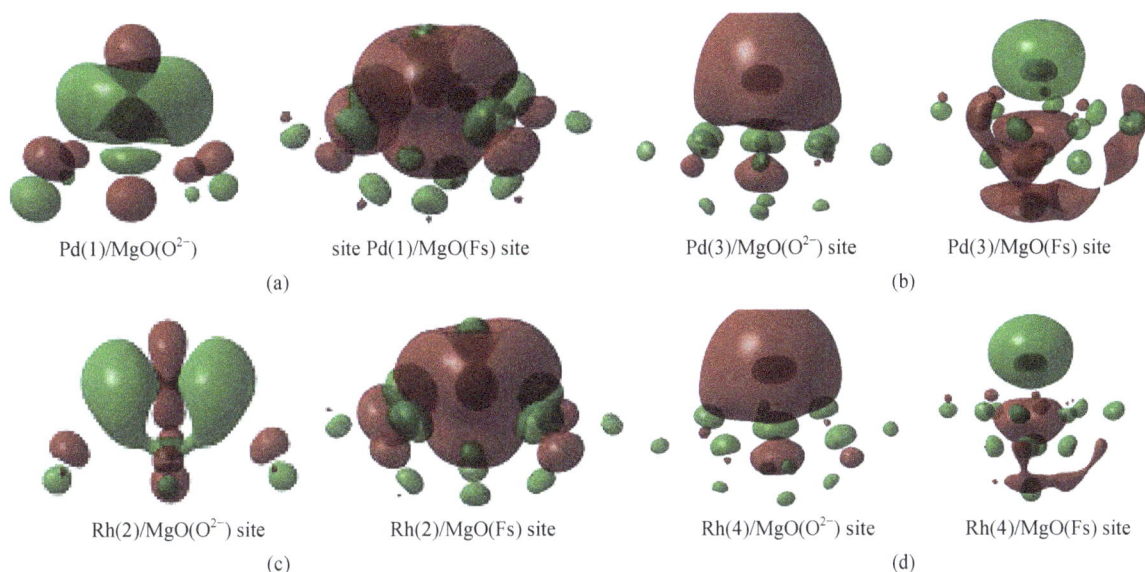

Pd(1)/MgO(O^{2-})   site Pd(1)/MgO(Fs) site   Pd(3)/MgO(O^{2-}) site   Pd(3)/MgO(Fs) site

(a)   (b)

Rh(2)/MgO(O^{2-}) site   Rh(2)/MgO(Fs) site   Rh(4)/MgO(O^{2-}) site   Rh(4)/MgO(Fs) site

(c)   (d)

Figure 1. Schematic representation of the highest occupied molecular orbitals (HOMOs) of Pd and Rh atoms with (a) low spin state and (b) high spin state deposited on MgO(O^{2-}) and MgO(Fs) sites using the embedded cluster model.

gurations, parallel and perpendicular to the surface plane have been considered. The best optimized geometries of Rh$_2$, Pd$_2$ and PdRh supported particles anchored on terrace sites of MgO(001) are with the molecular axis almost parallel to the surface and the supported two atoms of the dimer nearly on the top of two O$_{5c}$ anions, **Figure 2**. The results are agreement with [19] [56] [66].

Because of the spin polarization, the corresponding values of binding, nucleation, trapping and charges transfer for the deposition of the Rh$_2$, Pd$_2$ and PdRh particles on the regular oxygen site and Fs center have been summarized in **Table 2** by using B3LYP/CEP-121G at various spin states in order to find the most stable spin state for each dimer. It is interesting to note that, the ground state of Rh$_2$, Pd$_2$ is singlet, at variance with gas-phase of Pd$_2$ and Rh$_2$ which has a triplet $^3\Sigma_u^+$ [67] [68] and quintet $^5\Sigma_u^+$ [69] ground state, respectively, but in agreement with previous studies [5] [16]. Thus, the interaction with the substrate induces a change in the electronic structure of Pd$_2$ and Rh$_2$ dimers. The adsorption energies of Rh$_2$, Pd$_2$ and PdRh in the low spin states are stronger than those in the high spin states. Hence, the dimer-support interactions stabilize the low spin states of the adsorbed Rh$_2$, Pd$_2$ and PdRh dimers.

From these results, it is observed that the interaction of Rh$_2$, Pd$_2$ and PdRh on Fs site is characterized by stronger binding energy with shorter equilibrium adsorption distance than on the surface O^{2-} site. Although the binding energy is noticeably affected by the support, the nucleation energy is weakly affected; for both sites the values of E$_{nucl}$ are less significantly changed (0.001 - 0.246 eV). This can mean that (a) the regular metal oxide surface is always an appropriated place to form Rh$_2$, Pd$_2$ and Pd-Rh(b) the dimers formation is independent of the adsorption site (regular or diamagnetic Fs site). However the dimer formation will be favored on the Fs center due to the trapping energy, consistently with [3].

On a terrace site, the addition of second TM atom leads to a nucleation energy E$_{nucl.}$ = 1.716 eV, 2.465 eV, and 2.354 eV that are 0.466 eV, 1.165 eV and 1.053 eV higher than the adsorption energies of the TM atom on a terrace site. Consequently, the dimer formation of Rh$_2$, Pd$_2$ and PdRh are preferred with respect to two isolated atoms adsorbed on O^{2-} anions (**Table 2**), indicating that dimerization should be possible even on the MgO (001) terraces and the dimer nucleation is a thermodynamic favored process at O^{2-} anions. Although this result is in contrast with the results reported by Bogicevic and Jennison [70] who reported almost no difference in stability between the dimer and two isolated atoms on the MgO (001) terraces, it is agreement with [5] [66].

The elongation of the Pd-Pd, Rh-Rh and Pd-Rh distances with respect to the gas-phase is explained by the fact that the dimer is oriented towards two nearest neighbor O^{2-} anions on the surface to maximize the bonding with the O^{2-} anions. The Pd-Pd bond length becomes close to 2.98Å, is only 0.22 Å longer than in the free molecule, in agreement with [71].

Concerning the bimetallic PdRh particle, the ground state geometry of the bimetallic is significantly modified

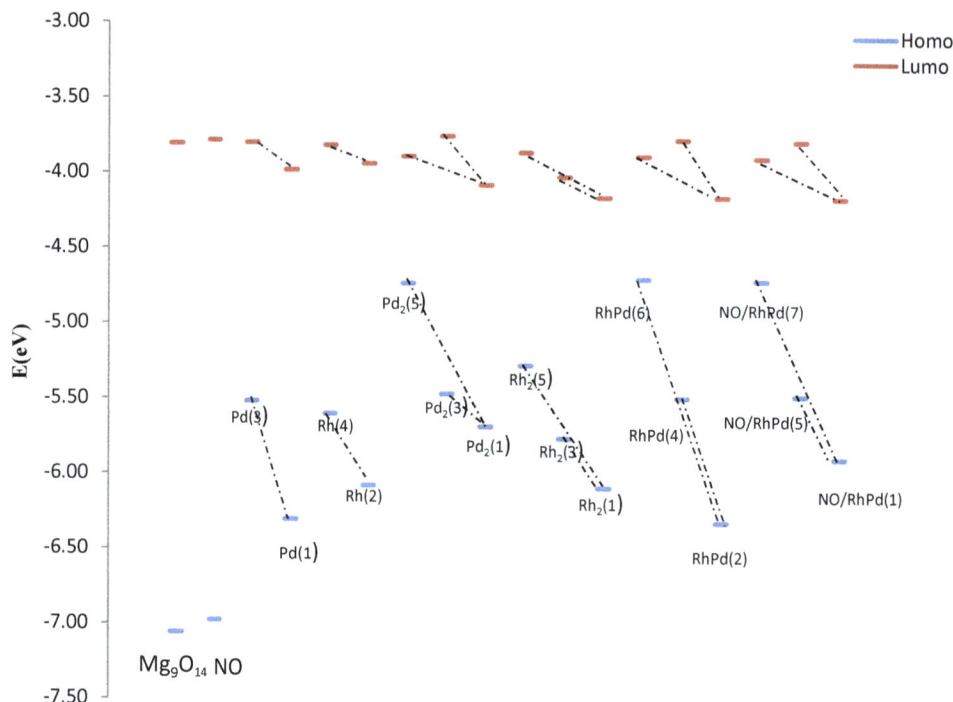

Figure 2. Frontier orbitals of regular surfaces Mg_9O_{14}, free NO molecule, Pd/Mg_9O_{14}, Rh/Mg_9O_{14}, Pd_2/Mg_9O_{14}, Rh_2/Mg_9O_{14}, $PdRh/Mg_9O_{14}$, $NO/Pd_2/Mg_9O_{14}$, $NO/Rh_2/Mg_9O_{14}$ and $NO/PdRh/Mg_9O_{14}$ at their high and low spin states.

Table 2. Geometrical parameters, binding, nucleation, trapping energies and atomic charges for the adsorption of Pd_2, Rh_2 and PdRh dimers with various spin multiplicities at regular (O^{2-}) site and defect center (Fs) site of the MgO (001) surface. Energies are expressed in eV, the corresponding equilibrium distances (d) in (Å), and charges in electron units.

	MgO (O^{2-}) site									MgO (Fs) site								
	Pd_2			Rh_2			PdRh			Pd_2			Rh_2			PdRh		
	M = 1	M = 3	M = 5	M = 1	M = 3	M = 5	M = 2	M = 4	M = 6	M = 1	M = 3	M = 5	M = 1	M = 3	M = 5	M = 2	M = 4	M = 6
E_a (M$_2$)	3.681	2.828	2.252	3.116	2.647	1.767	3.038	2.164	1.894	5.616	4.155	3.644	4.875	4.399	2.795	4.728	3.688	3.585
E_{nucl}	1.716	1.412	-0.894	2.465	2.653	2.502	2.354	2.139	4.647	1.717	0.804	-1.438	2.263	2.443	1.567	2.108	1.694	4.403
E_b	0.466	0.161	-2.145	1.165	1.352	1.201	1.053	1.071	2.119	0.466	-0.447	-2.688	0.962	1.142	0.267	0.807	0.659	-1.183
$E_{trap.}$	-	-	-	-	-	-	-	-	-	2.489	1.328	1.392	3.116	2.647	1.767	3.038	2.164	1.894
q(M1)	−0.069	0.03	−0.006	−0.064	−0.05	0.081	−0.01	0.008	−0.082	0.022	0.028	0.173	0.008	0.261	0.053	−0.001	−0.782	−0.842
q(M2)	−0.029	0.069	0.233	−0.097	0.087	0.112	0.001	0.107	0.334	−0.78	−0.727	−0.674	−0.999	−1.005	−0.751	−0.803	0.09	0.312
q(M$_2$)	−0.098	0.099	0.227	−0.161	0.037	0.193	−0.009	0.115	0.252	−0.758	−0.699	−0.501	−0.991	−0.744	−0.698	−0.804	−0.692	−0.530
d(M$_1$-S)	2.14	2.26	2.36	2.02	2.14	2.26	2.19	2.19	2.25	2.18	2.24	2.34	1.98	2.02	2.18	2.1	2.24	2.24
d(M$_2$-S)	2.23	2.28	2.41	2.06	2.1	2.12	2.1	2.19	2.23	1.54	1.62	1.76	1.58	1.62	1.62	1.57	1.59	1.67

q(M1), q(M2): atomic charges at each metal of the dimer. d(M$_1$-S), d(M$_2$-S): optimal distances between adsorbed metals of the dimer and surface site of MgO.

after deposition. The electronic density of states analysis reveals that after deposition, Pd-Rh favors doublet spin multiplicity as the lowest energy configuration, **Table 2**. In consequence, the binding, nucleation and dimer binding energies for Rh-Rh and Pd-Rh are very close. So that, such a comparison is of interest as many recent experimental studies on the supported palladium model catalytic systems [39] address the question whether the palladium can be used as an alternative to the expensive rhodium in the reaction of reduction of NO [72].

In this section, the stability trends of the Pd_2, Rh_2 and PdRh dimers for ground state structures are analyzed in terms of the energy gaps between HOMO and LUMO. A large HOMO-LUMO gap has been considered as an important requisite for the chemical stability of transition metal clusters [68]. The calculated HOMO-LUMO gaps for the ground states of Pd_2, Rh_2 and PdRh dimers are presented in **Table 2**. It can be seen from this table that, the calculated HOMO-LUMO gaps of Pd_2, Rh_2 and PdRh dimers follow the trend $Pd_2 < Rh_2 < PdRh$. Since a large energy gap corresponds to higher stability therefore, the PdRh dimers that deposited on the MgO (O^{2-}) and MgO (Fs) sites have the highest chemical stabilities.

Indeed, the molecular orbital, MO, interaction is controlled by the level of the frontier orbitals. Therefore, the relations between spin quenching of supported Rh, Pd, Pd_2, Rh_2 and PdRh dimers at MgO surface and energy gaps between frontier orbitals are established. In **Figure 3**, the HOMOs and LUMOs of the supported metals and dimers at the defect free surface of MgO are presented. While spin preservation occurs for Pd complex, spin quenching occurs for Rh, Rh_2, Pd_2 and RhPd complexes; this is agreement with **Tables 1-3**. This is clearly correlated with frontier orbital energies where the HOMO energy levels of Rh, Rh_2, Pd_2 and Pd-Rh low spin complexes are lower than their high spin counterparts. This extra stability allows for stronger interaction with the surface hence, the interaction in this case is strong enough to quench the spin.

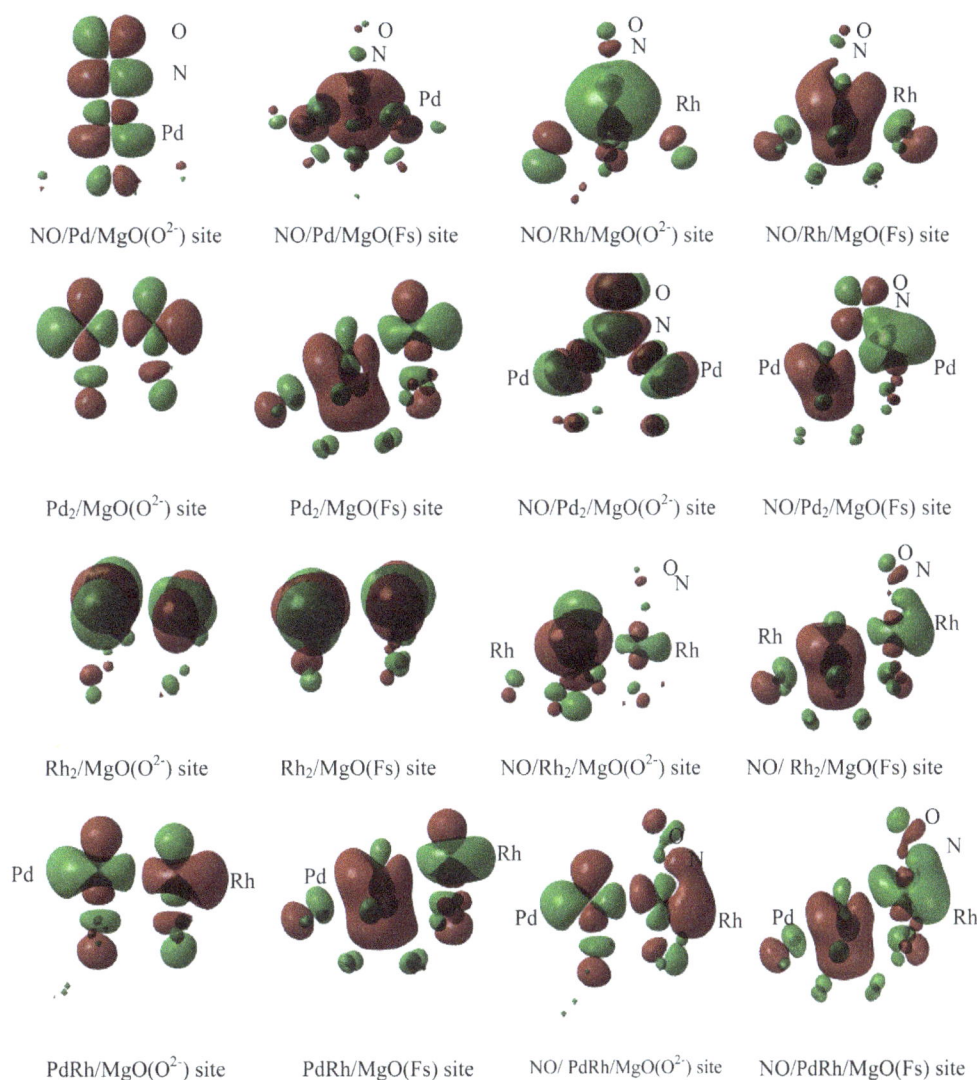

Figure 3. Schematic representation of the HOMO's of Pd_2/MgO, Rh_2/MgO, PdRh/MgO, NO/Pd/MgO, NO/Rh/MgO, NO/Pd_2/MgO, NO/Rh_2/MgO and NO/PdRh/MgO deposited on MgO(O^{2-}) and MgO(Fs) sites using the embedded cluster model.

Table 3. Transition energy, $\Delta E_{complex}^{H-L}$, required to excite the supported dimers, Pd_2, Rh_2 and PdRh, from the high- to low-spin states. A positive sign indicates that the ground state is provided by the low-spin coupling. $\Delta d_{M_1-s}^{H-L}$ and $\Delta d_{M_2-s}^{H-L}$ are the change in the equilibrium distance of each TM to the surface; $\Delta q_{M_1-M_2}^{H-L}$ is the change in charges at the dimers that supported to the regular (O^{2-}) site and oxygen vacancy (Fs) site at MgO (001) going from high- to low-spin state. A positive value indicates that d and q are larger in the high-spin state. Energies are expressed in eV, the corresponding equilibrium distances (d) in (Å), and charges in electron units H: High spin. L: Low spin.

	Pd_2		Rh_2		PdRh	
	Pd_2/MgO (O^{2-}) site	Pd_2/MgO (Fs) site	Rh_2/MgO (O^{2-}) site	Rh_2/MgO (Fs) site	PdRh/MgO (O^{2-}) site	PdRh/MgO (Fs) site
$(\Delta E_{complex}^{H-L})^a$	2.611	3.155	-0.036	0.695	2.223	2.222
$(\Delta E_{complex}^{H-L})^b$	2.306	2.242	0.151	0.875	2.009	1.842
$(\Delta d_{M_1-s}^{H-L})^a$	0.22	0.16	0.24	0.2	0.06	0.14
$(\Delta d_{M_1-s}^{H-L})^b$	0.1	0.1	0.12	0.16	0.00	0.00
$(\Delta d_{M_2-s}^{H-L})^a$	0.18	0.22	0.06	0.04	0.13	0.1
$(\Delta d_{M_2-s}^{H-L})^b$	0.13	0.14	0.02	0.00	0.04	0.08
$(\Delta q_{M_1-M_2}^{H-L})^a$	0.325	0.257	0.354	0.293	0.261	0.274
$(\Delta q_{M_1-M_2}^{H-L})^b$	0.13	0.19	0.156	0.046	0.137	0.162

*a: High spin multiplicity is (5) for Pd_2 Rh_2 and (6) for PdRh; *b: High spin multiplicity is (3) for Pd_2 Rh_2 and (4) for PdRh; Where low spin state is (1) for Pd_2 Rh_2 and (2) for PdRh

3.3. Interaction of NO with Supported Pd, Rh, Rh_2, Pd_2 and PdRh on MgO (001)

The binding energy for the interaction of NO molecule on different spin states of Pd, Rh, Rh_2, Pd_2 and PdRh that supported on MgO can be calculated as $E_a(NO) = -[E(NO/M_x/MgO\_site) - E(M_x/MgO\_site) - E(NO)]$ where x = 1 or 2 [55]. In all cases, the positive values correspond to exothermic processes. It is observed that NO adsorbs much more strongly at the Pd with low spin state and Rh with different spin state that deposited on the MgO(O^{2-}) site than on the MgO(Fs), **Table 4**. The reason of this behavior will be discussed later.

In **Tables 5-7**, the adsorption energies, optimal distances and charge transfer for the interaction of NO molecule on different spin states of Rh_2, Pd_2 and PdRh that supported at MgO (O^{2-}) and MgO (Fs) through its N atom at various spin states have been calculated. Again, the NO molecule adsorbs much more strongly at the Pd_2, with low spin state, Rh_2 and PdRh with different two spin state that deposited on the MgO (O^{2-}) site than on the MgO (Fs), **Table 5**. As well as, the NO has been adsorbed at the center of the Pd-Pd bond in a bridge position, which is energetically the most stable in agreement with previous works [37] [65] [73]. The Pd-Pd distance is about 2.98 Å - 3.05 Å, depending on the site considered, **Table 5**. The large Pd-Pd bond distance of adsorbed Pd_2 is in close agreement with the STM measurements [20]. Globally, the binding energy calculated for the adsorption of NO on the palladium dimers,1.808 eV, is in the range of the experimental value that measured in thermal desorption by Ramsier *et al.* [74].

At the high spin states a dramatic change is found when the NO which bind to the supported Pd and Pd_2 where there is an increase in the binding energies at the MgO (Fs) site than on the MgO (O^{2-}). The different behavior of Pd and Pd_2 with high spin states that adsorbed on Fs centers towards NO can be explained as follows. At the low spin states of supported Pd and Pd_2 the delocalization of the trapped electrons into the 5 s level leads to an increased Pauli repulsion with the NO molecule and in a strong weakening of the bond. On contrary, at the supported Pd and Pd_2 with high spin states this effect is smaller because of the presence of an incomplete d shell. **Figure 3** confirmed the results in **Table 4** and **Table 5** and the adsorption properties that discussed above.

Since the high occupied molecular orbital of NO is π^* *anti*-bonding orbital with an unpaired electron therefore, the charge transferred from Pd_2/MgO, Rh_2/MgO and PdRh/MgO to the NO will occupy the π^* orbital and weak the NO bond strength. The NO bond length is elongated after the adsorption of NO on the particular dimers, this

Table 4. Adsorption properties of NO interacting with regular (O^{2-}) site and oxygen vacancy (Fs) site of the MgO (001) surface supported Pd and Rh atoms at the low and high spin state.

	NO/Pd				NO/Rh			
	MgO (O^{2-}) site		MgO (Fs) site		MgO (O^{2-}) site		MgO (Fs) site	
Spin multiplicity	2	4	2	4	1	3	1	3
Spin state	L	H	L	H	L	H	L	H
E_a (NO)	1.499	1.038	0.702	1.261	3.048	2.115	1.449	1.026
E_a (MNO)	1.516	1.327	2.655	2.563	1.771	1.582	2.134	2.455
$\Delta E_{complex}^{H-L}$	1.695		1.598		0.934		0.423	
qMNO	−0.137	−0.126	−0.816	−0.936	−0.182	−0.063	−0.819	−0.742
d(M-S)	2.13	2.21	1.60	1.56	1.98	2.18	1.76	1.80
d(N-M)	1.90	1.96	2.06	2.02	1.74	1.82	1.78	1.96
d(N-O)	1.228	1.288	1.208	1.248	1.207	1.207	1.207	1.207

q(M): atomic charges at the adatom. q(NO): molecular charge at NO molecule. d(M-S): optimal distances between adatom and surface site of MgO. d(N-O): Equilibrium N-O distances. d(N-M): optimal distances between adatom and nitrogen atom.

Table 5. Adsorption properties of NO interacting with regular (O^{2-}) site and oxygen vacancy (Fs) site of the MgO (001) surface supported Pd_2 dimer at the low and high spin state.

	MgO (O^{2-}) site		MgO (Fs) site	
Spin multiplicity	2	4	2	4
Spin state	L	H	L	H

E_a (NO)	1.808	1.052	1.231	1.342
Ea (Pd_2NO)	1.887	2.352	3.245	3.969
Total q(Pd_2)	0.415	0.525	−0.493	−0.049
q(N)	−0.248	−0.473	−0.071	−0.460
q(O)	−0.27	−0.225	−0.199	−0.247
q(NO)	−0.518	−0.697	−0.27	−0.707
q(Pd_2NO)	−0.103	−0.172	−0.763	−0.756

Red spheres: O^{2-}; yellow spheres: Mg^{2+}; light blue spheres: Pd atom; dark blue sphere: N atom.

Table 6. Adsorption properties of NO interacting with regular (O^{2-}) site and oxygen vacancy (Fs) site of the MgO (001) surface supported Rh_2 dimer at the low and high spin state.

Spin multiplicity	2	4	2	4
Spin state	L	H	L	H
	MgO (O^{2-}) site		MgO (Fs) site	

E_a (NO)	2.909	2.506	2.971	2.097
Ea (Rh_2NO)	1.786	2.542	3.607	3.885
q(Rh_2)	−0.164	0.01	−0.912	−0.639
q(N)	0.22	0.093	0.219	0.070
q(O)	−0.185	−0.222	−0.188	−0.231
q(NO)	0.035	−0.129	0.031	−0.161
q(Rh_2NO)	−0.129	−0.119	−0.881	−0.799

Red spheres: O^{2-}; yellow spheres: Mg^{2+}; blue spheres: Rh atom; dark blue sphere: N atom.

Table 7. Adsorption properties of NO interacting with regular (O^{2-}) site and oxygen vacancy (Fs) site of the MgO (001) surface supported PdRh dimer at the low and high spin state.

Spin multiplicity	1	3	1	3
Spin state	L	H	L	H
	MgO (O^{2-}) site		MgO (Fs) site	

E_a (NO)	2.666	1.315	2.872	1.329
Ea (PdRhNO)	2.057	2.319	3.953	3.856
q(PdRh)	−0.201	0.539	−0.901	−0.109
q(N)	0.22	−0.413	0.206	−0.359
q(O)	−0.204	−0.248	−0.205	−0.236
q(NO)	0.016	−0.661	0.001	−0.595
q(PdRhNO)	−0.184	−0.122	−0.901	−0.704

Red spheres: O^{2-}; yellow spheres: Mg^{2+}; light blue spheres: Pd atom; blue spheres: Rh atom; dark blue sphere: N atom.

is consistent with the electron transfer direction [57], **Tables 5-7**. The higher electron transfer from Pd_2/MgO, Rh_2/MgO and PdRh/MgOto NO can explain the larger stretching of the NO interatomic distance on the oxygen anion and oxygen vacancies sites.

The adsorption energy of NO on Rh_2/MgO is larger than on Pd_2/MgO (2.909 vs. 1.808 eV), possibly due the decrease of the d electrons on the Rh which can lead to a decrease of the σ-σ repulsion. The metal-nitrogen bond is shorter for Rh than Pd (1.74 vs. 2.01 Å), which also indicates a strong bonding between NO and Rh_x/MgO. The larger M-N-O angle for Rh_2 than Pd_2 (180° vs. 132.2°) indicates that the 5σ orbital is much more involved in the adsorption at Rh than Pd. Although, the charge of Rh_2NO and Pd_2NO supported on the (O^{2-}) and (Fs) site is practically the same than that of the supported Rh_2 and Pd_2 at the same sites, the MgO(Fs) site acquires a much more significant negative charge, **Tables 5-7**. These results are confirmed by **Figure 3**.

Indeed, the different behavior of rhodium and palladium supported at MgO(001) towards NO can be explained as follows. On palladium the delocalization of the trapped electrons into the 5s level leads to an increased Pauli repulsion with the NO molecule and in a strong weakening of the bond, whereas on Rh this effect is smaller because of the presence of an incomplete d shell and the easier mixing of the 5s with the 4d orbitals to form new hybrid orbitals [62]. The high reactivity of Rh relative to Pd was also found by a periodic DFT calculation for NO adsorbed on (001) surfaces of Rh and Pd. Therefore, Rh appears to be more efficient than Pd for NO adsorption at Pd-Rhbimetallic and may justify its use as a catalyst in TWC. These observations agree well with previous theoretical calculations [75]. Therefore, it seems that NO prefers binding to Rh when both Rh and Pd sites co-exist in the bimetallic Pd-Rh. These results are confirmed by **Table 7** and **Figure 3**.

The optimized geometry of admolecule NO with the N-end to the Rh atom of bimetallic PdRh and the molecular axis of NO normal to the surface plane is presented in **Table 6**. The N-O distance elongates from its value of 1.158 Å for the free NO molecule to 1.207 Å for Rh_2 and PdRh at O^{2-} and Fs, respectively. Therefore, the results show that the adsorption energies of NO at Rh-Rh and Pd-Rh, with low spin state, that deposited at O^{2-} and Fs of MgO are very close, (2.909 vs. 2.666 eV) at oxygen anion and (2.971 vs. 2.872 eV) at Fs centers, **Table 6** and **Table 7**. **Figure 3** confirmed the results in **Tables 2-7** with the adsorption properties discussed above.

Interestingly, the interaction is assumed to mainly be a HOMO-LUMO type [76]. The differences in the adsorption energies reported for the interaction of NO on supported PdRh at the O^{2-} and Fs center can be due to the differences in energy between the HOMO of the surface and the LUMO of the NO molecule. The results show that, the formation of a vacancy on the MgO surface decreases the difference between the HOMO of PdRh supported at MgO (001) surface and LUMO of NO molecule by 0.367 eV. This result is in agreement with the greater strength of NO chemisorption on supported PdRh/MgO(Fs) in comparison with supported PdRh/MgO(O^{2-}), HOMO-LUMO = 2.174 eV and 2.541 eV, respectively. As the interaction of NO with PdRh occurs through a charge transfer from the HOMO of the surface to the LUMO of the adsorbed NO molecule, the smaller the value of the HOMO–LUMO gap the easier the charge transfer and consequently, the larger the adsorption energy, this is agreement with [77].

Recently, several authors were interested in studying the CO-induced modification of the metal-MgO interaction [78] [79] in the form of atoms and layers. For instance, it was reported that CO enhances the bonding between Pt or Pd atoms and the oxide, but for Au this effect is negligible [78]. To allow a similar analysis on NO, the adsorption of the MNO and M_2NO complexes on MgO(001) was considered. The corresponding binding energy can be defined in a similar way as that for M_x on MgO, $E_a(M_xNO) = -[E(NO/ M_x/MgO\_site) - E(MgO\_site) - E(M_xNO)]$ where x = 1 or 2. On terrace O_{5c} sites the $O_{5c}Rh_2$-NO and $O_{5c}PdRh$-NO bonds are definitely stronger than the O_{5c}-Rh_2NO and O_{5c}-PdRhNO. This means that an increase in temperature can lead to diffusion of the Rh_2NO and PdRhNO complexes before NO desorption occurs. Whereas on neutral F centers, the bonding of the Pd_2NO, Rh_2NO and PdRhNO unit to the surface is so strong that no diffusion of this species occurs once the complex is trapped at an oxygen vacancy. An increase in temperature will result in the loss of NO and TM atoms filling the vacancy. An important result is that, on Fs sites, the Fs-Pd_2NO, Fs-Rh_2NO and Fs-PdRhNO bonding is stronger than that of the $FsPd_2$-NO $FsRh_2$-NO and FsPdRh-NO, **Tables 5-7**.

The spin transition energies, $\Delta E_{complex}^{H-L}$, of ON·Pd_2·MgO, ON·Rh_2·MgO and ON·RhPd·MgO complexes have been considered, **Table 8**. The trend emerging from the present model calculations indicates that the metal-support interaction tends to stabilize the low-spin state with respect to the isolated atom thus, the low-spin state becoming the ground state at regular and F_S sites. Although, the combined effects of NO adsorbate and Pd_2, Rh_2 and RhPd supported at MgO were strong enough to quench the spin of Pd_2, Rh_2 and RhPd dimers (changes the sign of the spin transition energy), the low-spin states are preserved. Although, the interaction of ON.Rh_2·MgO (Fs)

Table 8. Transition energy, $\Delta E^{H-L}_{complex}$, required to excite ON·Pd$_2$·MgO, ON·Rh$_2$·MgO and ON·PdRh·MgO complexes from the high-to low-spin states. A positive sign indicates that the ground state is provided by the low-spin coupling. Δd^{H-L}_{M-N} is the change in optimal distances between TM and nitrogen atom. Δd^{H-L}_{N-O} is the change in optimal distances between N and O atom of NO molecule. $\Delta q^{H-L}_{M_2NO}$ is the change in charges at the M_2NO dimer that supported to the regular (O^{2-}) site and oxygen vacancy (Fs) site at MgO (001) going from high- to low-spin state. A positive value indicates that d is larger in the high-spin state. Energies are expressed in eV, the corresponding equilibrium distances (d) in (Å), and charges in electron units H: High spin. L: Low spin.

	NO/Pd$_2$/MgO (O^{2-}) site	NO/ Pd$_2$/MgO (Fs) site	NO/Rh$_2$/MgO (O^{2-}) site	NO/ Rh$_2$/MgO (Fs) site	NO/PdRh/MgO (O^{2-}) site	NO/PdRh/MgO (Fs) site
($\Delta E^{H-L}_{complex}$ (NO))	1.061	0.801	0.216	0.694	1.565	1.923
(Δd^{H-L}_{M-N})	0.14	0.27	0.06	0.04	0.427	0.35
(Δd^{H-L}_{N-O})	0.021	0.041	0	0.02	0.06	0.02
($\Delta q^{H-L}_{M_2NO}$)	-0.069	0.007	0.01	0.082	0.063	0.197

where low spin multiplicity is (2) for NOPd$_2$, NORh$_2$ and (1) for NOPdRh high spin multiplicity is (4) for NOPd$_2$, NORh$_2$ and (3) for NOPdRh.

and ON·RhPd·MgO (Fs) shows that the transition energy exhibits the largest increase, the interaction of ON·Pd$_2$·MgO (Fs)exhibits the largest decrease. Notice that, as expected, there is an inverse correlation between adsorption energy and equilibrium distance, the larger the former the shorter the later, **Tables 5-8**. In these cases, it is clear that the low-spin state is more favored because of the formation of a direct bond between the adsorbed transition metal dimer and the electronic levels corresponding to the oxygen vacancy electrons. The results show that the magnetic-spin states of transition metals atoms and clusters supported at metal oxide surface and the role of a precursor molecule on the considered magnetic and binding properties need to be explicitly taken into account.

3.4. Pairwise and Non-Pairwise Additivity

The concept of pairwise and non-pairwise additivity has been studied for atom clusters and insulators [80]-[82]. In studying a supported-metal catalyst system, it is very important to quantify the extent to which the support MgO (S) with regular and defective surface affects the interaction of the NO admolecule with the Pd, Rh, Pd$_2$, Rh$_2$ and PdRh particles. The interaction energy $E^{S-M_x-NO}_{(i)}$ among three subsystems; the support (S), (M$_x$) where x = 1 for Pd, Rh atom and x = 2 forPd$_2$, Rh$_2$, PdRh dimer, and the adsorbate (NO) molecule can be defined as:

$$E^{S-M_x-NO}_{(i)} = E^{S-M_x-NO} - E^S - E^{M_x} - E^{NO} \tag{5}$$

where every energy term on the right-hand side of Equation (5) is calculated using geometrical parameters corresponding to the equilibrium geometry of S-M$_x$-NO systems. The left-hand side represents the energy required to separate the three subsystems without altering any change in their geometrical parameters. Such energy can be divided into contributions from three-pairwise components and a non-additive term, ε^{nadd}, as follows:

$$E^{S-M_x-NO}_{(i)} = E^{S-M_x}_{(i)} + E^{S-NO}_{(i)} + E^{M_x-NO}_{(i)} + \varepsilon^{nadd} \tag{6}$$

where ε^{nadd} is a measure of cooperative interactions among the three subsystems [38] [81]. The four energy terms on the right-hand side of Equation (6) are calculated from the relations:

$$E^{S-M_x}_{(i)} = E^{S-M_x} - E^S - E^{M_x} \tag{7}$$

$$E^{S-NO}_{(i)} = E^{S-NO} - E^S - E^{NO} \tag{8}$$

$$E^{M_x-NO}_{(i)} = E^{M_x-NO} - E^{M_x} - E^{NO} \tag{9}$$

$$\varepsilon^{nadd} = E^{S-M_x-NO}_{(i)} - E^{S-M_x}_{(i)} - E^{S-NO}_{(i)} - E^{M_x-NO}_{(i)} \tag{10}$$

The total interaction energies, the pairwise energy components to the S-M_x-NO systems, and the non-additive energy term, ε^{nadd}, are presented in **Table 9**. As shown in this table, the total interaction energies of ON·Pd·$Mg_9O_{13}O^{2-}$, ON·Pd·$Mg_9O_{13}Fs$, ON·Rh·$Mg_9O_{13}O^{2-}$, ON·Rh·$Mg_9O_{13}Fs$, ON·Pd_2·$Mg_9O_{13}O^{2-}$, ON·Pd_2·$Mg_9O_{13}Fs$, ON·Rh_2·$Mg_9O_{13}O^{2-}$, ON·Rh_2·$Mg_9O_{13}Fs$, ON·PdRh·$Mg_9O_{13}O^{2-}$ and ON·PdRh·$Mg_9O_{13}Fs$ complexes are dominated by the pairwise additive components $E_{(i)}^{S-M}$ and $E_{(i)}^{S-M_2}$, respectively. On the other hand, the small values of $E_{(i)}^{S-NO}$ pairwise component that represent the interaction energy between support (S) and admolecule (NO) in the S-M_x-NO system may be attributed to the large separation between (S) and the NO admolecule. This result means that the binding of NO is mainly dominated by the $E_{(i)}^{M_1-NO}$ and $E_{(i)}^{M_2-NO}$ pairwise additive contributions of the considered complexes.

The non additivity term, ε^{nadd}, is a measure of cooperative interaction among the subsystems, decreases with surface defect-formation at ON·Pd·MgO, ON·Rh·MgO, ON·Pd_2·MgO and ON·Rh_2·MgO. Except at ON·PdRh·MgO complex, ε^{nadd} increases with surface defect-formation, **Table 9**. This suggests that the interaction of NO with Pd, Rh, Pd_2, Rh_2 and PdRh dimer is essentially affected by defect formation. This confirmed the adsorption properties of NO at supported Pd, Rh, Pd_2, Rh_2 and PdRh particles that discussed above. Finally, the role of metal oxide is not restricted only to supporting the metal, but influences the interaction of NO molecule with the Pd, Rh, Pd_2, Rh_2 and PdRh dimers.

4. Conclusions

An attempt has been made to understand the effect of surface site on the spin state for the interaction of NO with Pd_2, Rh_2 and PdRh nanoparticles that supported at regular and defective MgO (001) Surfaces. A spin-polarized treatment is considered to properly describe the ground-state electronic structure, adsorption energies and the low- to high-spin energy transition. The calculated results are compared with experimental data and previous theoretical studies as possible. The geometrical optimizations have been considered to represent the most stable structures for the adsorption of NO at the supported Pd_2, Rh_2 and PdRh and to investigate the changes induced by the oxide substrate in the chemisorption properties of the adsorbed particles.

Upon interaction with O^{2-} and Fs surface sites, the high to low spin transition energies of Pd atom is positive indicating that the spin states are preserved, and the low spin states are favored. Hence, the number of unpaired electrons in the adatom tends to be the same as in the gas phase and the ground state of Pd-MgO is spin singlet. However, the main contributions to the Rh atom, Pd_2, Rh_2 and PdRh at MgO are the polarization of the metal electrons induced by the ionic substrate and the small mixing between the s and d orbitals of the transition metal with the 2p orbitals of the surface oxygen. Consequently, the interaction of Rh atom, Pd_2, Rh_2 and PdRh dimers at MgO (001) surface induces a quenching of the magnetic moment, which results in a doublet ground state for Rh atom and PdRh as well as a singlet ground state for Pd_2 and Rh_2 at MgO (001) surface. As a consequence, the formation of the dimer in its singlet state, Rh_2 and Pd_2, and doublet state, Pd-Rh deposited at MgO (001), is

Table 9. Interaction energies of ON.Pd.MgO, ON.Rh.MgO, ON.Pd_2.MgO, ON.Rh_2.MgO, ON.PdRh.MgO complexes with the most stable spin states at O^{2-} and Fs sites, pairwise components and non additivity terms. All energies are given in eV.

Complex	$E_{(i)}^{S-M_x-NO}$		$E_{(i)}^{S-M_x}$		$E_{(i)}^{S-NO}$		$E_{(i)}^{M_x-NO}$		ε^{nadd}	
	O^{2-}	F	O^{2-}	F	O^{2-}	F	O^{2-}	F	O^{2-}	F
ON·Pd·MgO	−2.749	−3.888	−1.238	−3.188	−0.061	−0.092	−1.229	−1.128	−0.221	0.519
ON·Rh·MgO	−4.892	−4.713	−1.294	−2.938	−0.086	−0.080	−2.570	−2.537	−0.942	0.842
ON·Pd₂·MgO	−5.489	−6.847	−3.624	−5.614	0.035	−0.068	−3.529	−2.659	1.629	1.494
ON·Rh₂·MgO	−5.989	−7.846	−3.058	−4.870	−0.063	−0.120	−2.528	−2.528	−0.341	−0.328
ON·PdRh·MgO	−5.704	−7.600	−3.039	−4.727	−0.062	−0.117	−2.638	−2.619	0.035	−0.136

favored with respect to the presence of two isolated atoms on the surface. Notice that, as expected, there is an inverse correlation between adsorption energy and equilibrium distance, the larger the former the shorter the later. In any case, the extent of metal-metal bonding in supported dimer has been increased compared with the gas-phase unit. This leads to a considerable elongation of the metal-metal bond to maximize the metal-O interaction. Notice that the dimer as a unit adsorbs much more strongly on the MgO (Fs) site than on the MgO (O^{2-}) site. Moreover, the large enhancement in the activity of supported dimers is due mainly to the electron transfer from the cavity to the supported dimers. Theoretical calculations indicate that the formation of Rh_2, Pd_2 and PdRh dimer on an Fs center is favored by 0.466, 0.962, and 0.807 eV respectively with respect to a TM atom bound at the Fs center and other TM atom on a terrace site. The dimers deposited interact relatively strongly with the substrate oxide forming predominantly covalent bonds with the adsorbed sites. The interaction is not accompanied by a significant charge transfer at the interface. The PdRh bimetallic has larger HOMO-LUMO gap and is relatively more chemically stable than the Pd_2 and Rh_2 monometallic that deposited on the MgO (O^{2-}). The transition energy, $\Delta E^{H-L}_{complex}$, for the interaction of Pd-Rh with the oxygen anion and F_S center exhibits the largest increase, 2.223 eV and 2.222 eV respectively. In these cases, it is clear that the low-spin state is more favored because of the formation of a direct bond between the adsorbed bimetallic and the electronic levels corresponding to the oxygen anion and oxygen vacancy electrons. A molecular-scale understanding of the energetic and mechanisms for formation of such metal dimers on inert oxide surfaces can open new avenues to the design of catalysts with specific functions.

In summary, it seems that NO prefers to bound with Rh atoms when both Rh and Pd site co-exist in the Pd-Rh bimetallic. The electronic structures and N-O bond lengths of the chemisorbed systems are similar for $NO \cdot Rh_2 \cdot MgO$ and $NO \cdot PdRh \cdot MgO$ with the top geometries but show significant differences from bridge geometries, $NO/Pd_2/MgO$. Bridge-site adsorption causes the N-O bond to lengthen and soften due essentially to increase an electrostatic repulsion between both N and O atoms. In addition, the NO adsorbs much more strongly at the PdRh that is deposited on the MgO (Fs) than on the MgO (O^{2-}) site.

The transfer of electron charge density from such a defect to a dimer reinforces the metal-metal bonds. Therefore, color centers at the MgO surface not only reduce the diffusion of metal atoms and dimers, but also act as stabilizing agents for the whole structure. This point could be particularly important in the context of identifying methods to stabilize the support particles on an oxide substrate under chemical reaction conditions.

To summarize, the larger interaction of NO at bimetallic PdRh at oxygen anions and oxygen vacancies induces an enhancement of the energy required to switch from high spin to low spin 1.565 eV and 1.923 eV respectively. These results show that the spin state of adsorbed PdRh dimer on oxide supports tends to preserve the number of unpaired electrons as found in the case of the regular terrace sites.

Acknowledgements

My gratitude and deep thanks to Prof. Dr. A.S. Shalabi for his interest, and useful discussions.

References

[1]   Piccolo, L. and Henry, C.R. (2000) Reactivity of Metal Nanoclusters: Nitric Oxide Adsorption and CO+NO Reaction on Pd/MgO Model Catalysts. *Applied Surface Science*, **162-163**, 670-678.
        http://dx.doi.org/10.1016/S0169-4332(00)00267-1

[2]   Xu, C., Oh, W.S., Liu, G., Kim, D.Y. and Goodman, D.W. (1997) Characterization of Metal Clusters (Pd and Au) Supported on Various Metal Oxide Surfaces (MgO and TiO₂). *Journal of Vacuum Science & Technology A*, **15**, 1261.
        http://dx.doi.org/10.1116/1.580604

[3]   Florez, E., Mondragón, F., Truong, T.N. and Fuentealba, P. (2007) Density Functional Theory Characterization of the Formation of Copper Clusters on F_s and F_s^+ Centers on a MgO Surface. *Surface Science*, **601**, 656-664.
        http://dx.doi.org/10.1016/j.susc.2006.10.040

[4]   Wang, Y., Florez, E., Mondragón, F. and Truong, T.N. (2006) Effects of Metal-Support Interactions on the Electronic Structures of Metal Atoms Adsorbed on the Perfect and Defective MgO(100) Surfaces. *Surface Science*, **600**, 1703-1713. http://dx.doi.org/10.1016/j.susc.2005.12.062

[5]   Giordano, L., Di Valentin, C., Pacchioni, G. and Goniakowski, J. (2005) Formation of Pd Dimers at Regular and Defect Sites of the MgO(100) Surface: Cluster Model Calculations. *The Journal of Chemical Physics*, **309**, 41-47.

[6]   Inntam, C., Moskaleva, L.A., Neyman, K.M. and Nasluzov, V.A. (2006) Adsorption of Dimers and Trimers of Cu, Ag,

and Au on Regular Sites and Oxygen Vacancies of the MgO(001) Surface: A Density Functional study Using Embedded Cluster Models. *Applied Physics A*, **82**, 181-189. http://dx.doi.org/10.1007/s00339-005-3352-8

[7]　Brune, H. (1998) Microscopic View of Epitaxial Metal Growth: Nucleation and Aggregation. *Surface Science Reports*, **31**, 125-229. http://dx.doi.org/10.1016/S0167-5729(99)80001-6

[8]　Cinquini, F., Di Valentin, C., Finazzi, E., Giordano, L. and Pacchioni, G. (2007) Theory of Oxides Surfaces, Interfaces and Supported Nano-Clusters. *Theoretical Chemistry Accounts*, **117**, 827-845. http://dx.doi.org/10.1007/s00214-006-0204-3

[9]　Fernandez, S., Markovits, A. and Minot, C. (2008) Adsorption of the First Row of Transition Metals on the Perfect and Defective MgO(100) Surface. *Chemical Physics Letters*, **463**, 106-111. http://dx.doi.org/10.1016/j.cplett.2008.08.053

[10]　Markovits, A., Paniagua, J.C., Lopez, N., Minot, C. and Illas, F. (2003) Adsorption Energy and Spin State of First-Row Transition Metals Adsorbed on MgO(100). *Physical Review B*, **67**, 115417. http://dx.doi.org/10.1103/PhysRevB.67.115417

[11]　Neyman, K.M., Innatam, C., Nasluzov, V.A., Kosarev, R. and Rosch, N. (2004) Adsorption of *d*-Metal Atoms on the Regular MgO(001) Surface: Density Functional Study of Cluster Models Embedded in an Elastic Polarizable Environment. *Applied Physics A*, **78**, 823-828. http://dx.doi.org/10.1007/s00339-003-2437-5

[12]　Markovits, A., Skalli, M.K., Minot, C., Pacchioni, G., Lopez, N. and Illas, F. (2001) The Competition between Chemical Bonding and Magnetism in the Adsorption of Atomic Ni on MgO(100). *The Journal of Chemical Physics*, **115**, 8172. http://dx.doi.org/10.1063/1.1407824

[13]　Sousa, C., de Graaf, C., Lopez, N., Harrison, N.M. and Illas, F. (2004) *Ab Initio* Theory of Magnetic Interactions at Surfaces. *Journal of Physics*: *Condensed Matter*, **16**, S2557-S2574. http://dx.doi.org/10.1088/0953-8984/16/26/027

[14]　Paulus, U.A., Endruschat, U., Feldmeyer, G.J., Schimidt, T.J., Bonnemann, H. and Behm, R.J. (2000) New PtRu Alloy Colloids as Precursors for Fuel Cell Catalysts. *Journal of Catalysis*, **195**, 383-393. http://dx.doi.org/10.1006/jcat.2000.2998

[15]　Jalili, S., Isfahani, A.Z. and Habibpour, R. (2012) Atomic Oxygen Adsorption on Au (100) and Bimetallic Au/M (M = Pt and Cu) Surfaces. *Computational and Theoretical Chemistry*, **989**, 18-26. http://dx.doi.org/10.1016/j.comptc.2012.02.033

[16]　Sicolo, S. and Pacchioni, G. (2008) Charging and Stabilization of Pd Atoms and Clusters on an Electron-Rich MgO Surface. *Surface Science*, **602**, 2801-2807. http://dx.doi.org/10.1016/j.susc.2008.07.005

[17]　Kukovecz, Á., Pótári, G., Oszkó, A., Kónya, Z., Erdőhelyi, A. and Kiss, J. (2011) Probing the Interaction of Au, Rh and Bimetallic Au-Rh Clusters with the TiO2 Nanowire and Nanotube Support. *Surface Science*, **605**, 1048-1055. http://dx.doi.org/10.1016/j.susc.2011.03.003

[18]　Rassoul, M., Gaillard, F., Garbowski, E. and Primet, M. (2001) Synthesis and Characterisation of Bimetallic Pd-Rh/Alumina Combustion Catalysts. *Journal of Catalysis*, **203**, 232-242. http://dx.doi.org/10.1006/jcat.2001.3328

[19]　Ferrari, A.M. (1999) Pd and Ag Dimers and Tetramers Adsorbed at the MgO(001) Surface: A Density Functional Study. *Physical Chemistry Chemical Physics*, **1**, 4655-4661. http://dx.doi.org/10.1039/a904813h

[20]　Shinkarenko, V.G., Anufrienko, V.F., Boreskov, G.K., Ione, K.G. and Yureva, T.M. (1975) *Doklady Akademii Nauk SSSR*, **223**, 410.

[21]　Neyman, K.M. and Illas, F. (2005) Theoretical Aspects of Heterogeneous Catalysis: Applications of Density Functional Methods. *Catalysis Today*, **105**, 2-16. http://dx.doi.org/10.1016/j.cattod.2005.04.006

[22]　Nasluzov, V.A., Rivanenkov, V.V., Gordienko, A.B., Neyman, K.M., Birkenheuer, U. and Rösch, N. (2001) Cluster Embedding in an Elastic Polarizable Environment: Density Functional Study of Pd Atoms Adsorbed at Oxygen Vacancies of MgO(001). *The Journal of Chemical Physics*, **115**, 8157. http://dx.doi.org/10.1063/1.1407001

[23]　Matveev, A.V., Neyman, K.M., Yudanov, I.V. and Rösch, N. (1999) Adsorption of Transition Metal Atoms on Oxygen Vacancies and Regular Sites of the MgO(001) Surface. *Surface Science*, **426**, 123-139. http://dx.doi.org/10.1016/S0039-6028(99)00327-1

[24]　Berthier, G. (2001) Simulation of *Ab Initio* Results for Palladium and Rhodium Clusters by Tight-Binding Calculations. *International Journal of Quantum Chemistry*, **82**, 26-33. http://dx.doi.org/10.1002/1097-461X(2001)82:1<26::AID-QUA1018>3.0.CO;2-O

[25]　Lombardi, J.R. and Davis, B. (2002) Periodic Properties of Force Constants of Small Transition-Metal and Lanthanide Clusters. *Chemical Reviews*, **102**, 2431-2460. http://dx.doi.org/10.1021/cr010425j

[26]　Wu, Z.J. (2005) Theoretical Study of Transition Metal Dimer AuM (M = 3d, 4d, 5d Element). *Chemical Physics Letters*, **406**, 24-28. http://dx.doi.org/10.1016/j.cplett.2005.02.083

[27]　Wu, Z.J. (2004) Density Functional Study OF the Second Row Transition Metal Dimmers. *Chemical Physics Letters*,

383, 251-255. http://dx.doi.org/10.1016/j.cplett.2003.11.023

[28] Negreiros, F.R., Barcaro, G., Kuntová, Z., Rossi, G., Ferrando, R. and Fortunelli, A. (2011) Structures of AgPd Nanoclusters Adsorbed on MgO(100): A Computational Study. *Surface Science*, **605**, 483-488. http://dx.doi.org/10.1016/j.susc.2010.12.002

[29] Wang, M.Y., Liu, X.J., Meng, J. and Wu, Z.J. (2007) Interaction of H_2 with Transition Metal Homonuclear Dimers Cu_2, Ag_2, Au_2 and Heteronuclear Dimers PdCu, PdAg and PdAu. *Journal of Molecular Structure: THEOCHEM*, **804**, 47-55. http://dx.doi.org/10.1016/j.theochem.2006.10.007

[30] Gómez, T., Florez, E., Rodriguez, J.A. and Illas, F. (2010) Theoretical Analysis of the Adsorption of Late Transition-Metal Atoms on the (001) Surface of Early Transition-Metal Carbides. *The Journal of Physical Chemistry C*, **114**, 1622-1626. http://dx.doi.org/10.1021/jp910273z

[31] Die, D., Kuang, X.Y., Guo, J.J. and Zheng, B.X. (2009) First-Principle Study of Au_nFe (n = 1–7) Clusters. *Journal of Molecular Structure: THEOCHEM*, **902**, 54-58. http://dx.doi.org/10.1016/j.theochem.2009.02.009

[32] Chin, Y.H., King, D.L., Roh, H.S., Wang, Y. and Heald, S.M. (2006) Structure and Reactivity Investigations on Supported Bimetallic Au-Ni Catalysts Used for Hydrocarbon Steam Reforming. *Journal of Catalysis*, **244**, 153-162. http://dx.doi.org/10.1016/j.jcat.2006.08.016

[33] Liu, F.L., Zhao, Y.F., Li, X.Y. and Hao, F.Y. (2007) *Ab Initio* Study of the Structure and Stability of M_nTl_n (M = Cu, Ag, Au; n = 1, 2) Clusters. *Journal of Molecular Structure: THEOCHEM*, **809**, 189-194. http://dx.doi.org/10.1016/j.theochem.2007.01.018

[34] Vesecky, S.M., Rainer, D.R. and Goodman, D.W. (1996) Basis for the Structure Sensitivity of the CO+NO Reaction on Palladium. *Journal of Vacuum Science & Technology A*, **14**, 1457. http://dx.doi.org/10.1116/1.579969

[35] Rainer, D.R., Vesecky, S.M., Koranne, M., Oh, W.S. and Goodman, D.W. (1997) The CO+NO Reaction over Pd: A Combined Study Using Single-Crystal, Planar-Model-Supported, and High-Surface-Area Pd/Al_2O_3 Catalysts. *Journal of Catalysis*, **167**, 234-241. http://dx.doi.org/10.1006/jcat.1997.1571

[36] Viñes, F., Desikusumastuti, A., Staudt, T., Gorling, A., Libuda, J. and Neyman, K.N. (2008) A Combined Density-Functional and IRAS Study on the Interaction of NO with Pd Nanoparticles: Identifying New Adsorption Sites with Novel Properties. *The Journal of Physical Chemistry C*, **112**, 16539-16549. http://dx.doi.org/10.1021/jp804315c

[37] Grybos, R., Benco, L., Bučko, T. and Hafner, J. (2009) Interaction of NO Molecules with Pd Clusters: *Ab Initio* Density-Functional Study. *Journal of Computational Chemistry*, **30**, 1910-1922. http://dx.doi.org/10.1002/jcc.21174

[38] Abbet, S., Riedo, E., Brune, H., Heiz, U., Ferrari, A.-M., Giordano, L. and Pacchioni, G. (2001) Identification of Defect Sites on MgO(100) Thin Films by Decoration with Pd Atoms and Studying CO Adsorption Properties. *Journal of the American Chemical Society*, **123**, 6172-6178. http://dx.doi.org/10.1021/ja0157651

[39] Mineva, T., Alexiev, V., Lacaze-Dufaure, C., Sicilia, E., Mijoule, C. and Russo, N. (2009) Periodic Density Functional Study of Rh and Pd Interaction with the (100)MgO Surface. *Journal of Molecular Structure: THEOCHEM*, **903**, 59-66. http://dx.doi.org/10.1016/j.theochem.2009.01.025

[40] Lopez, N. and Illas, F. (1998) *Ab Initio* Modeling of the Metal-Support Interface: The Interaction of Ni, Pd, and Pt on MgO(100). *The Journal of Physical Chemistry B*, **102**, 1430-1436. http://dx.doi.org/10.1021/jp972626q

[41] D'Ercole, A., Giamello, E. and Pisani, C. (1999) Embedded-Cluster Study of Hydrogen Interaction with an Oxygen Vacancy at the Magnesium Oxide Surface. *The Journal of Physical Chemistry B*, **103**, 3872-3876. http://dx.doi.org/10.1021/jp990117d

[42] Abdel Halim, W.S., Abdel Aal, S. and Shalabi, A.S. (2008) CO Adsorption on Pd Atoms Deposited on MgO, CaO, SrO and BaO Surfaces: Density Functional Calculations. *Thin Solid Films*, **516**, 4360-4365. http://dx.doi.org/10.1016/j.tsf.2008.01.009

[43] Becke, A.D. (1993) Density-Functional Thermochemistry. III. The Role of Exact Exchange. *The Journal of Chemical Physics*, **98**, 5648. http://dx.doi.org/10.1063/1.464913

[44] Lee, C., Yang, W. and Parr, R.G. (1988) Development of the Colle-Salvetti Correlation-Energy Formula into a Functional of the Electron Density. *Physical Review B*, **37**, 785-789. http://dx.doi.org/10.1103/PhysRevB.37.785

[45] Lopez, N., Illas, F., Rösch, N. and Pacchioni, G. (1999) Adhesion Energy of Cu Atoms on the MgO(001) Surface. *The Journal of Chemical Physics*, **110**, 4873. http://dx.doi.org/10.1063/1.478373

[46] Moreira, I.P.R., Illas, F. and Martin, R.L. (2002) Effect of Fock Exchange on the Electronic Structure and Magnetic Coupling in NiO. *Physical Review B*, **65**, Article ID: 155102. http://dx.doi.org/10.1103/PhysRevB.65.155102

[47] Siegbahn, P.E. and Crabtree, R.H. (1997) Mechanism of C-H Activation by Diiron Methane Monooxygenases: Quantum Chemical Studies. *Journal of the American Chemical Society*, **119**, 3103-3113. http://dx.doi.org/10.1021/ja963939m

[48] Illas, F., Moreira, I.P.R., Graaf, C. and Barone, V. (2000) Magnetic Coupling in Biradicals, Binuclear Complexes and

Wide-Gap Insulators: A Survey of *Ab Initio* Wave Function and Density Functional Theory Approaches. *Theoretical Chemistry Accounts*, **104**, 265-272. http://dx.doi.org/10.1007/s002140000133

[49] Stevens, W., Krauss, M., Basch, H. and Jasien, P.G. (1992) Relativistic Compact Effective Potentials and Efficient, Shared-Exponent Basis Sets for the Third-, Fourth-, and Fifth-Row Atoms. *Canadian Journal of Chemistry*, **70**, 612-630. http://dx.doi.org/10.1139/v92-085

[50] Cundari, T.R. and Stevens, W.J. (1993) Effective Core Potential Methods for the Lanthanides. *The Journal of Chemical Physics*, **98**, 5555. http://dx.doi.org/10.1063/1.464902

[51] Larsen, G. (2000) A Performance Comparison between the CEP Effective Core Potential/Triple-Split Basis Set Approach and an All-Electron Computational Method with Emphasis on Small Ti and V Alkoxide Complexes. *Canadian Journal of Chemistry*, **78**, 206-211. http://dx.doi.org/10.1139/v99-225

[52] Henrich, V.E. and Cox, P.A. (1994) The Surface Science of Metal Oxides. Cambridge University Press, Cambridge.

[53] Grimes, R.W., Catlow, C.R.A. and Stoneham, A.M. (1989) Quantum-Mechanical Cluster Calculations and the Mott-Littleton Methodology. *Journal of the Chemical Society, Faraday Transactions II: Molecular and Chemical Physics*, **85**, 485-495. http://dx.doi.org/10.1039/f29898500485

[54] Frisch, M.J., Trucks, G.W., Schlegel, H.B., Scuseria, G.E., Robb, M.A., Cheeseman, J.R., *et al*. (1998) Gaussian 98. Gaussian Inc., Pittsburgh.

[55] Fuente, S.A., Ferullo, R.M., Domancich, N.F. and Castellani, N.J. (2011) Interaction of NO with Au Nanoparticles Supported on (100) Terraces and Topological Defects of MgO. *Surface Science*, **605**, 81-88. http://dx.doi.org/10.1016/j.susc.2010.10.003

[56] Giordano, L. and Pacchioni, G. (2005) Pd Nanoclusters at the MgO(100) Surface. *Surface Science*, **575**, 197-209. http://dx.doi.org/10.1016/j.susc.2004.11.024

[57] Silvia, A., Patricia, G., Ferullo, M. and Castellani, J. (2008) Adsorption of NO on Au Atoms and Dimers Supported on MgO(100): DFT Studies. *Surface Science*, **602**, 1669-1676. http://dx.doi.org/10.1016/j.susc.2008.02.037

[58] Yulikov, M., Sterrer, M., Heyde, M., Rust, H.-P., Risse, T., Freund, H.-J., Pacchioni, G. and Scagnelli, A. (2006) Binding of Single Gold Atoms on Thin MgO(001) Films. *Physical Review Letters*, **96**, Article ID: 146804. http://dx.doi.org/10.1103/PhysRevLett.96.146804

[59] Moseler, M., Häkkinen, H. and Landman, U. (2002) Supported Magnetic Nanoclusters: Soft Landing of Pd Clusters on a MgO Surface. *Physical Review Letters*, **89**, Article ID: 176103. http://dx.doi.org/10.1103/PhysRevLett.89.176103

[60] Xu, L., Henkelman, G., Campbell, C.T. and Jónsson, H. (2006) Pd Diffusion on MgO(100): The Role of Defects and Small Cluster Mobility. *Surface Science*, **600**, 1351-1362. http://dx.doi.org/10.1016/j.susc.2006.01.034

[61] Stirling, A., Gunji, I., Endow, A., Oumi, Y., Kubo, M. and Miyamoto, A. (1995) Γ-Point Density Functional Calculations on Theadsorption of Rhodium and Palladium Particles on MgO(001) Surface and Their Reactivity. *Journal of the Chemical Society, Faraday Transactions*, **93**, 1175-1178. http://dx.doi.org/10.1039/a604388g

[62] Giordano, L., Vitto, A.D., Pacchioni, G. and Ferrari, A.M. (2003) CO Adsorption on Rh, Pd and Ag Atoms Deposited on the MgO Surface: A Comparative *Ab Initio* Study. *Surface Science*, **540**, 63-75. http://dx.doi.org/10.1016/S0039-6028(03)00737-4

[63] Reed, A., Weinstock, R.B. and Weindhold, F. (1985) Natural Population Analysis. *The Journal of Chemical Physics*, **83**, 735. http://dx.doi.org/10.1063/1.449486

[64] Zhao, S., Ren, Y., Ren, Y., Wang, J. and Yin, W. (2011) Density Functional Study of NO Binding on Small Ag_nPd_m ($n + m \leq 5$) Clusters. *Computational and Theoretical Chemistry*, **964**, 298-303. http://dx.doi.org/10.1016/j.comptc.2011.01.009

[65] Dufaurea, C., Roques, J., Mijoule, C., Sicilia, E., Russo, N., Alexiev, V. and Mineva, T. (2011) A DFT Study of the NO Adsorption on Pd_n ($n = 1 - 4$) Clusters. *Journal of Molecular Catalysis A: Chemical*, **341**, 28-34. http://dx.doi.org/10.1016/j.molcata.2011.03.020

[66] Giordano, L., Valentin, C.D., Goniakowski, J. and Pacchioni, G. (2004) Nucleation of Pd Dimers at Defect Sites of the MgO(100) Surface. *Physical Review Letters*, **92**, Article ID: 096105. http://dx.doi.org/10.1103/PhysRevLett.92.096105

[67] Zhang, W., Ge, Q. and Wang, L. (2003) Structure Effects on the Energetic, Electronic, and Magnetic Properties of Palladium Nanoparticles. *The Journal of Chemical Physics*, **118**, 5793. http://dx.doi.org/10.1063/1.1557179

[68] Kumar, V. and Kawazoe, Y. (2002) Icosahedral Growth, Magnetic Behavior, and Adsorbate-Induced Metal-Nonmetal Transition in Palladium Clusters. *Physical Review B*, **66**, Article ID: 144413. http://dx.doi.org/10.1103/PhysRevB.66.144413

[69] Yang, J.X., Cheng, F.W. and Guo, J.J. (2010) Density Functional Study of Au_nRh ($n=1-8$) Clusters. *Physica B: Condensed Matter*, **405**, 4892-4896. http://dx.doi.org/10.1016/j.physb.2010.09.029

[70] Bogicevic, A. and Jennison, D.R. (2002) Effect of Oxide Vacancies on Metal Island Nucleation. *Surface Science*, **515**, L481-L486. http://dx.doi.org/10.1016/S0039-6028(02)02024-1

[71] Efremenko, I. (2001) Implication of Palladium Geometric and Electronic Structures to Hydrogen Activation on Bulk Surfaces and Clusters. *Journal of Molecular Catalysis A: Chemical*, **173**, 19-59. http://dx.doi.org/10.1016/S1381-1169(01)00144-3

[72] Piccolo, L. and Henry, C.R. (2001) NO-CO Reaction Kinetics on Pd/MgO Model Catalysts: Morphology and Support Effects. *Journal of Molecular Catalysis A: Chemical*, **167**, 181-190. http://dx.doi.org/10.1016/S1381-1169(00)00505-7

[73] Yamaguchi, A. and Iglesia, E. (2010) Catalytic Activation and Reforming of Methane on Supported Palladium Clusters. *Journal of Catalysis*, **274**, 52-63. http://dx.doi.org/10.1016/j.jcat.2010.06.001

[74] Ramsier, R.D., Gao, H.N.Q., Lee, K.W., Nooji, O.W., Lefferts, L. and Yates, J.T. (1994) NO Adsorption and Thermal Behavior on Pd Surfaces. A Detailed Comparative Study. *Surface Science*, **320**, 209-237. http://dx.doi.org/10.1016/0039-6028(94)90310-7

[75] Tsai, M.H. and Hass, K.C. (1995) First-Principles Studies of NO Chemisorption on Rhodium, Palladium, and Platinum Surfaces. *Physical Review B*, **51**, Article ID: 14616. http://dx.doi.org/10.1103/PhysRevB.51.14616

[76] Pacchioni, G. (1993) Physisorbed and Chemisorbed CO_2 at Surface and Step Sites of the MgO(100) Surface. *Surface Science*, **281**, 207-219. http://dx.doi.org/10.1016/0039-6028(93)90869-L

[77] Florez, E., Fuentealba, P. and Mondragón, F. (2008) Chemical Reactivity of Oxygen Vacancies on the MgO Surface: Reactions with CO_2, NO_2 and Metals. *Catalysis Today*, **133**, 216-222. http://dx.doi.org/10.1016/j.cattod.2007.12.087

[78] Sterrer, M., Yulikov, M., Risse, T., Freund, H.J., Carrasco, J., Illas, F., Valentin, C.D., Giordano, L., Pacchioni, G., Risse, T. and Freund, H.J. (2006) When the Reporter Induces the Effect: Unusual IR Spectra of CO on Au_1/MgO(001)/Mo(001). *Angewandte Chemie International Edition*, **45**, 2633-2635. http://dx.doi.org/10.1002/anie.200504473

[79] Grönbeck, H. and Broqvist, P. (2003) CO-Induced Modification of the Metal/MgO(100) Interaction. *The Journal of Physical Chemistry B*, **107**, 12239-12243.

[80] Abbeta, S., Heizb, U., Ferraric, A.M., Giordanod, L., Valentin, C.D. and Pacchioni, G. (2001) Nano-Assembled Pd Catalysts on MgO Thin Films. *Thin Solid Films*, **400**, 37-42. http://dx.doi.org/10.1016/S0040-6090(01)01444-4

[81] Abdel Halim, W.S., Assem, M.M., Shalabi, A.S. and Soliman, K.A. (2009) CO Adsorption on Ni, Pd, Cu and Ag Deposited on MgO, CaO, SrO and BaO: Density Functional Calculations. *Applied Surface Science*, **255**, 7547-7555. http://dx.doi.org/10.1016/j.apsusc.2009.04.026

[82] Shalabi, A.S., Nour, E.M. and Abdel Halim, W.S. (2000) Characterization of van der Waals Interaction Potentials D_{4h} and Td Configurations of He_4. *International Journal of Quantum Chemistry*, **76**, 10-22. http://dx.doi.org/10.1002/(SICI)1097-461X(2000)76:1<10::AID-QUA2>3.0.CO;2-1

13

Molecular Dynamics Simulations of Perovskites: The Effect of Potential Function Representation on Equilibrium Structural Properties

Kholmirzo T. Kholmurodov[1,2], Sagille A. Ibragimova[2], Pavel P. Gladishev[2], Anatoly V. Vannikov[3], Alexey R. Tameev[3], Tatyana Yu. Zelenyak[2]

[1]Frank Laboratory of Neutron Physics, JINR, Dubna, Russia
[2]Dubna International University, Dubna, Russia
[3]A.N. Frumkin Institute of Physical Chemistry and Electrochemistry, Russian Academy of Sciences, Moscow, Russia
Email: kholmirzo@gmail.com

Abstract

The perovskites with general formula ABX_3 have been widely used as for materials with their unique properties (ferroelectric, piezoelectric, dielectric, catalytic and so on). Hybrid organolead halide perovskites are a class of semiconductors with ABX_3 (X = Cl, Br, and I) structures consisting of lead cations in 6-fold coordination (B site), surrounded by an octahedron of halide anions (X site, face centered) together with the organic components in 12-fold cub octahedral coordination. These hybrid perovskites have a direct band gap, a large absorption coefficient as well as high charge carrier mobility that represent a very attractive characteristic of cost-effective solar cells. Basically, these crystals are inorganic solids of $CaTiO_3$ type held together by bonds that are either ionic or partially ionic and partially covalent. In spite of the partially covalent character of the Ti-O bond, the system is modeled by a two-body central force interatomic potential (the form of the Vashishta and Rahman interatomic potential), which has been used successfully for many materials with a perovskite structure. In the present work using molecular dynamics (MD) simulation method we investigate the dynamical and structural behavior of $CaTiO_3$ perovskite at normal pressure and temperature conditions. The MD calculations were performed on a system of 16,000 particles (3200Ca + 3200Ti + 96,000O), initially in an orthorhombic-Pbnm structure. The orthorhombic MD box had edges $L_x = 53.4$ Å, $L_y = 53.4$ Å and $L_z = 61.12$ Å, which provided a density matching the experimental value of $\rho = 4$ g/cm^3. Starting with this structure and using proposed interatomic potentials the MD system stabilizes at room temperature in its initial configuration. The aim of the present study to explore the effect of potential function representations on structural equilibrium properties for the perovskite models including hybrid halide ones outlined above. Concerning the perovskite equilibrium state we elucidate the role of potential function modification on the atomic pair correlation and structural re-organization. The details of the interatomic potential representation have to be crucially important for obtaining of correct analysis data in crystallic, liquid and amorphous phases including perovskite systems.

Keywords

Perovskites, Halides, Potential Functions, Structural Properties, MD Simulations

1. Introduction

The cost effective solar energy production and reproducible device performance are the important subjects in today nanotechnology research. In this respect, computer design and modeling of nanostructures aimed on developing of solar cells prototypes of greater efficiencies represent a great scientific interest. The computer methods based on atomic/molecular modeling approach may provide a lot of useful information concerning the crystal chemistry of solar cell materials, the dynamical and structural data, the thermodynamic properties and phase transitions, charge transfer and diffusion processes, and so on. The computer analysis would obviously be helpful for performing experiments with fewer resources thereby indicating the rationally modifying ways and finding the best structure design for the solar cells materials. In the present article the review of structural characterization of a number solar cell systems together with a novel molecular dynamics (MD) simulation data are presented [1]-[5].

Organic-inorganic hybrid solar cells. Recently, there are new opportunities for the development of photovoltaics. In 2012-2013 in the field of perovskite semiconductor photovoltaic cells were obtained principally new results for which it was achieved an efficiency of 16%. For today, the efficiency of organic-inorganic hybrid solar elements exceeded up 20% [6]. This relatively new type of photovoltaics has remarkable properties. The roots of these photovoltaic systems lie in the Grätzel' "liquid" dye-sensitized solar cells (DSSCs). In contrast, the perovskite photovoltaic cells are solid state system with all their advantages. The best results were obtained for the synthetic organic-inorganic perovskite crystals. Organic-inorganic hybrid solar cells that combine mesoporous framework are perovskite light absorber and the electron and hole transporter. Perovskite can be formed as flexible and semi-transparent solar cells. The elements of the perovskite inferior in efficiency while the silicon single-stage and other photoelectric converters, however, have a great advantage in cost and simplicity of manufacture. Some experts are already predicting new material displacement of silicon solar cells [4]-[7].

Organo-metallic perovskite semiconductor. Organo-metallic perovskite semiconductor structures are attractive because they have high charge carrier mobility and a large diffusion length, allowing the photogenerated electrons and holes effectively to travel long distances without losing energy. As a result, one can use solar cells with more thin layers, which absorb more light and hence, generate more electricity. In this regard, organo-metallic, in particular organo-lead perovskite halide solar cells has become one of the most promising candidates for next-generation solar cells [2]-[8].

Perovskite thin film solar cells. Until recently, the perovskite thin film solar cells have been used exclusively and organic polymeric materials with a relatively low mobility of the holes. In this respect the more promising as a hole conductive materials to use more stable inorganic structures, in particular copper iodide. Hole conductors based on copper previously successfully used in solar cells sensitized with dyes and quantum dots. Copper iodide is inexpensive hole conductive material that can serve as a possible alternative to spiro-OMeTAD. This may lead to the development of inexpensive, high-performance solar cell perovskite. In the future, for such solar cells can be increased voltage and efficiency, in particular by reducing the high rate of recombination. Best efficiency solar cells perovskite already competitive with modern commercial technology at a much lower cost price [1] [5]-[8].

Further improvement of photovoltaics. Further improvement of such devices is promising the most promising avenue of research in the field of photovoltaics. It is important to carry out a systematic study and optimization of organic-inorganic perovskite structure in order to improve the photoelectric conversion efficiency based on it. These problems will involve the precise calculation methods (as molecular dynamics and related techniques) for exploring the stability effect of substitution known organo-metallic perovskites of different ions to other similar ions. Parallel on the basis of quantum-mechanical calculations the spectral properties of various perovskite systems could be predicted. These results have to bring to determination and identification of the most stable crystal structures with the necessary spectral, photovoltaics, and other properties important for photoelectric conversion.

Stability of developed new perovskite solar cells. The formation of perovskite thin film photovoltaic cells with high efficiency involves their further studies from the point of view of the radiation and thermal stabilities. One believes that for the newly developed perovskite solar cells the problems of radiation damages will be on the top of the structural, spectroscopic and other physical methods research as future trends.

2. The Perovskites Structure Properties and Design Aspects

Understanding the basic properties and functioning of organo-metallic perovskites solar cells represent a great scientific and technological interest due to a number of their unique properties (ferroelectric, piezoelectric, dielectric, catalytic and so on). The radiation and thermal stability of perovskite developed solar cells in this respect are the problems of a great research interest as well. Synthesis of new optimal perovskite semiconductor systems inevitably demands the studies of their spectral, structural and photovoltaic properties. The formation of new organic-inorganic perovskite solar cells of high efficiency characteristics correlate with the theoretical research investigation. So far, the development of theoretical and computational methods, as like as molecular dynamics and quantum chemical calculations, are important issues for the search of optimizing perovskite structures, studying the melting and phase transitions, diffusion and conductivity phenomena, etc., that are aimed on the innovation of new photovoltaic systems unknown for today [1] [4] [9]. In the present study we have performed molecular dynamics (MD) simulation on $CaTiO_3$ system as a basic perovskite structure. The above mentioned system represents the most studied perovskite model. We have been aimed to compare the structural behavior of the $CaTiO_3$ perovskite under different intermolecular potential representations. Such analysis and comparison between different simulation approaches allow one to extend the studies on more complicated perovskite structures (as like as hybrid organolead ones) on stronger motivated basis.

The perovskites with general formula ABO_3 have been widely used for crystal materials. A typical representative of the class of modern crystal materials known as perovskites is calcium-titanate ($CaTiO_3$) The perovskite $CaTiO_3$ is feroelastic with an orthorhombic symmetry at room temperature (space group Pbnm) and undergoes two phases transitions at respectively $T_1 \sim 1398$ K (space group Mmcm) and $T_2 \sim 1523$ K (space group Pm3m). $CaTiO_3$ can be prepared by the combination of CaO and TiO_2 at temperatures >1300°C. The sol-gel processes have been used to make a more pure substance, as well as lowering the synthesis temperature. These compounds synthesized are more compressible due to the powders from the sol-gel process as well and bring it closer to its calculated density (~ 4.04 g·mol^{-1}). Below in **Figure 1** several typical perovskite structures are shown.

Another group of metallo-organic perovskites is a system of type $CH_3NH_3PbX_3$. For example, $[CH_3NH_3]^+Pb^{2+}X_3$, perovskite of mixed halide form, methylammonium lead iodide. For the above system, X represents halogens (Br or I), halogen mixes ($CH_3NH_3PbI_{3-x}Cl_x$) or more complicated one, $(R(CH_2)_2NH_3)_2PbX_4$, with R as phenyl or halogen derivatives (see **Figure 2**).

In the basic structure, $CaTiO_3$ (Ca^{2+}, Ti^{4+}, O^{2-}), or general representation ABX_3, alkali atoms occupy A sites, A (Cs, Rb, CH_3NH_3), Pb atoms occupy B sites, B (Pb), and halogen atoms occupy X sites, X (Cl, Br, I). In this regard, the perovskite system of $Cs^+Pb^{2+}F^-$ type represents the most interesting object. We emphasize the modern trend in developing of more complicate perovskites as $[Me\ NH_3]^+Pb^{2+}F^-$, $[Rn\ NH_3]^+Pb^{2+}F^-$ (n > 1), $[R_nR_mNH_3]^+Pb^{2+}F^-$, $[R_nR_mR_kNH]^+Pb^{2+}F^-$, $[R_nR_mR_kR_lN]^+Pb^{2+}F^-$ with halogens $F^- = (F^-, Cl^-, Br^-, I^-)$. **Figure 3** demonstrates schematic aspects of the perovskite structure models.

Basically, hybrid organolead perovskites are inorganic solids of $CaTiO_3$ type structure held together by the ionic or partially ionic and partially covalent bonds. In spite of the partially covalent character of the Ti-O bond, the system is modeled by a two-body central force interatomic potential, which has been used successfully for many ceramics materials of hybrid halide perovskites type. In the present work a series of the MD simulations were performed to investigate the effect of potential function representations on structural equilibrium properties for the $CaTiO_3$ model structure possessing similar behavior as hybrid halides and other complicated perovskites [3]-[10].

3. The MD Simulation and Interaction Potential

The MD calculations were performed on a system of 16,000 particles (3200Ca + 3200Ti + 9600O), initially in

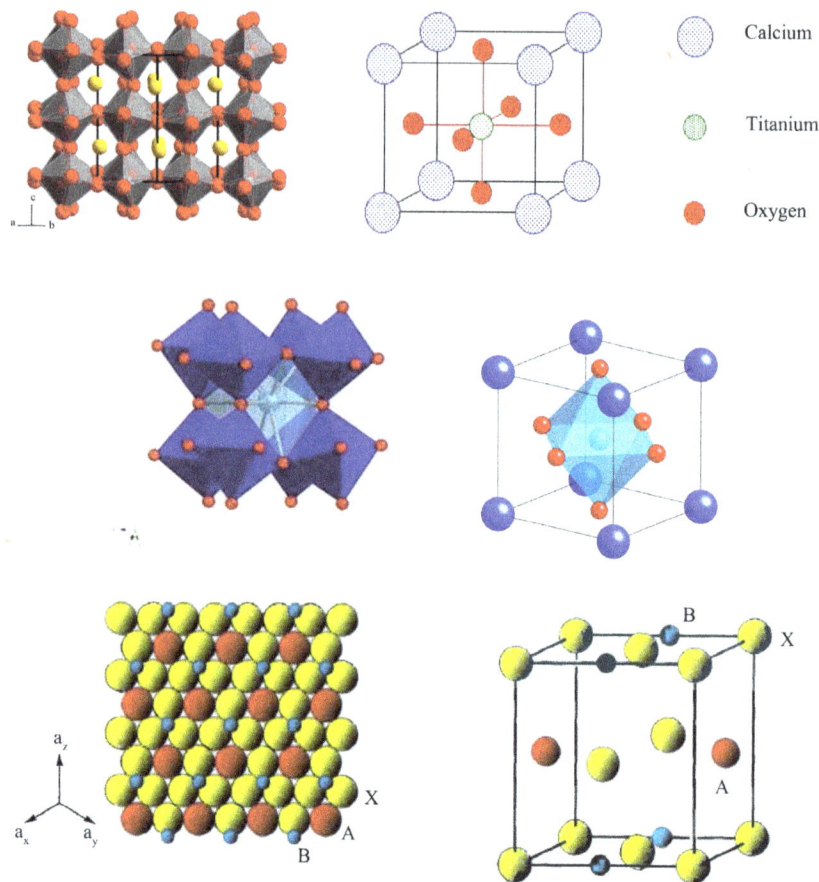

Figure 1. Several typical perovskite structures with a chemical formula ABX_3 are shown. The red spheres are X atoms (usually oxygens O_3^{2-}), the blue spheres are B-atoms (a smaller metal cation, such as Ti^{4+}), and the green spheres are the A-atoms (a larger metal cation, such as Ca^{2+}). (Refs: A public domain of the World Wide Web and https://en.wikipedia.org/wiki/Perovskite_(structure)).

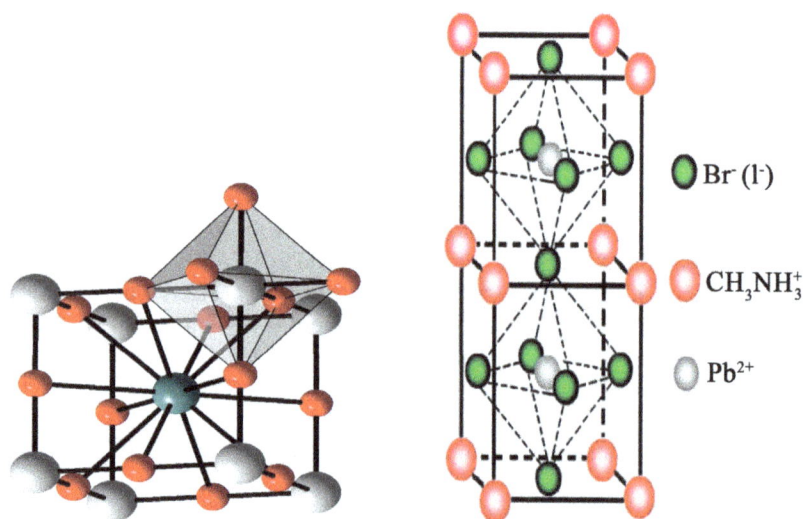

Figure 2. The structure of perovskite of mixed halide form, methylammonium lead iodide $[CH_3NH_3]^+Pb^{2+}X_3$ (Refs: A public domain of the World Wide Web).

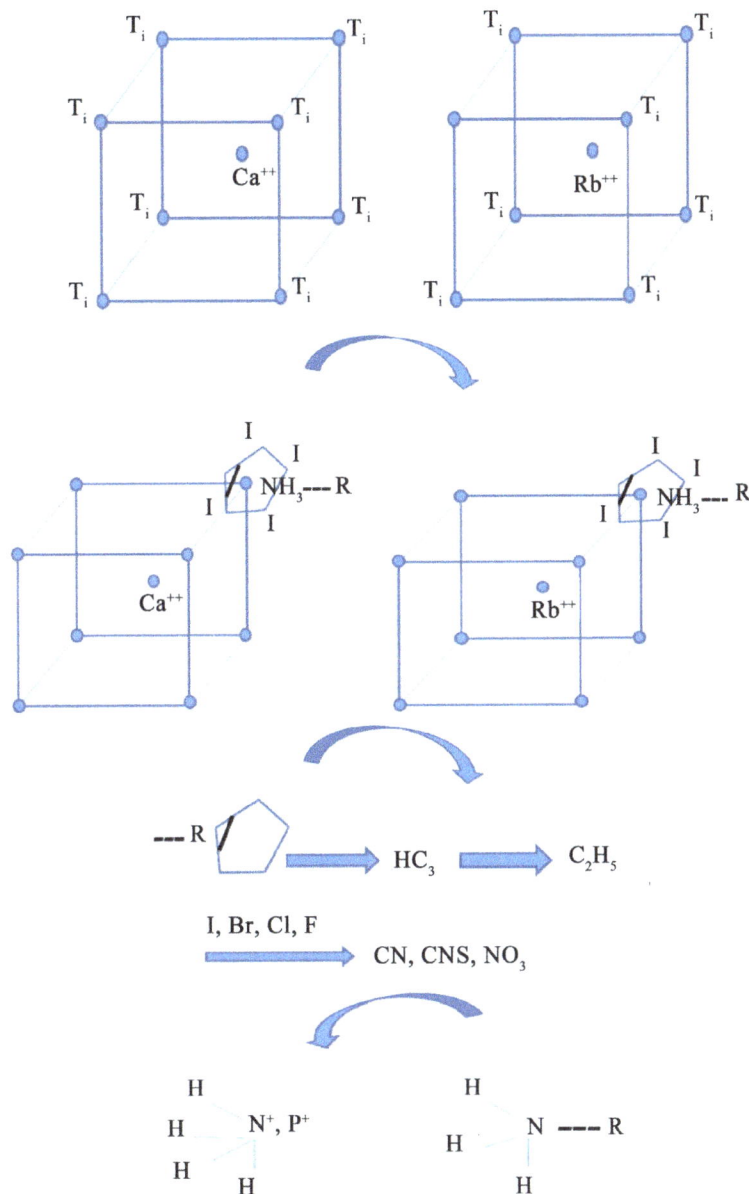

Figure 3. The structure representation of more complicate perovskite system.

an orthorhombic Pbnm structure. The structure was built using the data base of Institute of Experimental Mineralogy of Russian Academy of Sciences, http://database.iem.ac.ru/mincryst (Card No: 3594, PEROVSKITE CaTiO(3), Orthorhombic Pbnm, Z = 4; Ref.: Wyckoff R.W.G. (1963), Crystal Structures, 2, 410). The lattice parameters were as: a = 5.37 Å, alpha = 90o; b = 5.44 Å, beta = 90o; c = 7.64 Å, gamma = 90˚; unit cell volume = 223.19 Å^3; molar volume = 33.61 cm^3/mol.

Table 1 and **Table 2** demonstrate the atomic position details of the perovskite CaTiO$_3$ structure model.

The orthorhombic MD box had edges L_x = 53.4 Å, L_y = 53.4 Å and L_z = 61.12 Å, which provided a density matching experimental value of ρ = 4.0 g·cm^{-3} as in work [1]. (Comparing with [1] where the MD parameters were used for a system of 10,240 particles = 2048Ca + 2048Ti + 6144O of an orthorhombic-Pbnm structure within the box L_x = 43.022 Å, L_y = 43.494Å and L_z = 61.107 Å, providing the density ρ = 4 g/cm^3). Starting with the initial structure as described above and using the proposed interatomic potential, the MD system stabilizes at room temperature in the relaxed configuration. The MD simulation have been performed on the basis of the DL_POLY general-purpose code [11] [12]. Various combinations of the integration algorithm were used

Table 1. Co-ordinates, thermal parameters, occupation for atomic positions.

NoP	$x \cdot a^{-1}$	$y \cdot b^{-1}$	$z \cdot c^{-1}$	B(j)	Atom/occupation
1	0.0	0.03	0.25	0.0	Ca = 1.0
2	0.0	0.50	0.0	0.0	Ti = 1.0
3	0.037	0.482	0.25	0.0	O = 1.0
4	0.073	0.268	0.026	0.0	O = 1.0

Table 2. Co-ordinates for all atomic positions.

No	NoP	$x \cdot a^{-1}$	$y \cdot b^{-1}$	$z \cdot c^{-1}$
1	1	0.0	0.03	0.25
2	2	0.0	0.5	0.0
3	3	0.037	0.482	0.25
4	4	0.732	0.268	0.026
5	1	0.0	0.97	0.75
6	3	0.963	0.518	0.75
7	4	0.268	0.732	0.974
8	2	0.0	0.268	0.5
9	4	0.268	0.53	0.526
10	4	0.732	0.0	0.474
11	1	0.5	0.982	0.25
12	2	0.5	0.768	0.0
13	3	0.463	0.47	0.25
14	4	0.768	0.018	0.026
15	1	0.5	0.232	0.75
16	3	0.537	0.0	0.75
17	4	0.232	0.232	0.974
18	2	0.5	0.0	0.5
19	4	0.232	0.232	0.526
20	4	0.768	0.768	0.474

(NPT and NVT ensembles) controlling the pressure/temperature of the system with the standard termostat and barostat relaxation procedures (see, Appendices 1 and 2 for the FIELD and CONTROL files).

It should be noted that perovskite crystals are inorganic solids held together by bonds that are either ionic or partially ionic and partially covalent [1]. In spite of the partially covalent character of the Ti-O bond, the system is modeled by a two-body central force interatomic potential, based on the form of the Vashishta and Rahman (VR) interatomic potential, which has been used successfully for many different systems. The total potential energy of the system is written as ([1], see also **Figure 4** [1]):

The first term in the above formula is the Coulomb interaction potential between the ions Z_i, Z_j (in units of the electron charge $|e|$), $r_{ij} = r_i - r_j$ is he interatomic distance between ions i and j, and λ is the screening length for the Coulombic interactions. The second term represents the steric effects of the ions sizes, where H_{ij} and η_{ij} are the strength and exponent of steric repulsion, respectively (in [1] the following values were used $\eta_{ij} = 11$ (for Ca-Ca and Ti-Ti pairs), 9 (for Ca-Ti, Ca-O and Ti-O), and 7 (for O-O)). The third term represents the charge-induced dipole interaction, to include the electronic polarizabilities of the atoms, where D_{ij} is the strength of the charge-dipole attraction (O_2- is a highly polarizable ion). The last is the induced dipole-dipole potential

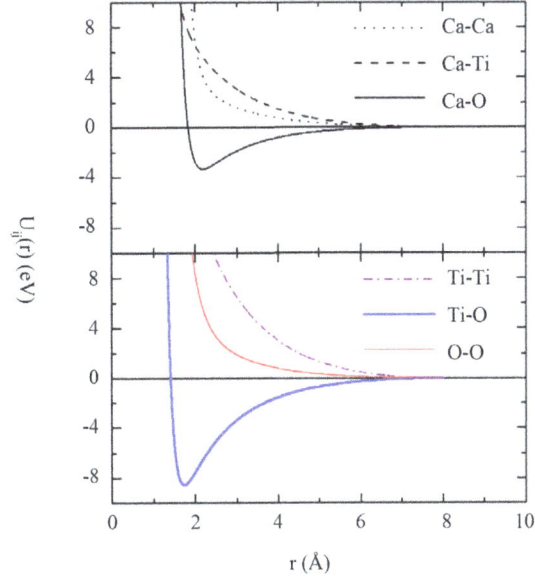

$$U = \sum_{i<j} U_{ij}$$

$$U_{ij} = \frac{Z_i Z_j e^2}{r} e^{-\frac{r}{\lambda}} + \frac{H_{ij}}{r^{\eta_{ij}}} - \frac{D_{ij}}{2r^4} e^{-\frac{r}{\xi}} - \frac{W_{ij}}{r^6}$$

Figure 4. Two-body interaction potential as a function of distance. The total interaction potential is the sum of all two-body potentials [1].

based on the van der Waals interaction, where W_{ij} is its strength. Parameter ξ is the screening length for charge-dipole interactions, respectively.

We neglect both screening terms for the Coulombic (*i.e.* $\exp(-r/\lambda) = 1$) and charge-dipole ($\xi \to 0$) interactions. Also for the ions steric repulsion we consider a fixed value $\eta_{ij} = 12$ for all interacting atomic pairs. In such approximation the above mentioned VR interatomic potential approximates the well-know Lennard-Jones (LJ) or (12 - 6) potential types that are being widely used for the MD simulations of the condensed molecular systems similar to perovskites [12]-[16]:

$$U(r) = 4\varepsilon\left[\left(\frac{\sigma}{r}\right)^{12} - \left(\frac{\sigma}{r}\right)^6\right], \quad U(r) = \frac{A}{r^{12}} - \frac{B}{r^6}.$$

Table 3 and **Table 4** show the atomic and potential parameters of the perovskite $CaTiO_3$ model.

4. Result and Discussion

Configuration snapshots. In **Figure 5** and **Figure 6** we present the computer generated $CaTiO_3$ perovskite structures at initial and relaxed (equilibrium) states. During the equilibrization the sample structure undergoes the recrystallization modification due to the cooling and melting processes. It is well know that in the single crystal the recrystallization occurs in the orthorhombic structure though the amorphous regions to be formed during the relaxation procedure. The present results obtained by present MD calculations agree with the simulation and experimental reported for the polycrystalline material similar to $CaTiO_3$ perovskite structure [1].

Pair distribution functions. The MD simulation results for the RDF (radial distribution function) $g_{\alpha\beta}(r)$ are summarized in the **Figures 7-10**. For the comparison we also present the results reported in [1] and obtained by the VR potential. In **Figure 7**. the graphs of RDF (radial distribution function) $g_{\alpha\beta}(r)$ vs. r are presented for the ionic pair Ca-Ca: (a) 12 - 6 potential and (b) VR potential [1]. The comparison indicate a similar RDF behave for both 12 - 6 and VR potentials. We see existing of the three largest $g_{\alpha\beta}(r)$ peaks for both (a) and (b) models. However, for the 12 - 6 potential the position of the $g_{\alpha\beta}(r)$ peaks are located closer to the origin of r axis. This means that the Ca-Ca ionic pair forms in 12 - 6 model relatively stronger bond than in the VR model. In **Figure 8**. the RDF graphs are presented for the ionic pair Ca-O. The comparison with the MD results of [1] indicate that the location of the $g_{\alpha\beta}(r)$ peak for both 12 - 6 and VR potentials are very close to each other. However, for the 12 - 6 potential we observe a very large $g_{\alpha\beta}(r)$ peak (the value of the $g_{\alpha\beta}(r)$ peak in model (a) is four times larger than in model (b)). This implies for the Ca-O a stronger ordering and ionic pair correlation for 12 - 6

Table 3. The atomic masses and charges of the perovskite system CaTiO$_3$.

Atom	m (in a.u. mass)	q (in proton charge)
Ca	40.0800	+0.9697
Ti	47.8800	+1.9394
O	15.9994	−0.9697

Table 4. Parameters of the two-body interaction potential used in the MD simulations.

Atomic pair	A (=H$_{ij}$) (eV·Å^{12})	B (=W$_{ij}$) (eV·Å^6)
Ca-Ca	8223.56	0.0
Ca-Ti	216.65	0.0
Ca-O	2365.84	242.64
Ti-Ti	25.22	0.0
Ti-O	374.99	0.0
O-O	684.09	0.0

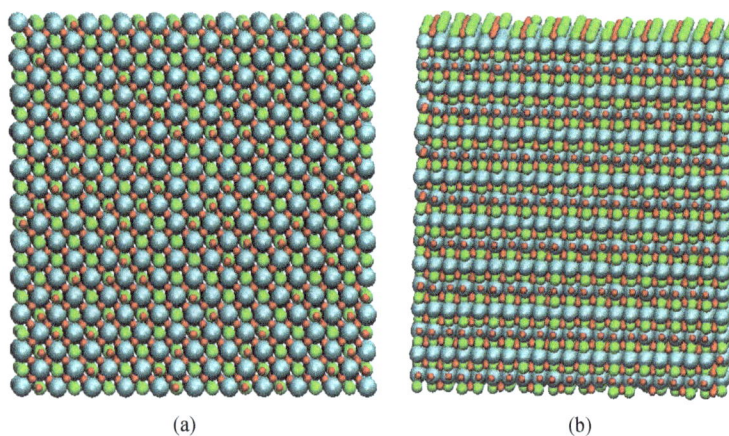

(a)   (b)

Figure 5. Computer generated perovskite structures are shown for the CaTiO$_3$ ((a): top view, (b): side view). The red spheres are oxygens (O$_3^{2-}$), the smaller green spheres are metal cation Ti^{4+}, the larger blue spheres are metal cation Ca^{2+}.

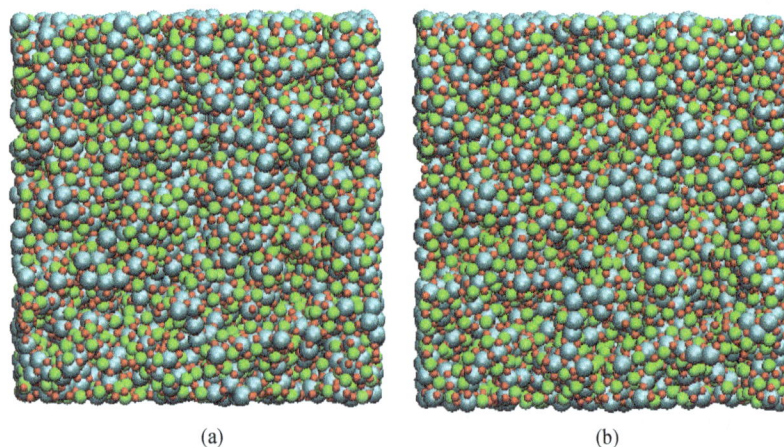

(a)   (b)

Figure 6. Computer generated two relaxed configurations of the CaTiO$_3$ perovskite structure are shown. The red spheres are oxygens (O$_3^{2-}$), the smaller green spheres are metal cation Ti^{4+}, the larger blue spheres are metal cation Ca^{2+}.

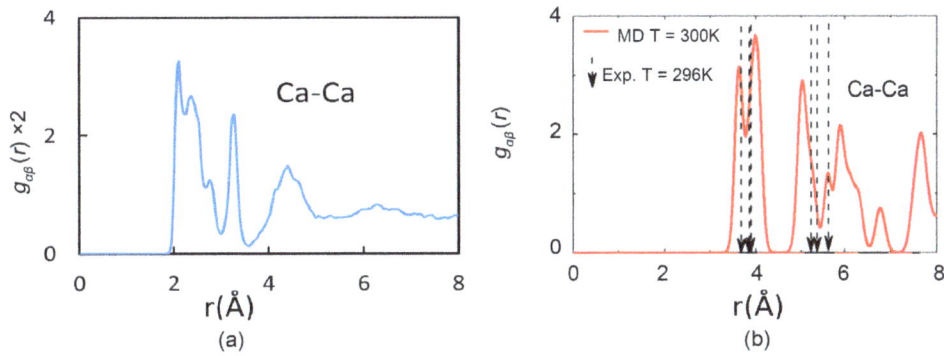

Figure 7. The RDF (radial distribution function) $g_{\alpha\beta}(r)$ vs. r for the ionic pair Ca-Ca from the MD simulations at T = 300 K obtained by (a) 12 - 6 potential and (b) VR potential [1].

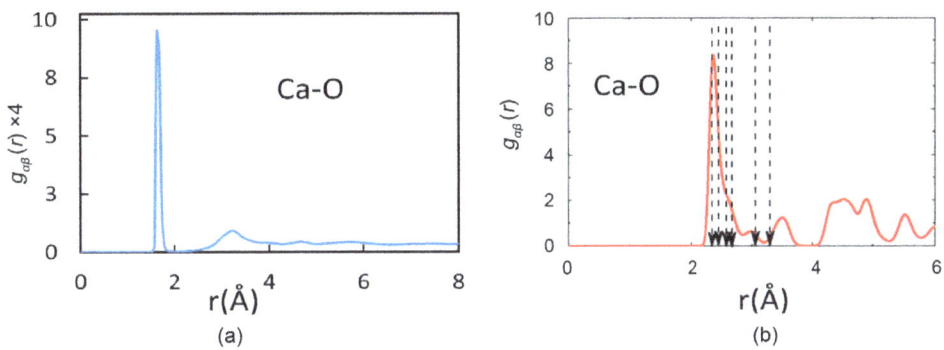

Figure 8. The RDF (radial distribution function) $g_{\alpha\beta}(r)$ vs. r for the ionic pair Ca-O from the MD simulations at T = 300 K obtained by (a) 12 - 6 potential and (b) VR potential [1].

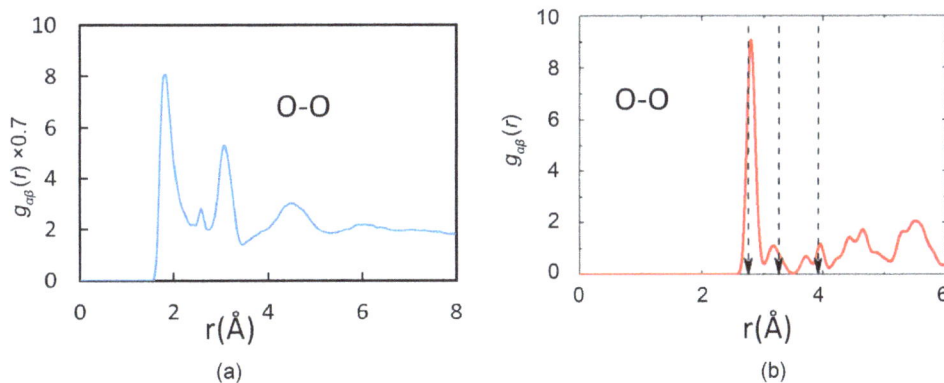

Figure 9. The RDF (radial distribution function) $g_{\alpha\beta}(r)$ vs. r for the ionic pair O-O from the MD simulations at T = 300 K obtained by (a) 12 - 6 potential and (b) VR potential [1].

potential comparing to the VR one. In **Figure 9** the RDF graphs are shown for the ionic pair O-O. The comparison of the left and right hand-side graphs indicate that the amplitudes of the $g_{\alpha\beta}(r)$ for both 12 - 6 and VR potentials are not so much differ from each other. Though even in (a) 12 - 6 model a secondary $g_{\alpha\beta}(r)$ peak seem to appear which does not exist in the (b) VR poteitial model. Thus, for the O-O pair interactions negleting the Coulombic and charge-dipole screening potentail terms do not effect O-O ordering even so the O^{2-} is a highly polarizable ion. In **Figure 10** the RDF graphs are shown for the ionic pair Ti-Ti. The results show that the Ti-Ti ordering and interaction for the (a) 12 - 6 potential visibly are weakened in comparison with the (b) VR potential.

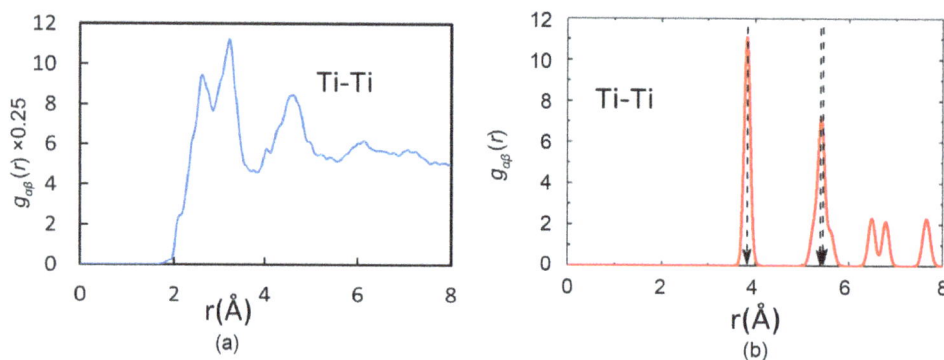

Figure 10. The RDF (radial distribution function) $g_{\alpha\beta}(r)$ vs. r for the ionic pair Ti-Ti from the MD simulations at T = 300 K obtained by (a) 12 - 6 potential and (b) VR potential [1].

5. Conclusion

The paper is aimed on molecular-dynamics (MD) simulation of the $CaTiO_3$ perovskite model concerning the effect of interatomic potential function modification on the relaxed equilibrium states. The proposed earlier in [8] the Vashishta and Rahman (VR) interatomic potential has shown to successfully work for many different systems including perovskites. In [1] the VR interaction potential was used for the MD simulation of $CaTiO_3$ system and it was proved that the VR potential to be very effective for the describing of structural phase transitions under the temperature change; the MD results obtained in [1] were the same as in experimental observations. In the present study work we have made modification of the VR potential, neglecting both screening terms for the Coulombic (*i.e.* $\exp(-r/\lambda) = 1$) and charge-dipole ($\xi \to 0$) interactions, thereby approaching the VR potential to the well-know Lennard-Jones (LJ) or (12 - 6) ones. With such modifications the MD generated configuration snapshot remain similar to existing experimental and simulation observations. Nevertheless, the behavior of the RDF (radial distribution function) $g_{\alpha\beta}(r)$ undergoes essential changes. For the ionic pair Ca-Ca the comparison indicate a similar RDF behave for both 12 - 6 and VR potentials, showing existing of three largest $g_{\alpha\beta}(r)$ peaks, but for the 12 - 6 potential the position of the $g_{\alpha\beta}(r)$ peaks are located closer to the origin of r axis. For the ionic pair Ca-O the $g_{\alpha\beta}(r)$ peak for both 12 - 6 and VR potentials are very close to each other, but for the 12 - 6 potential we observe a very large $g_{\alpha\beta}(r)$ peak where the value of the $g_{\alpha\beta}(r)$ in (12 - 6) model has to be four times larger than in VR model. That just implies the Ca-O stronger ordering and ionic pair correlation for 12-6 potential comparing to the VR one. On the contrary, for the ionic pair O-O the $g_{\alpha\beta}(r)$ for both 12 - 6 and VR potentials are not so much differ from each other, implying that for the O-O pair interactions neglecting the Coulombic and charge-dipole screening potentail terms do not effect O-O ordering even so the O^{2-} is a highly polarizable ion. Also for the ionic pair Ti-Ti our MD results indicate that the Ti-Ti ordering and interaction for the (12 - 6) potential have to weaken in comparison with the VR potential. In conclusion, the details of the atomic pair re-organization due to interatomic potential modifications elucidate on the role of potential function representation for cystallic, liquid and amorphous phases including the perovskite systems.

Acknowledgements

The authors acknowledge the financial support from RFBR (Russian Foundation for Basic Research), grant No. 14-43-03544 (the research ideas and calculation schemes for studies organic-inorganic perovskites crystallic structures) and RSCF (Russian SCience Foundation), grant No. 15-13-00170 (the performance of molecular-dynamics research for organic-inorganic perovskites of *Pbnm* structure type and generalization of the results).

References

[1]   Souza, J.A. and Rino, J.P. (2011) A Molecular Dynamics Study of Structural and Dynamical Correlations of $CaTiO_3$. *Acta Materialia*, **59**, 1409-1423. http://dx.doi.org/10.1016/j.actamat.2010.11.003

[2]   Taibi-Benziada, L., Mezroua, A. and Von Der Mühll, R. (2004) CaTiO3 Related Materials for Resonators. *Ceramics-Silikáty*, **48**, 180-184.

[3]   Calleja, M., Dove, M.T. and Salje, E.K.H. (2003) Trapping of Oxygen Vacancies on Twin Walls of $CaTiO_3$: A Com-

puter Simulation Study. *Journal of Phyics*: *Condensed Matter*, **15**, 2301-2307.

[4]   Cai, B., Xing, Y.D., Yang, Z., Zhang, W.-H. and Qiu, J.S. (2013) High Performance Hybrid Solar Cells Sensitized by Organolead Halide Perovskitest. *Energy & Environmental Science*, **6**, 1480.

[5]   Haruyama, J., Sodeyama, K., Han, L.Y. and Tateyama, Y. (2014) Termination Dependence of Tetragonal $CH_3NH_3PbI_3$ Surfaces for Perovskite Solar Cells. *Journal of Physical Chemistry Letters*, **5**, 2903-2909.

[6]   Jishi, R.A., Ta, O.B. and Sharif, A.A. (2014) Modeling of Lead Halide Perovskites for Photovoltaic Applications. *Journal of Physical Chemistry C*, **118**, 28344-28349. http://dx.doi.org/10.1021/jp5050145

[7]   Mashiyama, H., Kawamura, Y., Kasano, H., Asahi, T., Noda, Y. and Kimura, H. (2007) Disordered Configuration of Methylammonium of $CH_3NH_3PbBr_3$. *Determined by Single Crystal Neutron Diffractometry, Ferroelectrics*, **348**, 182-186. http://dx.doi.org/10.1080/00150190701196435

[8]   Vashishta, P. and Rahman, A. (1978) Ionic Motion in α-AgI. *Physical Review Letters*, **40**, 1337-1340.

[9]   Voronov, V.N. (2006) Ion Mobility and Properties of Perovskite ABX_3 Type Systems. Preprint № 000F. Physics Institute SB RAS, Krasnoyarsk, 64 p. http://test.kirensky.ru/zdoc/vvn_preprint.pdf

[10]   Razumov, V.F. and Klyuev, M.V., Eds. (2015) Organic and Hybrid Nanomaterials: Synthesis and Application Aspects: Monograph. Ivanovo State University, Ivanovo, 426-556.

[11]   Smith, W. and Forester, T.R. (1996) DL_POLY_2.0: A General-Purpose Parallel Molecular Dynamics Simulation Package. *Journal of Molecular Graphics*, **14**, 136-141.

[12]   Smith, W., Forester, T.R. and Todorov, I.T. (2008) The DL POLY 2 User Manual. STFC Daresbury Laboratory Daresbury, Warrington, Version 2.19.

[13]   Kholmurodov, K., Ed. (2007) Molecular Simulation Studies in Material and Biological Sciences. International Workshop. Nova Science Publishers Ltd., New York, 196 p.

[14]   Kholmurodov, K., Ed. (2009) Molecular Simulation in Material and Biological Research. Nova Science Publishers Ltd., New York, 155 p.

[15]   Kholmurodov, K., Ed. (2011) Molecular Dynamics of Nanobistructures. Nova Science Publishers Ltd., New York, 210 p.

[16]   Kholmurodov, K., Ed. (2013) Models in Bioscience and Materials Research: Molecular Dynamics and Related Techniques. Nova Science Publishers Ltd., New York, 208 p.

Appendix A

CaTiO₃ (PEROVSKITE) WITH LENNARD-JONES (lj) POTENTIAL

```
UNITS      eV
MOLECULES 1
CaTiO₃
NUMMOL 180
ATOMS    20
Ca              40.0800      +0.9697    1
Ti              47.8800      +1.9394    1
O               15.9994      −0.9697    2
Ca              40.0800      +0.9697    1
O               15.9994      −0.9697    2
Ti              47.8800      +1.9394    1
O               15.9994      −0.9697    2
Ca              40.0800      +0.9697    1
Ti              47.8800      +1.9394    1
O               15.9994      −0.9697    2
Ca              40.0800      +0.9697    1
O               15.9994      −0.9697    2
Ti              47.8800      +1.9394    1
O               15.9994      −0.9697    2
FINISH
VDW 6
Ca      Ca      12-6         8223.56    0.00
Ca      Ti      12-6         216.65     0.00
Ca      O       12-6         2365.84    242.64
Ti      Ti      12-6         25.22      0.00
Ti      O       12-6         374.99     0.00
O       O       12-6         684.09     0.00
CLOSE
```

Appendix B

```
CaTiO₃ (PEROVSKITE) STRUCTURE No.3594
integrate leapfrog verlet
#pressure                1.00
temperature              300.0
#ensemble npt hoover     0.1 0.05
ensemble nvt berendsen      0.01

steps                    20000
equilibration            1000

timestep                 0.001

scale                    100
print every              100
stats                    100
rdf                      100

cutoff           10.0000
delr width       0.5000
rvdw cutoff      7.5000

ewald precision    1.0E−04
quaternion tolerance   1.0E−4

print rdf

trajectory         1         100       0

job time                 10000.0
close time               1000.0
finish
```

14

Some Ideas about the Thermal Equilibrium in the Biosphere and the Entropy Variation Ascribed to Changes in the Radiations Wavelengths

Jaime González Velasco

Universidad Autónoma de Madrid, Facultad de Ciencias, Cantoblanco, Madrid, Spain
Email: jaime.gonzalez@uam.es

Abstract

An explanation is given for the thermal equilibrium in the biosphere, which is based in the equality between the thermal energy received from the sun and the thermal energy reemitted from the atmosphere to the space. In order to understand the origin of the energy that gives rise to the processes and phenomena taking place in the biosphere, it is necessary to take into account the free energy represented by the product of temperature times the change in entropy, TΔS, whose magnitude can be attributed to the variation experimented by the wavelengths (or, consequently, the frequencies) of the radiations composing the radiation spectrum received from the sun compared with the radiation spectrum reemitted from the biosphere into the space. A simple discussion allows to predict that the entropy increase driving the processes is connected with a spontaneous conversion of high frequency radiations (with lower "content" of entropy) in radiations of lower frequencies (with higher "content" of entropy). A consequence of this is that high frequency radiations would correspond to more ordered states and, therefore, to less probable states than those corresponding to radiations of lower frequencies.

Keywords

Biosphere, Thermal Equilibrium, Entropy, Radiation Frequency and Entropy

1. Introduction

The discussion on the conditions that should be accomplished for the establishment of the thermal equilibrium in the biosphere leads to an apparent paradox. The thermal equilibrium which allows an average temperature of 15°C in the biosphere [1] is based in the reemission to the space, in the form of radiating energy, with a power spectrum corresponding, approximately, to the emission spectrum of a blackbody at a temperature of 15°C, of the same amount of thermal energy that it is absorbed from the sun by the terrestrial atmosphere. This is the process, according to which the thermal equilibrium in the biosphere is admitted to take place. Nonetheless, before the radiation is reemitted, a large number of processes and phenomena take place in the biosphere, whose development requires the degradation of the energy to another, lower degree. The lowering of degree in the energy is a measure of the amount of free energy that is necessary to give rise to all the processes required to maintain the functioning of the biosphere, and this free energy is represented by the term $T\Delta S$, $i.e.$, by the variation in entropy and temperature which experiments the incident solar radiation with respect to the spectrum of radiations reemitted to the space surrounding the Earth.

The electromagnetic radiation reaching the outskirts of the terrestrial atmosphere presents a distribution of power by surface and frequency unit which is very similar to the radiation spectrum corresponding to a black body at 6000 K and arrives at a rhythm of around 1360 $W \cdot m^{-2}$. On the other hand, in order that the temperature in the biosphere remains constant along the time, the energy should be irradiated from it to the surrounding space at the same rhythm at which it is absorbed, $i.e.$, it is necessary to reach a thermal equilibrium. Since the average temperature of the biosphere is estimated to be 15°C, the spectrum of the energy irradiated is similar to the corresponding to a black body at a temperature of 15°C, which is characterized by a distribution of radiation of much lower frequencies (frequencies located in the range of the far infrared) than the spectrum of the electromagnetic radiation coming from the Sun, whose frequencies are located in the ultraviolet and visible ranges. Thus, the phenomena occurring in the biosphere must be the consequence of the lowering of the frequency which takes place in the electromagnetic radiation and the change in entropy responsible for the degradation of the electromagnetic energy should be attributed to the change in the frequency experimented by the different radiations composing the solar spectrum reaching the Earth, in comparison with the lower frequencies (or larger wavelengths) corresponding to the radiation spectrum emitted by a black body at 15°C.

2. Thermodynamic Interpretation of the Transformations Taking Place in the Biosphere

In order to explain how a lowering of the frequency of a radiation is able to give rise to the multiples transformations taking place in the biosphere it is necessary to resort to the Thermodynamics. According to the arguments exposed in the former paragraph, at a high frequency radiation should correspond a low value of the entropy, whereas at a radiation of low frequency, $i.e.$, of high wavelength, would correspond a high value of the entropy. According to this supposition, under conditions in which the enthalpy of any transformation remains constant, there would be a spontaneous tendency of the entropy to grow, which means that a spectrum composed by radiations of high frequency would present a spontaneous tendency to convert itself in another of lower frequency radiations. The free energy term associated to this transformation (which can be identified with the amount of free energy dissipated through all the changes induced in the biosphere) can be represented by the product of the absolute temperature, T, multiplied by the entropy change occurring in it, ΔS ($T\Delta S = h\Delta\nu$, where $\Delta\nu$ denotes the change in frequency associated with any process induced by the changes in the characteristics of the radiation).

The thermodynamic formulation of the energy change which gives rise to every phenomenon occurring in the biosphere can be expressed according to the following equation [2]:

$$\Delta G = \Delta H - T\Delta S = h\Delta\nu = \Delta E \qquad (1)$$

where ΔE represents the free energy dissipated as consequence of all the transformations which take place in the biosphere.

In the case of an infinitesimal change of the magnitudes produced under conditions of constant enthalpy ($dH = 0$), the above equation reduces to the following expression:

$$dG = -TdS = hd\nu = dE \qquad (2)$$

The result of the differentiation of the last equation with respect to the time (at constant temperature) gives rise to the rate of change of the free energy with the time (or the rate of free energy dissipation), which is given by the following expression.

$$\frac{dG}{dt} = -T\frac{dS}{dt} = h\frac{dv}{dt} = \frac{dE}{dt} \tag{3}$$

Remembering that $c = \lambda v$; $v = c/\lambda$, y: $dv = (-c/\lambda^2) \times d\lambda$. By substitution of this equivalence in Equation (3) one obtains the following expression:

$$\frac{dG}{dt} = -T\frac{dS}{dt} = h\frac{dv}{dt} = -\frac{hc}{\lambda^2}\frac{d\lambda}{dt}. \tag{4}$$

Finally, the rate of generation of entropy, measured by (dS/dt) would be given by the following expression:

$$\frac{dS}{dt} = -\frac{1}{T}\frac{dG}{dt} = \left(\frac{hc}{T\lambda^2}\right)\left(\frac{d\lambda}{dt}\right) = \left(\frac{hv}{T}\right)\left(\frac{d\lambda}{dt}\right) = (PC)\times\left(\frac{d\lambda}{dt}\right). \tag{5}$$

According to Equation (5) the rate of generation of entropy is proportional to the rate of change of the wavelength with the time. The coefficient of proportionality (denoted as PC) is always positive and is given by the following expression:

$$(PC) = \left(\frac{hc}{T\lambda^2}\right) = \left(\frac{hv}{T\lambda}\right) \tag{6}$$

This coefficient is an inverse function of the absolute temperature and would be measured in following units in the SI:

$$\left(\frac{hc}{T\lambda^2}\right) = \frac{(J\cdot s)\times\left(kg\cdot m^2\cdot s^{-2}\times J^{-1}\right)\times\left(m\cdot s^{-1}\right)}{K\cdot m^2}$$

$$= J\cdot K^{-1}\cdot m^{-1}\cdot s^{-2} = \left(kg\cdot m\cdot s^{-2}\right)\cdot K^{-1} = N\cdot K^{-1}$$

The fact that $(PC) > 0$ implies that a growth of the entropy with the time, which is indicated by $(dS/dt) > 0$ and expresses the spontaneous evolution of the entropy with the time in any system considered, would be the result of a growth of the wavelength with the time, i.e., would be the consequence of the positive value of the rate of change of the wavelength with the time, $(d\lambda/dt) > 0$. In other words, it can be said that the higher the wavelength of a radiation the higher would be its content in entropy.

On the other hand, the proportionality coefficient in Equation (6) $[(PC) = (hc/T\lambda^2) = (hv/T\lambda)]$ is an inverse function of the square of the wavelength and of the temperature. Thus, the most energetic radiations component of the solar light (i.e., those radiations of higher frequency or of lower wavelength), would be those that would give rise to a more rapid free energy dissipation or a higher rate of generation of entropy. Likewise, according to Equation (5) the rate of generation of entropy grows proportionally to the inverse of the temperature.

Finding dS from Equation (5) one comes to following expression:

$$dS = \left(\frac{hc}{T\lambda^2}\right)d\lambda = \left(\frac{hv}{T\lambda}\right)d\lambda \tag{7}$$

By integrating this equation it is possible to calculate the change of entropy ($\Delta S = S_f - S_{in}$), associated with a wavelength variation from an initial value of $\lambda = \lambda_{in}$ to a final value λ_f:

$$\Delta S = S_f - S_{in} = \int_{S_{in}}^{S_f} dS = \left(\frac{hc}{T}\right)\int_{\lambda_{in}}^{\lambda_f}\frac{d\lambda}{\lambda^2} = \frac{hc}{T}\left(\frac{1}{\lambda_{in}} - \frac{1}{\lambda_f}\right) \tag{8}$$

$$= \Delta S = S_f - S_{in} = \frac{h}{T}\left(v_{in} - v_f\right). \tag{9}$$

According to this expression $\Delta S > 0$ whenever $\lambda_{in} < \lambda_f$ or, in the case that the entropy change would be expressed as a function of the frequencies, whenever $v_f < v_{in}$.

These expressions allow to perform calculations about the entropy changes which accompany to the conversion of an Avogadro number of photons of visible radiation of wavelength corresponding to the blue ($\lambda_{blue} = 430$ nm) to an Avogadro number of photons of another radiation corresponding to the maximum in the far infrared spectrum corresponding to a black body at $15°C = 288$ K, which is situated at about $\lambda_{máx} = 20000$ nm.

3. Conclusions

According to the ideas above developed, it could be said that the "arrow of the time" defined by Clausius [3] could be interpreted as the irremediable spontaneous conversion along the time of radiations of high frequency, *i.e.*, of shorter wavelengths, in radiations of lower frequencies, *i.e.*, of longer wavelengths, which are those emitted by a black body at 15°C, as the biosphere can be supposed to be. Therefore, the energy necessary to maintain all the processes taking part in the biosphere should be attributed to the free energy changes proceeding from the term T∆S, *i.e.*, from a change in the radiation spectrum corresponding to the sun surface (around 6000 K) to the radiation spectrum of a black body at 288 K. In this case, in which it can be supposed that the enthalpy variation is equal to null, the driving force for the spontaneous transformation of short-length-radiations in radiations of longer wavelengths is the increase of entropy associated with such a change.

It can be said that the "quality" or grade of the energy transported by the electromagnetic radiation is inversely proportional to the wavelength, or, directly proportional to the frequency of the radiation considered. According to the second principle of thermodynamics, there is a natural tendency to the energy to spontaneously degrade and to convert itself from a higher into a lower grade or "noblesse" [4]. This degradation in the "noblesse" of the energy explains the origin of the work amount necessary for driving the multiple processes and phenomena taking place in the biosphere, whereas no thermal energy change can be observed.

According to the arguments above exposed, also the so called cosmic microwave radiation background that according to S. Weinberg [5] could be considered as a black-body radiation with a temperature of 3 K, which could be the consequence of the spontaneous degradation of electromagnetic radiations of higher frequencies produced along the time.

References

[1]   Twidell, J. and Weir, T. (2006) Renewable Energy Resources. 2nd Edition, Taylor and Francis, London, New York.

[2]   Sears, F.W. (1952) An Introduction to Thermodynamics, the Kinetic Theory of Gases, and Statistical Mechanics. Addison Wesley Publishing Company, Massachusetts.

[3]   Castellan, G.W. (1971) Physical Chemistry. Addison Wesley Publishing Company, Reading.

[4]   González Velasco, J. (2009) Energías Renovables. Ed. Reverté, Barcelona.

[5]   Weinberg, S. (1980) The First Three Minutes. A Modern View of the Origin of the Universe. Bentam Books, New York.

A Theoretical Study of β-Amino Acid Conformational Energies and Solvent Effect

Victor F. Waingeh[1*], Felix N. Ngassa[2*], Jie Song[3]

[1]School of Natural Sciences, Indiana University Southeast, New Albany, IN, USA
[2]Department of Chemistry, Grand Valley State University, Allendale, MI, USA
[3]Department of Chemistry and Biochemistry, University of Michigan-Flint, Flint, MI, USA
Email: [*]vwaingeh@ius.edu, [*]ngassaf@gvsu.edu, jiesong@umflint.edu

Abstract

The conformations of four β-amino acids in a model peptide environment were investigated using Hartree-Fock (HF) and density functional theory (DFT) methods in gas phase and with solvation. Initial structures were obtained by varying dihedral angles in increments of 45° in the range 0° - 360°. Stable geometries were optimized at both levels of theory with the correlation consistent double-zeta basis set with polarization functions (cc-pVDZ). The results suggest that solvation generally stabilizes the conformations relative to the gas phase and that intramolecular hydrogen bonding may play an important role in the stability of the conformations. The β^3 structures, in which the R-group of the amino acid is located on the carbon atom next to the N-terminus, are somewhat more stable relative to each other than the β^2 structures which have the R-group on the carbon next to the carbonyl.

Keywords

Density Functional Theory, β-Amino Acids, Conformational Search

1. Introduction

The functions of numerous biological systems depend on RNA and proteins. The conformation adopted by these biopolymers is linked to their functions in biological systems. As a consequence, there has been an increasing need to identify synthetic polymer backbones that adopt discrete and predictable conformations ("foldamers") to mimic natural biological systems. Such backbones can serve as tools to probe the functions of large-molecule interactions, such as protein-protein and protein-RNA interactions. In foldamer design, β-amino acids are highly

---

[*]Corresponding author.

attractive building blocks because the additional carbon confers conformational flexibility to β-amino acids compared to their α-amino acid counterparts.

Early attempts at realizing ordered peptide structures with β-amino acids dated from the end of the 1960s and were continued in the 1970s [1]-[3]. In 1994, Dado and Gellman studied the intramolecular hydrogen bonding properties of β- and γ-amino acid derivatives [4]. They hypothesized that the formation of secondary structures by oligomers of α-amino acids was due in part to the formation of short range hydrogen bonds, such as intramolecular 5- or 7-membered hydrogen bonded rings which were energetically unfavorable among α-amino acid oligomers (**Figure 1**). Based on their hypothesis, Dado and Gellman reasoned that other oligomers in which the formation of short range hydrogen bonds was unfavorable might also be predisposed to form secondary structures. To test this hypothesis further, Dado synthesized various derivatives of β-alanine as well as γ-isobutyric acid using IR spectroscopy, and examined the intramolecular hydrogen bonding of these molecules. The results demonstrated that while short range hydrogen bonds (7- and 9-membered rings) were common among the diamides of γ-amino-butyric acid, the same was not true for diamides of β-amino acids (**Figure 1**). Therefore, based on these studies, Dado and Gellman concluded that oligomers of β-amino acids (β-peptides) could potentially form stable secondary structures. The studies carried out by Dado and Gellman did not take into account the substitution on α-, β-, or γ-carbons which could potentially had some constraining effects.

Although, it was predicted that β-peptides were capable of forming stable and compact secondary structures, investigations of the NMR solution structures of poly-β-alanine indicated that the polymer had no folded conformation [2]. In contrast, IR data in the solid state indicated the formation of sheets [5]. Various research groups have studied the polymers of α-isobutyl-L-aspartate by X-ray diffraction, CD, and IR spectroscopy. In 1978, Yuki *et al.* reported that the polymer of α-isobutyl-L-aspartate formed extended sheets with hydrogen bonding between the strands [6]. In the mid 1980s, Fernández-Santín and co-workers reported that in the solid state, the polymer of α-isobutyl-L-aspartate formed two helical conformations characterized by intramolecular 16- and 20-membered hydrogen bonded rings [7] [8]. In 1995, López-Carrasquero *et al.* contrasted initial reports by Fernández-Santín and co-workers and instead proposed that the helical conformations formed by the polymer of α-isobutyl-L-aspartate was characterized by 14- and 18-membered hydrogen bonded rings [9].

The biggest breakthrough in the synthesis β-peptides of defined sequence that allowed crystallographic and high-resolution NMR data to be obtained was achieved by the research groups of Gellman [10]-[15] and Seebach [16]-[19]. Gellman and co-workers have synthesized and characterized the oligomers of *trans*-2-aminocyclo-pentanecarboxylic acid (ACPC) [12], *trans*-2-aminocyclohexanecarboxylic acid (ACHC) [11] and *trans*-3-aminopyrrolidine-4-carboxylic acid (ACP) [15]. Gellman *et al.* showed that while ACHC adopted a 14-helical conformation in organic solvent, ACPC adopted a 12-helical conformation. The Oligomers containing both ACPC and ACP residues were also shown to form 12-helical conformations in aqueous solution [15]. Seebach and co-workers showed that β-peptides composed of acyclic residues with side chains derived from α-amino acids also formed 14-helical conformations in organic solvent [16].

In polypeptides of α-amino acids, the backbone conformations are defined by three sets of torsion angles and (**Figure 2**) [20]. Generally in peptides and proteins, the torsion angle about the peptide bond is restricted to a *trans* geometry ($\beta = 180$). Backbone conformation of β-amino acid residues in peptides is determined by four main chain torsional variables θ and using the convention of Balaram (**Figure 2**) [20]. By this convention, the

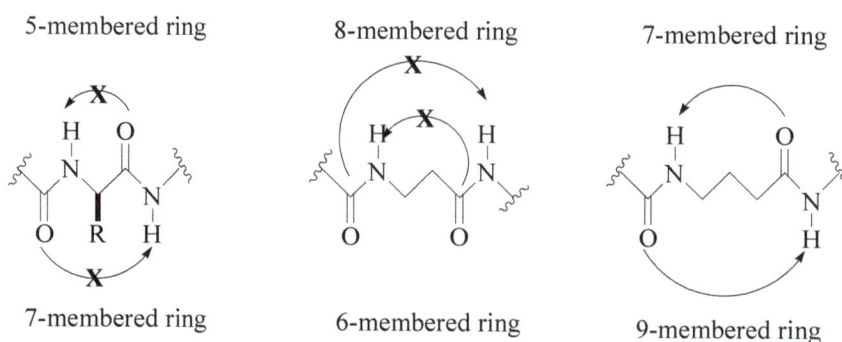

Figure 1. Favorable and non-favorable intramolecular hydrogen bonding in α-, β-, and γ-amino acid peptides.

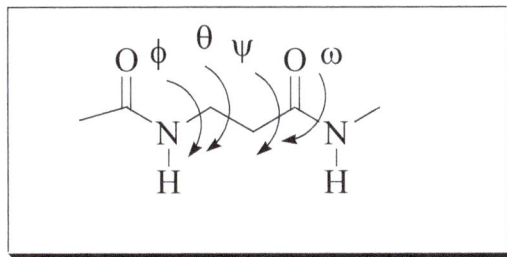

Figure 2. Backbone conformation of β-amino acid.

backbone conformation is read sequentially from the N- to the C-terminus with the torsional variables defined as $(N-C^\beta)$, $\theta (C^\beta-C^\alpha)$, $(C^\alpha-CO)$, and $(CO-N)$ (**Figure 2**). In β-amino acids, the torsion angle about the C-C bond (θ) can lie close to the gauche $(\theta = \pm 60°)$ and *trans* $(\theta = 180°)$ conformations [20]. A predominance of gauche conformations can easily lead to the formation of folded structures.

The interactions between oligomers and solvent can affect the conformations adopted by oligomers. The ability for an oligomer to fold can be favored or disfavored by such solvent properties as dielectric constant, solubility, and hydrogen bonding capability. Solvent effects on the conformations of phenylacetylene oligomers have been reported by Nelson and co-workers [21]. Based on NMR and UV spectroscopic studies, they concluded that a well-ordered conformation was observed in deuterated acetonitrile, but not in deuterated chloroform. The difference in behavior of the oligomer in the different solvents could be a result of difference in solubility. The oligomer was found to be completely soluble in chloroform irrespective of the number of residues and had many conformations. In acetonitrile, however, the oligomer became less soluble as the chain length increased resulting in a single, well-defined conformation.

Like natural peptides and proteins, β-peptides can adopt folded conformations including common secondary structures such as helices [11]-[13] [22]-[28], turns [18] [29] and sheets [14] [30]. Compared to α-peptides, β-peptides have the advantage of increased conformational stability in an aqueous environment. [15] The β-peptide backbone compared to natural peptides is resistant to protease degradation and has the potential for a great variety of substitution patterns [31]. In addition β-peptides form more stable helices in solution compared to α-peptides; β-peptides can form secondary structures with as few as four to six residues in solution [13] [16] [19], compared to over 30 residues needed for stability of the natural analogs. The stability of β-peptides, important for biological activity, makes them good candidates for useful drugs. [18] It has been shown that foldamers comprising a mixture of α-, β-, and γ-amino acid residues are not degraded by proteases which bodes well for biological application [26] [27].

Investigating the 3-D structure of β-peptides is critical to understanding their biological functions. Although crystal structures and NMR spectroscopy can provide adequate structural information for β-peptides, these methods still have some limitations. Crystallography only provides solid-state structural information and in most cases obtaining a good crystal structure for X-ray crystallography is often difficult. NMR spectroscopy can provide structural information corresponding to solution structures. However, limitations exist in the size of the peptide; NMR works best for relatively small peptides.

Computational modeling can provide structural information, at the atomic level, for β-peptides from the sequence of β-amino acids [32]. A good computational method depends on the ability to reproduce the structures and energies of β-amino acid conformations in a target molecule. Herein, the conformations of some selected β-peptides resulting from rotation along their backbones are studied theoretically. The conformations are first calculated in the gas phase at the HF and DFT level to determine their relative stabilities. Then, solvent effect is included by applying the continuum solvation model and comparisons of the two models are made.

2. Computational Details

Several conformations of selected β-amino acids were computed. To mimic the environment in longer peptide chains, the amino-end of the each amino acid was capped with an acetyl group and the carboxylic end was capped with methylamine. The resulting structure template and torsion angle labels are as shown in **Figure 3**.

Selected β-amino acids for this study include β-alanine (R = −CH₃), β-cysteine (R = −CH₂SH), β-leucine

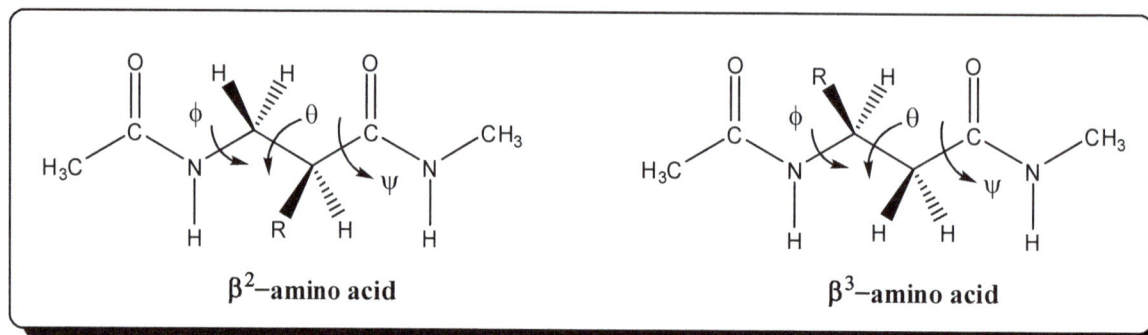

Figure 3. Structure of β-amino acid in peptide environment.

(R= $-CH_2CH(CH_3)_2$), and β-serine (R = $-CH_2SH$). In the β^2 structure, the R-group of the amino acid is on the carbon attached to the carbonyl function at the C-end while in β^3 the R-group is closer to the N-end. The β^3 conformation is included in this study because studies have shown that β^3-peptides can populate a secondary structure known as a 14-helix, which is characterized by 14-membered ring hydrogen bonds between the amide at potion i and the carbonyl at position $i + 2$, a left-handed helical twist with three discrete faces [33]-[35].

Initial conformations were obtained by rotating θ angles by 45° increments in the range 0° to 360°. The structures were then optimized at HF/cc-pVDZ and DFT/cc-pVDZ levels of theory in the gas phase, without any restraints. Calculations were then repeated to include solvation effects by employing the Polarizable Continuum Model (PCM) for water at the DFT level. For comparisons with gas-phase conformational energies, calculations with solvation were all done with a constant dielectric of 1.0. In all instances, the estimation of relative conformational energies was concluded by performing single point energy calculations at both the HF and DFT levels for each conformation. All calculations were performed using GAMESS suite program [36] and Avogadro [37] was used for visualization.

3. Results and Discussions

HF and DFT were used to study the conformations of selected β-amino acids in both the gas phase and with solvation (DFT only). Stable conformations were identified and the relative energies calculated as the difference in energy between each identified conformation and the lowest energy conformation.

3.1. β-Alanine

Gas phase calculations for β-alanine predicted 5 stable conformations for both HF and DFT. The optimized dihedral angles and relative energies of conformers are given in **Table 1**. The minimum energy conformation occurred at a dihedral angle of 57° in the HF calculations and at −64° in the DFT calculations (**Figure 4**).

The highest energy conformation in both cases is one with an optimized dihedral angle closer to 180° and is about 17 kJ/mol higher in energy than the minimum. When solvation is included, calculations yield seven stable conformations, with a 17 kJ/mol energy difference between the most and the least stable conformations. The β^3 structure of alanine shows a lot more variation in relative stabilities of its conformations with the lowest energy conformation at least 5 kJ/mol lower that the second lowest conformation as opposed to the smaller differences of 1 - 2 kJ/mol seen with the β^2 structures. Solvation seems to stabilize the β^3 structure to some extent as the relative energies of the identified conformations above the minimum are all within 2 kJ/mol of each other.

3.2. β-Cysteine

The presence of the $-CH_2SH$ side chain in cysteine would be expected to introduce additional degrees of freedom and conformational flexibility when compared to alanine. The calculated relative energies (**Table 2**) show a wide variation ranging from 0 - 29 kJ/mol. As was the case with β-alanine, solvation seems to introduce to form of stabilization for both the β^2 and β^3 structures. **Figure 5** shows the minimum energy conformations in gas phase.

Unlike in the corresponding β-amino acid, in β-cysteine, the positioning of $-CH_2SH$ group also allows for the

Table 1. *β*-Alanine conformations and relative energies.

	HF			DFT			DFT-Solvation	
Structure	Optimized Dihedral Angle	Relative Energy (KJ/mol)	Structure	Optimized Dihedral Angle	Relative Energy (KJ/mol)	Structure	Optimized Dihedral Angle	Relative Energy (KJ/mol)
	57.39	0.00		−64.45	0.00		−63.15	0.00
	−66.89	1.33		64.96	0.27		−127.20	0.47
	66.92	1.34		56.62	1.90		63.76	0.80
β^2-ala	−50.56	11.80	β^2-ala	−56.57	2.24	β^2-ala	−58.47	2.28
	170.86	17.90		180.00	16.75		−180.00	6.60
							177.26	7.74
							115.67	17.61
	−62.89	0.00		−62.75	0.00		−65.69	0.00
	5689	6.04		54.09	5.24		−60.07	1.39
	167.96	11.31		134.56	5.57		62.33	5.61
β^3-ala	64.04	17.88	β^3-ala	166.18	13.71	β^3-ala	66.91	6.18
				63.17	20.48		48.24	6.59
							168.25	7.62
							130.88	8.40

Table 2. *β*-Cysteine conformations and relative energies.

	HF			DFT			DFT-Solvation	
Structure	Optimized Dihedral Angle	Relative Energy (KJ/mol)	Structure	Optimized Dihedral Angle	Relative Energy (KJ/mol)	Structure	Optimized Dihedral Angle	Relative Energy (KJ/mol)
	−47.47	0.00		−51.89	0.00		−55.80	0.00
	−136.11	0.44		−131.22	5.46		−128.16	0.91
β^2-cys	80.37	5.68	β^2-cys	52.85	27.25	β^2-cys	55.79	11.04
	161.87	24.81		−161.15	29.00		−62.53	12.50
	−155.83	28.46		161.20	29.55		−158.40	15.09
	−65.23	0.00		−63.88	0.00		−64.34	0.00
	57.29	6.05		54.64	5.74		66.64	7.54
	52.85	13.36		135.30	6.63		50.50	7.57
β^3-cys	171.15	23.03	β^3-cys	72.70	23.27	β^3-cys	72.43	13.29
				62.21	23.93		176.85	16.08
							168.54	16.30
							−177.76	16.55

possibility of weak intramolecular hydrogen bonding between the side chain –SH and the backbone –C=O groups.

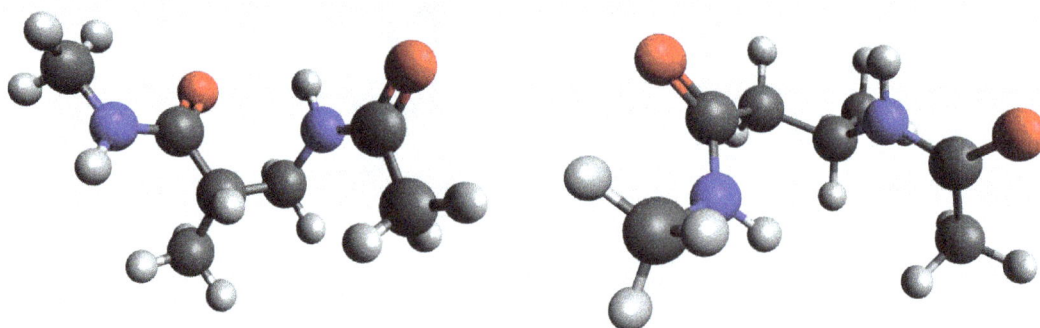

Figure 4. Most stable gas-phase conformers of β^2-alanine (left) and β^3-alanine (right).

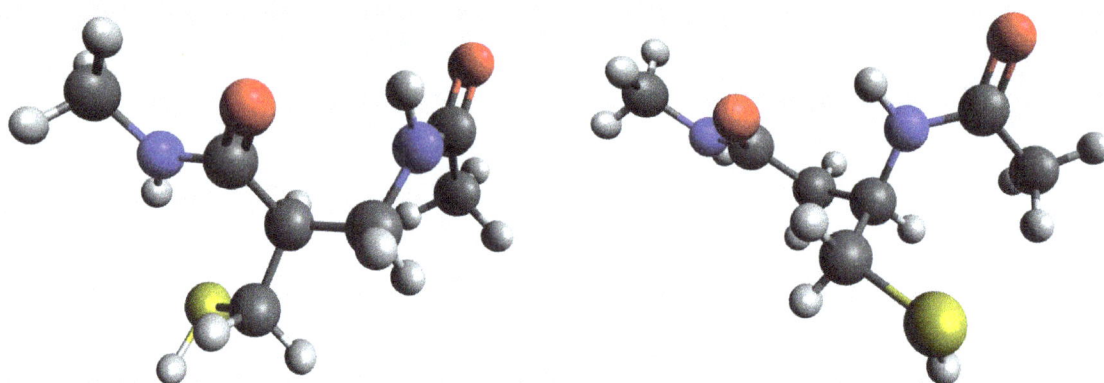

Figure 5. Most stable conformers of β^2-cysteine (left) and β^3-cysteine (right).

3.3. β-Serine

The –CH$_2$OH side chain in serine is similar to that of cysteine, and would introduce the same conformational flexibility. However, the replacement of the sulfur in cysteine with the oxygen in serine results in less polarizability and more hydrogen bonding capability. The conformational search and single point energy calculations resulted in three stable conformations in the gas phase for both the β^2 and β^3 structures in both HF and DFT. The β^3 structures are more stable relative to each other than are the β^2 structures as evident from the calculated relative energies as seen in **Table 3**.

In both HF and DFT, the next lowest energy conformation is at least 14 kJ/mol higher than the minimum, however for the β^3 structures, the higher energy conformations are all within ~1 kJ/mol of each other. This would suggest that the possibility of favorable and strong intramolecular hydrogen bonding in the β^3 structure helps to restrict the molecule. **Figure 6** shows the optimized structures of the most stable conformers of β^2 and β^3-serine. When solvation is included in the calculations, there is evidence of some substantial stabilization in solution with a mean deviation of about 22.9 ± 2.5 kJ/mol for the β^2 structure and 11.2 ± 1.9 kJ/mol for β^3 structures.

3.4. β-Leucine

In the gas phase, the lowest energy conformation for β^2-leucine occurs at a optimized dihedral angle of 58° in both HF and DFT with relative energies of higher energy conformations ranging from 13 - 26 kJ/mol in HF and 17 - 29 kJ/mol in DFT. For β^3-leucine, the most favorable conformation occurs at −63° (**Figure 7**) with relative energies in the range 6 - 23 kJ/mol in HF and 8 - 23 kJ/mol in DFT. As was the case with other β-amino acids, solvation stabilizes the conformations relative to each other, with relative energies for the β^2-leucine conformation in the range 3 - 18 kJ/mol and those for β^3-leucine in the 2 - 11 kJ/mol range (**Table 4**). In general, the β^3

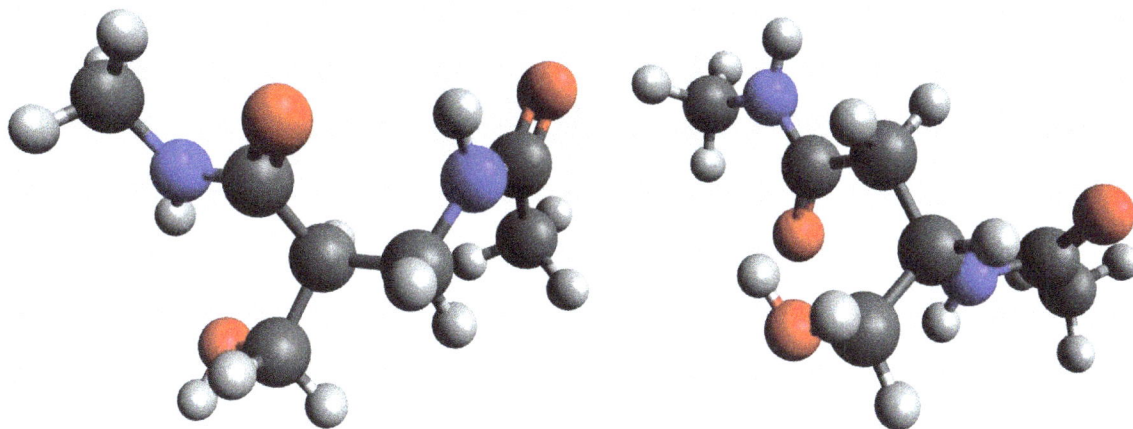

Figure 6. Most stable conformers of β^2-serine (left) and β^3-serine (right).

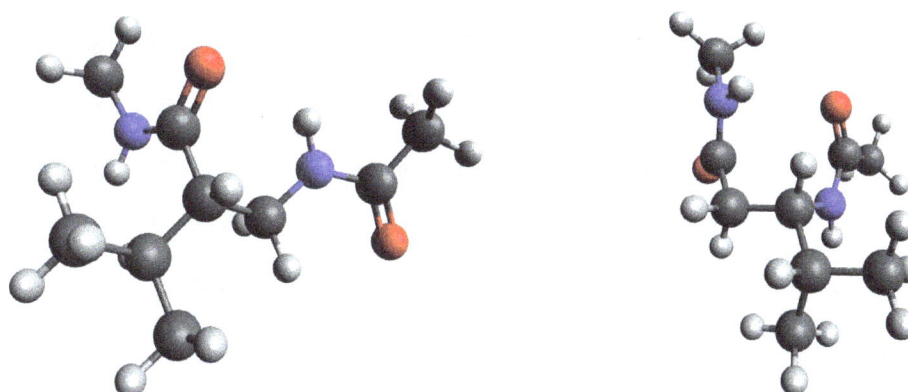

Figure 7. Most stable conformers of β^2-leucine (left) and β^3-leucine (right).

Table 3. β-Serine conformations and relative energies.

	HF			DFT			DFT-Solvation	
Structure	Optimized Dihedral Angle	Relative Energy (KJ/mol)	Structure	Optimized Dihedral Angle	Relative Energy (KJ/mol)	Structure	Optimized Dihedral Angle	Relative Energy (KJ/mol)
	−65.58	0.00		−63.98	0.00		−167.90	0.00
	141.02	13.80		134.19	21.74		69.93	19.70
β^2-ser	75.94	32.00	β^2-ser	78.50	42.62	β^2-ser	128.61	21.90
							170.23	22.17
							61.53	25.25
							−65.11	25.58
	57.29	0.00		123.14	0.00		54.88	0.00
	−62.50	18.54		160.22	23.28		160.83	9.12
β^2-ser	163.25	19.79	β^2-ser	107.29	23.58	β^2-ser	−60.93	11.69
							−64.50	12.88

Table 4. Optimized dihedral angles and relative energies of β-Leucine conformations.

Structure	HF Optimized Dihedral Angle	HF Relative Energy (KJ/mol)	Structure	DFT Optimized Dihedral Angle	DFT Relative Energy (KJ/mol)	Structure	DFT-Solvation Optimized Dihedral Angle	DFT-Solvation Relative Energy (KJ/mol)
	58.05	0.00		58.40	0.00		−66.43	0.00
β^2-leu	−56.39	13.11	β^2-leu	−56.65	17.47	β^2-leu	58.35	3.47
	151.77	24.62		174.40	27.93		−176.53	16.31
	173.20	26.11					−75.03	18.01
	−62.83	0.00		−62.74	0.00		−61.46	0.00
	165.11	6.03		52.13	8.40		166.26	2.14
β^3-leu	53.91	8.09	β^3-leu	163.70	12.46	β^3-leu	69.61	8.02
	61.16	19.12		153.44	22.65		42.89	11.02
	152.70	22.96		62.38	22.77		150.74	11.12

conformations seem to show slightly more solvent stabilization than the β^2 conformations.

4. Conclusion

In this study, the conformations of β-amino acids were investigated in the gas phase by HF and DFT calculations. The PCM water model was employed with DFT to investigate the solvation effects. Relative energies were computed for both gas phase and solution structures. The results suggest that solvation generally stabilizes the conformations relative to the gas phase, with smaller energy differences between the conformations in solution than in gas phase. It is also likely that intramolecular hydrogen bonding may play an important role in the stability of the conformations. The β^3 structures, in which the R-group of the amino acid is located on the carbon atom next to the N-terminus, are somewhat more stable relative to each other than the β^2 structures which have the R-group on the carbon next to the carbonyl. These results provide insight to the conformational structures of β-amino acids and may be useful in establishing the potential use of β-amino acids in the backbone of polypeptide chain that will be less susceptible to degradation. However, further work will need to be done, with a larger collection of β-amino acids. Also the use of a continuum solvent model to describe solvation of these systems is reasonable and provides a basis for qualitative comparison. However, to obtain a more quantitatively accurate description of the system, future work may employ explicit solvation models.

References

[1] Bestian, H. (1968) Poly-β-Amides. *Angewandte Chemie International Edition in English*, **7**, 278-285. http://dx.doi.org/10.1002/anie.196802781

[2] Glickson, J.D. and Applequist, J. (1971) The Conformation of Poly-β-Alanine in Aqueous Solution from Proton Magnetic Resonance and Deuterium Exchange Studies. *Journal of the American Chemical Society*, **93**, 3276-3281. http://dx.doi.org/10.1021/ja00742a030

[3] Kovacs, J., Ballina, R., Rodin, R.L., Balasubramanian, D. and Applequist, J. (1965) Poly-β-L-Aspartic Acid. Synthesis through Pentachlorophenyl Active Ester and Conformational Studies. *Journal of the American Chemical Society*, **87**, 119-120. http://dx.doi.org/10.1021/ja01079a022

[4] Dado, G.P. and Gellman, S.H. (1994) Redox Control of Secondary Structure in a Designed Peptide. *Journal of American Chemical Society*, **115**, 12609-12610.

[5] Narita, M., Doi, M., Kudo, K. and Terauchi, Y. (1986) Conformations in the Solid State and Solubility Properties of Protected Homooligopeptides of Glycine and Beta-Alanine. *Bulletin of Chemical Society of Japan*, **59**, 3553-3557.

[6] Yuki, H., Okamoto, Y., Taketani, Y., Tsubota, T. and Marubayshi, Y. (1978) Poly(β-Amino Acid)s. IV. Synthesis and Conformational Properties of Poly(α-Isobutyl-L-Aspartate). *Journal of Polymer Science Part A: Polymer Chemistry*, **16**, 2237-2251. http://dx.doi.org/10.1002/pol.1978.170160913

[7] Fernandez-Santin, J.M., Aymami, J., Rodrigues-Galan, A., Munoz-Guerra, S. and Subirana, J.A. (1984) Pseudo α-Helix from Poly(α-Isobutyl-L-aspartate), a Nylon-3 Derivative. *Nature*, **311**, 53-54. http://dx.doi.org/10.1038/311053a0

[8] Fernandez-Santin, J.M., Munoz-Guerra, S., Rodrigues-Galan, A., Aymami, J., Lloveras, J., Subrina, J.A., Giralt, E. and Ptak, M. (1987) Helical Conformations in Polyamide of the Nylon-3 Family. *Macromolecules*, **20**, 62-68. http://dx.doi.org/10.1021/ma00167a013

[9] Lopez-Carrasquero, F., Aleman, C. and Munoz-Guerra, S. (1995) Conformational Analysis of Helical Poly(β-L-Aspartate)s by IR Dichroism. *Biopolymers*, **36**, 263-271. http://dx.doi.org/10.1002/bip.360360302

[10] Appella, D.H., Barchi Jr., J.J., Durell, S.R. and Gellman, S.H. (1999) Formation of Short, Stable Helices in Aqueous Solution by β-Amino Acid Hexamers. *Journal of American Chemical Society*, **121**, 2309-2310. http://dx.doi.org/10.1021/ja983918n

[11] Appella, D.H., Christianson, L.A., Karle, I.L., Powell, D.R. and Gellman, S.H. (1999) Synthesis and Characterization of *Trans*-2-Aminocyclohexanecarboxylic Acid Oligomers: An Unnatural Helical Secondary Structure and Implications for β-Peptide Tertiary Structure. *Journal of the American Chemical Society*, **121**, 6206-6212. http://dx.doi.org/10.1021/ja990748l

[12] Appella, D.H., Christianson, L.A., Klein, D.A., Richards, M.R., Powell, D.R. and Gellman, S.H. (1999) Synthesis and Structural Characterization of Helix-Forming β-Peptides: *Trans*-2-Aminocyclopentanecarboxylic Acid Oligomers. *Journal of the American Chemical Society*, **121**, 7574-7581. http://dx.doi.org/10.1021/ja991185g

[13] Barchi Jr., J.J., Huang, X., Appella, D.H., Christianson, L.A., Durell, S.R. and Gellman, S.H. (2000) Solution Conformations of Helix-Forming β-Amino Acid Homooligomers. *Journal of the American Chemical Society*, **122**, 2711-2718. http://dx.doi.org/10.1021/ja9930014

[14] Krauthauser, S., Christianson, L.A., Powell, D.R. and Gellman, S.H. (1997) Antiparallel Sheet Formation in β-Peptide Foldamers: Effects of β-Amino Acid Substitution on Conformational Preference. *Journal of the American Chemical Society*, **119**, 11719-11720. http://dx.doi.org/10.1021/ja9730627

[15] Wang, X., Espinosa, J.F. and Gellman, S.H. (2000) 12-Helix Formation in Aqueous Solution with Short β-Peptides Containing Pyrrolidine-Based Residues. *Journal of the American Chemical Society*, **122**, 4821-4822. http://dx.doi.org/10.1021/ja000093k

[16] Seebach, D., Abele, S., Gademann, K., Guichard, G., Hintermann, T., Juan, B., Mathews, J.L. and Schreiber, J.V. (1998) Beta[2]- and Beta[3]-Peptides with Proteinaceous Side Chains: Synthesis and Solution Structures of the Constitutional Isomers, a Novel Helical Secondary Structure and the Influence of Solvation and Hydrophobic Interactions on Folding. *Helvetica Chimica Acta*, **81**, 932-982. http://dx.doi.org/10.1002/hlca.19980810513

[17] Seebach, D., Ciceri, P., Overhand, M., Juan, B., Rigo, D., Oberer, L., Hommel, U., Amstutz, R. and Widmer, H. (1996) Probing the Helical Secondary Structure of Short-Chain-Beta-Peptides. *Helvetica Chimica Acta*, **79**, 2043-2066. http://dx.doi.org/10.1002/hlca.19960790802

[18] Seebach, D. and Mathews, J.L. (1997) Beta-Peptides: A Surprise at Every Turn. *Chemical Communications*, No. 21, 2015-2022. http://dx.doi.org/10.1039/a704933a

[19] Seebach, D., Schreiber, J.V., Abele, S., Daura, X. and van Gunsteren, W.F. (2000) Structure and Conformation of β-Oligopeptide Derivatives with Simple Proteinogenic Side Chains: Circular Dichroism and Molecular Dynamics Investigations. *Helvetica Chimica Acta*, **83**, 34-57. http://dx.doi.org/10.1002/(SICI)1522-2675(20000119)83:1<34::AID-HLCA34>3.0.CO;2-B

[20] Banerjee, A. and Balaram, P. (1997) Stereochemistry of Peptides and Polypeptides Containing Omega Amino Acids. *Current Science*, **73**, 1067-1077.

[21] Nelson, J.C., Saven, J.G., Moore, J.S. and Wolynes, P.G. (1997) Solvophobically Driven Folding of Nonbiological Oligomers. *Science*, **277**, 1793-1796. http://dx.doi.org/10.1126/science.277.5333.1793

[22] Gellman, S.H. (1998) Foldamers: A Manifesto. *Accounts of Chemical Research*, **31**, 173-180. http://dx.doi.org/10.1021/ar960298r

[23] Hayen, A., Schmitt, M.A., Ngassa, F.N., Thomasson, K.A. and Gellman, S.H. (2004) Two Helical Conformations from a Single Foldamer Backbone: "Split Personality" in Short Alpha/Beta-Peptides. *Angewandte Chemie*, **43**, 505-510. http://dx.doi.org/10.1002/anie.200352125

[24] Hill, D.J., Mio, M.J., Prince, R.B., Hughes, T.S. and Moore, J.S. (2001) A Field Guide to Foldamers. *Chemical Reviews*, **101**, 3893-4012. http://dx.doi.org/10.1021/cr990120t

[25] Porter, E.A., Wang, X., Lee, H.S., Weisblum, B. and Gellman, S.H. (2000) Non-Haemolytic Beta-Amino-Acid Oligomers. *Nature*, **404**, 13. http://dx.doi.org/10.1038/35003742

[26] Porter, E.A., Weisblum, B. and Gellman, S.H. (2002) Mimicry of Host-Defense Peptides by Unnatural Oligomers: An-

timicrobial Beta-Peptides. *Journal of the American Chemical Society*, **124**, 7324-7330. http://dx.doi.org/10.1021/ja0260871

[27] Raguse, T.L., Porter, E.A., Weisblum, B. and Gellman, S.H. (2002) Structure-Activity Studies of 14-Helical Antimicrobial Beta-Peptides: Probing the Relationship between Conformational Stability and Antimicrobial Potency. *Journal of the American Chemical Society*, **124**, 12774-12785. http://dx.doi.org/10.1021/ja0270423

[28] Werder, M., Hauser, H., Abele, S. and Seebach, D. (1999) Beta-Peptides as Inhibitors of Small-Intestinal Cholesterol and Fat Absorption. *Helvetica Chimica Acta*, **82**, 1774-1783. http://dx.doi.org/10.1002/(SICI)1522-2675(19991006)82:10<1774::AID-HLCA1774>3.0.CO;2-O

[29] DeGrado, W.F., Schneider, J.P. and Hamuro, Y. (1999) The Twists and Turns of Beta-Peptides. *The Journal of Peptide Research: Official Journal of the American Peptide Society*, **54**, 206-217. http://dx.doi.org/10.1034/j.1399-3011.1999.00131.x

[30] Gung, B.W., Zou, D. and Miyahara, Y. (2000) Synthesis of a Hybrid Peptide with Both α- and β-Amino Acid Residues: Toward a New β-Sheet Nucleator. *Tetrahedron*, **56**, 9739-9746. http://dx.doi.org/10.1016/S0040-4020(00)00881-4

[31] Hintermann, T. and Seebach, D. (1997) The Biological Stability of Beta-Peptides: No Interactions between Alpha- and Beta-Peptide Structures. *Chimia*, **51**, 244-247.

[32] Kaminsky, J. and Jensen, F. (2007) Force Field Modeling of Amino Acid Conformational Energies. *Journal of Chemical Theory and Computation*, **3**, 1774-1788. http://dx.doi.org/10.1021/ct700082f

[33] Cheng, R.P., Gellman, S.H. and DeGrado, W.F. (2001) Beta-Peptides: From Structure to Function. *Chemical Reviews*, **101**, 3219-3232. http://dx.doi.org/10.1021/cr000045i

[34] Seebach, D., Beck, A.K. and Bierbaum, D.J. (2004) The World of Beta- and Gamma-Peptides Comprised of Homologated Proteinogenic Amino Acids and Other Components. *Chemistry & Biodiversity*, **1**, 1111-1239. http://dx.doi.org/10.1002/cbdv.200490087

[35] Qiu, J.X., Petersson, E.J., Matthews, E.E. and Schepartz, A. (2006) Toward Beta-Amino Acid Proteins: A Cooperatively Folded Beta-Peptide Quaternary Structure. *Journal of the American Chemical Society*, **128**, 11338-11339. http://dx.doi.org/10.1021/ja063164+

[36] Schmidt, M.W., Baldridge, K.K., Boatz, J.A., Elbert, S.T., Gordon, M.S., Jensen, J.H., Koseki, S., Matsunaga, N., Nguyen, K.A., Su, S., Windus, T.L., Dupuis, M. and Montgomery, J.A. (1993) General Atomic and Molecular Electronics Structure System. *Journal of Computational Chemistry*, **14**, 1347-1363. http://dx.doi.org/10.1002/jcc.540141112

[37] Hanwell, M.D., Curtis, D.E., Lonie, D.C., Vandermeersch, T., Zurek, E. and Hutchison, G.R. (2012) Avogadro: An Advanced Semantic Chemical Editor, Visualization, and Analysis Platform. *Journal of Cheminformatics*, **4**, 17. http://dx.doi.org/10.1186/1758-2946-4-17

16

The Crevice Corrosion of 316L SS Alloy in NaCl Solution at Different Applied Potentials

Sanaa T. Arab[1], Mohammed I. Abdulsalam[2], Huda M. Alghamdi[1], Khadijah M. Emran[3]

[1]Department of Physical Chemistry, KAU, Jeddah, KSA
[2]Department of Chemical and Materials Engineering, KAU, Jeddah, KSA
[3]Department of Chemistry, Taibah University, Al-Madinah Al-Monawarah, KSA
Email: dr.s.arab@hotmail.com, starab@kau.edu.as, hudaalghamdi@hotmail.com, kabdalsamad@taibahu.edu.sa

Abstract

316L SS alloy was tested under different applied potentials to study the susceptibility of this alloy to crevice corrosion. XPS measurements have been carried out to detect and define the products which formed on the surface of 316L SS in 3.5% NaCl at room temperature at applied potential = 200 mV_{SCE}. The formation of Fe, Cr and Mo compounds were found, and these compounds play a great role in protecting the alloy which was found. The boundaries of the corroded area under washer teeth are in agreement with IR drop. The potentiodynamic technique is also studied to examine the overall corrosion behavior of 316L SS.

Keywords

Crevice Corrosion, 316L SS, Applied Potentials, Potentiostatic Measurements, Potentiodynamic Measurements, XPS Measurement, IR Drop

1. Introduction

316L SS is an austenitic stainless steel alloy containing at least 17% Cr with some alloying elements additions in varying concentrations. It has a high corrosion resistance due to the passive film existing on the alloy. Its passivity depends on the environment conditions. The common type of corrosion which causes failure in 316L SS is the localized corrosion which leads to breakdown the passive film. One of the localized corrosions is the crevice corrosion which occurs in narrow openings or spaces between two metal surfaces or between metals and non metal surfaces [1]. The susceptibility of $316L$ SS to localized corrosion depends on applied potential at the outer surface, electrolyte resistivity, crevice length, crevice opening dimension, pH value, oxygen in the crevice electrolyte and temperature [2].

In the case of applied potential at the outer surface, the cathodic reaction rate did not limit the anodic reaction rate due to the potentiostat used to control the applied potential at the outer surface.

There are two models explaining the crevice corrosion process: *CCS model* and *IR drop*. *CCS model* assumes that the sole cause for the localized attack is related to compositional aspects, e.g., the acidification or the migration of aggressive ions into the crevice solution. These changes in the solution composition can cause breakdown the passivation film and promote acceleration and auto catalyzation of crevice corrosion [3]-[5]. The *CCS* theory predicts that the most severe attack will occur at the deepest part of the crevice, *i.e.*, the most occluded portion [6].

IR drop assumes that crevice corrosion is caused by the potential drop which placed the local electrode potential existing on the crevice wall in the active peak region of the polarization curve. Separation between the anodic and cathodic reactions is necessary for the occurrence of crevice corrosion by the *IR* drop mechanism [3] [7]-[9]. *IR-drop* theory predicts the most severe attack at intermediated distance from the crevice mouth [10]-[12].

The aim of this study is to determine the susceptibility of $316L$ SS alloy at different applied potentials in 3.5% NaCl. *XPS* technique was used to identify the constituent species of corrosion products. $316L$ SS alloy was also studied using potentiodynamic technique. The morphology of the corroded surface under the washer tooth was discussed.

2. Experimental

For potentiostatic study, a rectangular-shaped specimen with 60 mm × 30 mm × 3 mm exposed areas was used. The specimen has a hole in the center (7 mm diameter). This hole was used for the attachment of the multiple crevice test assembly. Another smaller hole on the top of the specimen was used for electrical contact using a titanium wire. The specimen was finally polished to 1000-silicon carbide paper then it was cleaned and degreased by ultrasound using acetone. The compositions of $316L$ SS alloy are listed in **Table 1**.

A multiple crevice test assembly made of Teflon from *Metal Samples Co.* Each washer has 20 teeth. Circular disc of 25 mm diameter was cleaned and degreased by ultrasound using double distilled water. The multiple crevice washers were bolted to both sides of each specimen using screw and nuts (made of *Titanium*). The torque value which used in this study is 2.5 *in lbf*. The screw bolt was inserted in a plastic tube to protect it from the solution (**Figure 1(a)**).

For potentiodynamic study, the sample was in square plate chap of (1.4 × 1.4 cm). It was finally polished to 1200 *silicon carbide* papers, cleaned and degreased using double distilled water. A copper wire was attached to the specimen from the back by soldering. This wire was connected to the potentiostat. The sample was placed in a cylindrical mold then epoxy resin was poured inside the mold over the sample. After the resin had dried, the specimen was polished from the front side to expose the alloy sample. The crevice between metal-non-metal in the front side was masked using a resin (from *Shin Etsu*) as shown in (**Figure 1(b)**). The samples were also, cleaned using de-ionized water.

Corrosion cell (**Figure 2**) containing 1000 ml of aerated NaCl solution (3.5%) was equipped with carbon counter electrode. Salt bridge with Luggin capillary was used to connect the reference electrode to the corrosion cell. Test solution was used in the salt bridge to avoid chloride concentration changes in the test solution during the study. The potential was measured using a Saturated Calomel Electrode (*SCE*). The corrosion cell was connected to *PC*14/750 *Gamry* potentiostat for the potentiostatic and potentiodynamic studies. Susceptibility of $316L$ SS alloy at different applied potentials was tested using potentiostatic measurment. Effect of applied potential was tested at values −200, −50, 0, 50 and 200 mV_{SCE}. Chemical compositions on 316 SS surface after the study were defined by *XPS* using a multi-technique surface analysis system (MAX200, Leybold).

The potentiodynamic technique was used to examine the overall corrosion behavior of $316L$ SS. Potentiody-

Table 1. Chemical composition (%wt) of *the studied* $316L$ SS alloy.

C	Cr	Cu	Mn	Mo	Ni	N	P	Si	S	Fe
0.030	16.0 - 18.0	0.5 - 1	2.0	2.0 - 3.0	10.0 - 14.0	0.10	0.045	0.75	0.03	Balance

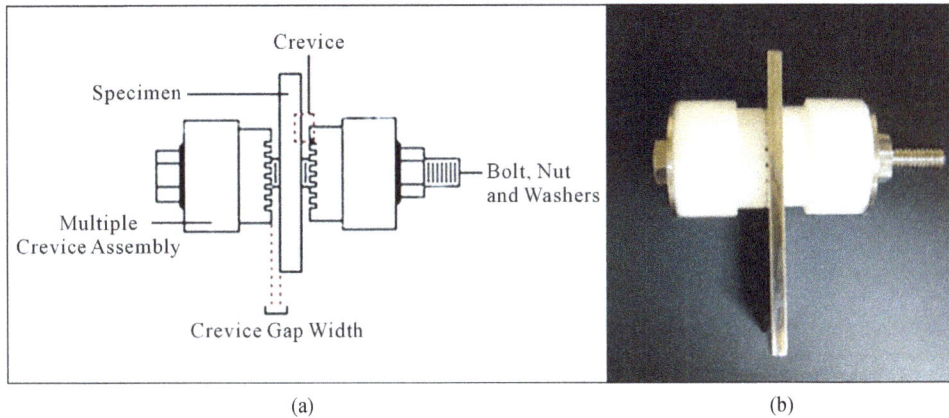

Figure 1. Schematic and photograph of the crevice assembly used in the study.

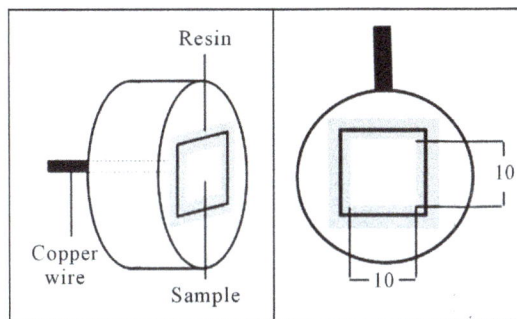

Figure 2. Schematic of the specimen used in potentiodynamic experiments and its dimensions in mm.

namic potential sweep rate (0.1667 mV/s) was applied according to *ASTM* standard *G5*-94 [13] with change in potential from −600 to +1200 mV_{SCE}. Potentiostatic and potentioynamic Tests were run at room temperature (25°C).

3. Results and Discussion

Potentiostatic Measurements

Crevice corrosion susceptibility results from potentiostatic tests in 3.5% NaCl solution are summarized in **Table 2**.

316L SS sample did not show any crevice attack on the surface under the environmental conditions in 3.5% NaCl at room temperature (25°C) at applied potentials of −200, −50 and 0.0 mV_{SCE}, or under the crevice washers after three days of the experiment. But at an applied potential 50 mV_{SCE}, 316L SS showed a localized corrosion attack under the crevice washers. Thirty three out of forty (33/40) crevice sets showed the crevice corrosion attack as shown in (**Figure 2**). The 316L SS sample was removed after ~8 h and 19 min of the experiment.

In test No. 5, the applied potential was increased from 50 to 200 mV_{SCE}, 316L SS showed a localized corrosion attack under the crevice washers as shown in (**Figure 3**) Thirty three out of forty (33/40) crevice sites showed crevice corrosion attack. The number of corrosion sets in 316L SS at applied potential 50 and 200 mV_{SCE} is similar but the crevice corrosion attack on 316L SS sample at 50 mV_{SCE} took more time to form which means the corrosion rate value for 316L SS at 200 mV_{SCE} is higher than the value of corrosion rate for the same alloy at applied potential 50 mV_{SCE}.

Figure 4 shows the effect of different applied potentials on the behavior of the crevice corrosion current for 316L SS in 3.5% NaCl at room temperature (25°C). It is observed that the current is almost constant with the time at applied potentials: 0.0, −50 and −200 mV_{SCE} which means that there is no crevice corrosion attack for 316L SS in this range of applied potentials. The crevice corrosion in the samples at applied potentials: 0.0, −50

Table 2. Crevice corrosion susceptibility results for 316L SS at different applied potentials in 3.5% NaCl solution at room temperature (25˚C) at torque value = 2.5 *in lbf*.

Test No.	Applied potential (mV_{SCE})	Time (ks)	Corrosion attacks
1	−200	259.2	No
2	−50	259.2	No
3	0.0	259.2	No
4	50	30.24	Yes* (33/40**)
5	200	2.220	Yes (33/40)

*Yes = corrosion; **n/40 = number of crevice corrosion sets for the total number of sets.

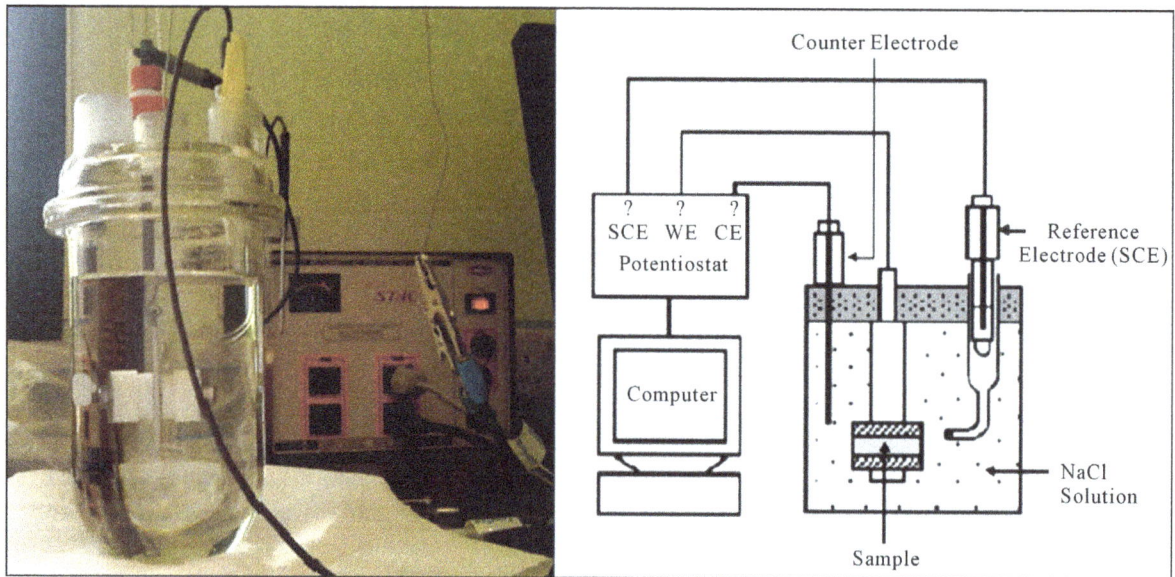

Figure 3. Schematic and photograph of corrosion cell used in experiments.

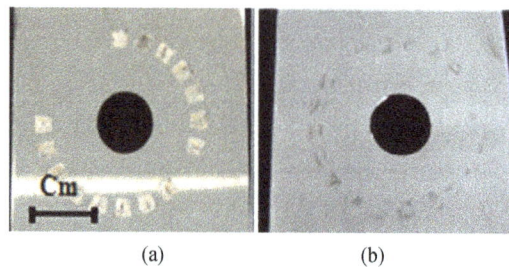

(a)   (b)

Figure 4. Crevice corrosion attack for 316L SS sample in 3.5% NaCl at room temperature (25˚C) after the experiment (a) at applied potential = 50 mV_{SCE}, (b) at applied potential = 200 mV_{SCE}.

and −200 mV_{SCE} weren't observed which means that alloy 316L SS is still passive in the bulk solution and was held several tens millivolts below the crevice potential [14]. According to *IR*-drop theory, during the induction period the inverse of equation $IR > \Delta\varphi^*$ to $IR < \Delta\varphi^*$ is hold.

At applied potentials = 50 and 200 mV_{SCE}, the current increased with time. This is in agreement with Cho *et al.* [15] who found that the susceptibility to crevice corrosion decreases as E_{appl} becomes nobler.

The magnitude of the *IR* voltage between the bottom and opening of the crevice is greater than $\Delta\emptyset^*$ upon application of E_{appl}. Increasing the applied potential from 50 mV_{SCE} to 200 mV_{SCE} leads to decrease the induction period as shown in **Table 3**.

4. *XPS* Measurement for 316*L SS* after Corrosion

XPS measurements have been carried out for identifying the constituent species of corrosion products, which formed on the surface of 316*L SS* in 3.5% NaCl, at room temperature (25˚C) at 200 mV_{SCE}. The formation of Fe_2O_3, CrO, MoO_3, $MoCl_3$ and NiO compounds were found as shown in (**Figure 5**). It is expected that the presence of molybdenum can prevent the raising of corrosion speed through forming MoO_3 oxide layers preventing flux of OH^- and Cl^- ions into films and pits [16].

5. The Morphology of the Corroded Surface under the Washer Tooth

Three different regions can be seen at corrosion area under each washer teeth as shown in (**Figure 6**). The *first* region (the passive region) extended from the crevice mouth to the x_{pass} boundary. The *second* region (active region) started at $x > x_{pass}$ and extended to x_{lim}. The *third* region (etched region) extended from x_{lim} to the crevice set center. The x_{lim} boundary separating the latter two regions is diffuse and, unlike the x_{pass} boundary, is not readily identified in the photograph. These three regions were observed on 316*L SS* at applied potential 50 and 200 mV_{SCE}. This distribution of the corrosion is in agreement with *IR-drop* theory which predicts the most severe attack at intermediated distance from the crevice mouth [10]-[12].

6. Potentiodynamic Polarization Measurement

Potentiodynamic polarization curve based on *ASTM G5*-94 [13] was recorded, with a scan rate of 0.1667 mV/s, on 316*L SS* in 3.5% NaCl solution at room temperature using a potential range from −600 to 1200 mV_{SCE} as shown in (**Figure 7**). The figure shows three distinguished regions, passive, active and transpassive.

In the *active region*, the behavior of the metal shows an increase in the applied potential causes a correspondingly rapid increase in corrosion rate. When the applied potential is increased sufficiently, the corrosion rate suddenly decreases. This behavior corresponds to the beginning of the *passive region*. Fe_2O_3, CrO, MoO_3, and NiO compounds play roles to protect 316*L SS*. In the passive region, there is a fluctuation in the current density

Table 3. *Induction period* values for 316*L SS* at different applied potentials in 3.5% NaCl at room temperature (25˚C).

Applied potential (mV_{SCE})	Induction period (s)
50	525
200	60

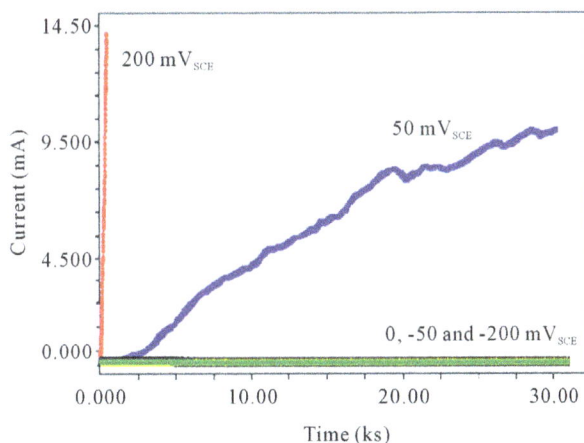

Figure 5. Measured current with time for 316 *SS* sample at different applied potential in 3.5% NaCl at room temperature (25˚C).

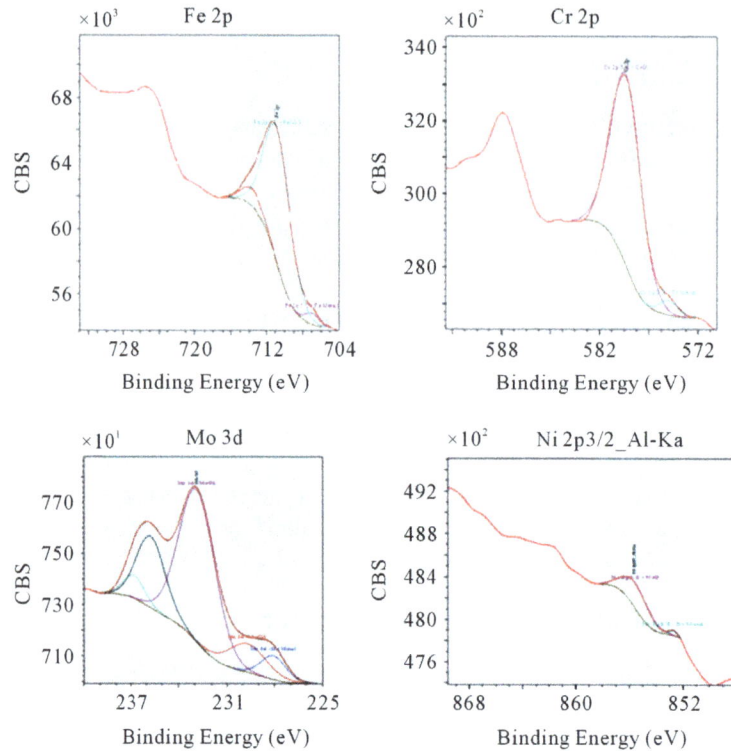

Figure 6. X-ray spectra obtained for Fe, Cr, Mo and Ni on 316 *SS* surface after corrosion in 3.5% NaCl at 200 mV_{SCE} at room temperature (25˚C).

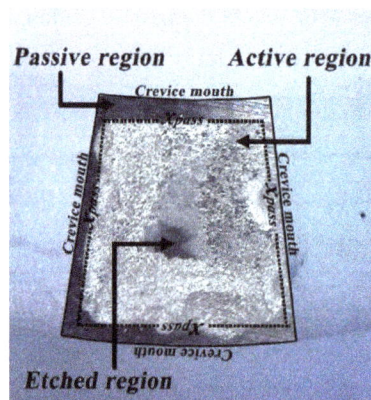

Figure 7. Photograph of the crevice set after the experiment of 316 *stainless steel* alloy in 3.5% NaCl at $E_{appl} =$ 50 mV_{SCE} at room temperature (25˚C) (Magnification: ×70).

values from 77.38 mV up to 476.2 mV and there is more than one corrosion potential appeared indicating the unstable passive system. The stability of the film which formed on the alloy surface decreases with the applied potential increasing. A competition between corrosion and re-passive processes happened. At a very high applied potentials, the corrosion rate begins to increase with increasing potentials in a region called the *transpassive region*, where the protective passive film is thermodynamically unstable.

Included in the plot of (**Figure 7**) are various terms used to define the behavior of 316*L SS* alloy. Corrosion potential (E_{corr}) and corrosion current (i_{corr}) values were extracted from the potentiodynamic polarization plot. The average values E_{corr}, I_{corr}, β_a and β_c, from polarization curves are presented in **Table 4**. The value of β_a is greater than β_c which means that the alloy 316L SS prone to passivity [17].

Effect of Scan Direction on 316L SS

Figure 8 shows the polarization curves for 316L SS sample after scanning in two directions, from active to passive and vice versa, using scan rate = 0.1667 mV/s in 3.5% NaCl at room temperature. The polarization curves are not similar for scanning from active to passive and passive to active directions for 316L SS sample.

At scan direction from active to passive **Figure 9**, more than one corrosion potential appeared this is an indication of an unstable passive system. The stability of the film formed decreased with increase in the applied potentials. At scan direction from passive to active, the value of $E_{corr} = -530$ mV. This value is lower than E_{corr} for the alloy scanning from active to passive (-135.4 mV). This may be attributed to the instability of the passive film.

7. Conclusions

- Crevice corrosion and the value of current with time for 316L SS increased with applied potential increasing.
- The values of induction period decreased with applied potential increasing, which means the growth of crevice corrosion becomes faster with applied potential increasing.
- *XPS* measurements show that the Cr, Fe and Mo compounds resulting from crevice corrosion play an important role in protecting 316L SS alloy.

Table 4. Electrochemical data resulted from polarization curve for 316L SS immersed in 3.5% NaCl at room temperature.

E_{corr} (mV)	I_{corr} (A/cm^2)	β_a (V/decade)	β_c (V/decade)	E_{pass} (mV)	Corrosion Rate (mpy)
-135.4	90.84e$-$9	0.338	0.138	250	41.12e$-$3

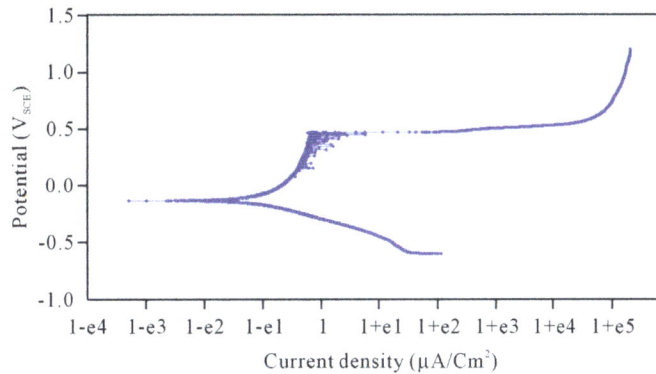

Figure 8. Potentiodynamic polarization curve for 316 SS with 3.5% NaCl at room temperature.

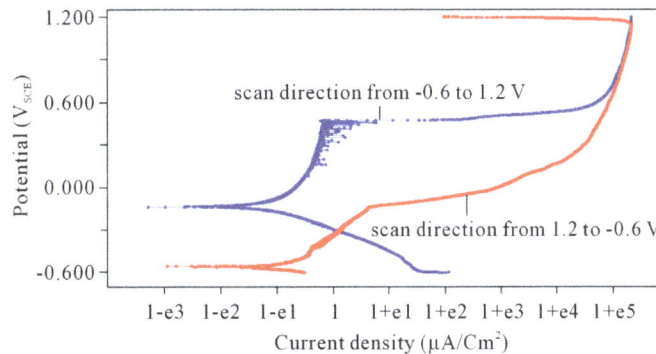

Figure 9. Effect of potential scanning direction on the potentiodynamic polarization curve of 316L SS sample in 3.5% NaCl at room temperature (25°C).

- The crevice corrosion occurred in the intermediate area under the crevice in agreement with *IR* theory.

Acknowledgements

This work was financially supported by KACST, Saudi Arabia. The authors wish to thank Prof. Hamad A. Al-Turaif.

References

[1] Nicastro, D.H. (1997) Failure Mechanisms in Building Construction. ACSE. http://dx.doi.org/10.1061/9780784402832

[2] Al-Zahrani, A.M. and Pickering, H.W. (2005) Electrochim. *Acta*, **50**, 3420.

[3] Cho, K. and Pickering, W. (1990) Demonstration of Crevice Corrosion in Alkaline Solution without Acidification. *Journal of the Electrochemical Society*, **137**, 3313-3314. http://dx.doi.org/10.1149/1.2086211

[4] Fontana, M.G. (1986) Corrosion Engineering. 3rd Edition, McGraw-Hill Book Co., New York.

[5] Sundararajan, T., Akiyama, E. and Tsuzaki, K. (2005) Hydrogen Mapping across a Crevice: Effect of Applied Potential. *Scripta Materialia*, **53**, 1219-1223. http://dx.doi.org/10.1016/j.scriptamat.2005.08.016

[6] Watson, M. and Postlethwaite, J. (1990) Numerical Simulation of Crevice Corrosion of Stainless Steels and Nickel Alloys in Chloride Solutions. *Corrosion*, **46**, 522-530. http://dx.doi.org/10.5006/1.3585142

[7] Cho, K. and Pickering, H.W. (1991) The Role of Chloride Ions in the IR > IR* Criterion for Crevice Corrosion in Iron. *Journal of the Electrochemical Society*, **138**, L56-L58. http://dx.doi.org/10.1149/1.2085386

[8] Pickering, H.W. (2003) Important Early Developments and Current Understanding of the IR Mechanism of Localized Corrosion. *Journal of the Electrochemical Society*, **150**, K1-K13. http://dx.doi.org/10.1149/1.1565142

[9] Abdulsalam, M.I. (2005) Behaviour of Crevice Corrosion in Iron. *Corrosion Science*, **47**, 1336-1351. http://dx.doi.org/10.1016/j.corsci.2004.08.001

[10] Pickering, H.W., Cho, K. and Nystrom, E. (1993) Microscopic and Local Probe Method for Studying Crevice Corrosion and Its Application to Iron and Stainless Steel. *Corrosion Science*, **35**, 775-781. http://dx.doi.org/10.1016/0010-938X(93)90215-3

[11] Xu, Y., Wang, M. and Pickering, H.W. (1993) On Electric Field Induced Breakdown of Passive Films and the Mechanism of Pitting Corrosion. *Journal of the Electrochemical Society*, **140**, 3448-3457. http://dx.doi.org/10.1149/1.2221108

[12] Xu, Y. and Pickering, H.W. (1993) The Initial Potential and Current Distributions of the Crevice Corrosion Process. *Journal of the Electrochemical Society*, **140**, 658-668. http://dx.doi.org/10.1149/1.2056139

[13] ASTM G 5-94 (2009) Standard Reference Test Method for Making Potentiostatic and Potentiodynamic Anodic Polarization Measurements.

[14] Kelly, R.G. and Stewart, K.C. (2001) Combining the Ohmic Drop and Critical Crevice Solution Approaches to Rationalize Intermediate Attack in Crevice Corrosion. In: Luo, J.L. and Rodda, J.R., Eds., *Passivity of Metals and Semiconductors*, The Electrochemical Society, Pennington, 547.

[15] Cho, K., Abdulsalam, M.I. and Pickering, H.W. (1998) The Effect of Electrolyte Properties on the Mechanism of Crevice Corrosion in Pure Iron. *Journal of the Electrochemical Society*, **145**, 1862-1869. http://dx.doi.org/10.1149/1.1838568

[16] Razavi, G.R., Gholami, H., Zirepour, G.R., Zamani, D., Saboktakin, M. and Monajati, H. (2011) Study Corrosion of High-Mn Steels with Mo in 3.5% NaCl Solution. 2011 *International Conference on Advanced Materials Engineering IPCSIT*, **15**, 36.

[17] Mareci, D., Romaş, M., Căilean, A. and Sutiman, D. (2011) Electrochemical Studies of Cobalt-Chromium-Molybdenum Alloys In Artificial Saliva. *Revue Roumaine de Chimie*, **56**, 697-704.

17

Matrix Isolation and Computational Study on the Photolysis of CHCl₂COCl

Nobuaki Tanaka

Department of Environmental Science and Technology, Faculty of Engineering, Shinshu University, Nagano, Japan

Email: ntanaka@shinshu-u.ac.jp

Abstract

UV light photolysis of dichloroacetyl chloride ($CHCl_2COCl$) has been investigated by infrared spectroscopy in cryogenic Ar, Kr, Xe, and O_2 matrices. The formation of $CHCl_3$ and CO was found to be the dominant process over the ketene formation. The C-C bond cleaved products $CHCl_2$ and COCl were also observed. As the number of the chlorine atom substitution to methyl group of acetyl chloride increased, the C-C bond cleaved product yield in the triplet state increased, which can be attributed to an internal heavy-atom effect where the intersystem crossing rate was enhanced.

Keywords

Dichloroacetyl Chloride, Photolysis, Cryogenic Matrix

1. Introduction

Dichloroacetyl chloride ($CHCl_2COCl$) is known to be produced in the oxidation of chlorinated ethenes [1]-[4]. In the chlorine atom initiated oxidation of chlorinated ethenes, relatively high product yields of chlorinated acetyl chloride were reported by Hasson and Smith [5]. Conformations of $CHCl_2COCl$ were studied by vibrational spectroscopy [6]-[9], electron diffraction [10], and theoretical method [11]. Two conformers exist in the $CHCl_2$ internal rotation potential: *syn* conformer having an H-C-C=O dihedral angle of 0° and *gauche* conformer having a non-zero value of the dihedral angle. As for the photolysis of chlorinated acetyl chloride in rare gas matrix, one chlorine atom substitution to methyl group of acetyl chloride opened the additional reaction paths in the T_1 state [12] [13]. Without chlorination the ketene···HCl complex was exclusively produced in the S_0 state after the internal conversion from the S_1 state [14] [15]. In the CCl_3COCl photolysis in an Ar matrix, the C-C bond cleavage was found to be the major reaction path [16].

In the present study, the UV light photolysis of $CHCl_2COCl$ was investigated in cryogenic Ar, Kr, Xe, and O_2

matrices with the aid of the calculation using the B3LYP and MP2 methods to clarify how the two chlorine atom substitutions affect the reaction mechanism.

2. Experimental

Light irradiation was performed using a low pressure mercury arc lamp (HAMAMATSU L937-04, $\lambda > 253.7$ nm). IR spectra were measured in the range 4000 - 700 cm^{-1} with 1.0 cm^{-1} resolution by a SHIMADZU 8300A Fourier transform IR spectrometer with a liquid-nitrogen-cooled MCT detector. Each spectrum was obtained by scanning over 128 times. A closed-cycle helium cryostat (Iwatani M310/CW303) was used to control the temperature of the matrix.

Argon (Nippon Sanso, 99.9999%), krypton (Taiyo Sanso), xenon (Nippon Sanso), and O_2 (Okaya Sanso) were used without further purification. Dichloroacetyl chloride (Wako Pure Chemicals) was used after freeze-pump-thaw cycling at 77 K. Chloroform (Wako Pure Chemicals) was used as an authentic sample for product identification. Samples were deposited on a CsI window at 6 K.

For product identification and energetic consideration, molecular orbital calculation was utilized. Geometry optimizations were performed using the second-order Møller-Plesset theory (MP2) and density functional theory (B3LYP [17] [18], CAM-B3LYP [19], and M06-2X [20]) with the 6-311++G(3df,3pd) and aug-cc-pV(T+d)Z basis sets. Harmonic vibrational frequency calculation was performed to confirm the predicted structures as local minima and to elucidate zero-point vibrational energy corrections (ZPE). The vertical transition energy was calculated at the SAC-CI/D95+(d,p) level based on the structures optimized at the CCSD/D95+(d,p) level. All calculations were performed using Gaussian 09 [21].

3. Results and Discussion

3.1. CHCl$_2$COCl/Ar

A mixture of CHCl$_2$COCl/Ar was deposited on a CsI window with a ratio of CHCl$_2$COCl/Ar = 1/1000. In the infrared spectrum obtained after deposition, two conformers, *gauche-* and *syn*-CHCl$_2$COCl were distinguished by the C=O stretching vibration bands at 1816 and 1784 cm^{-1}, respectively [8] [9]. **Figure 1(a)** shows the infrared difference spectrum obtained upon $\lambda > 253.7$ nm irradiation of a matrix CHCl$_2$COCl/Ar for 60 min. The positive and negative bands indicate the growth and depletion, respectively, during the irradiation period. **Table 1** lists the observed wavenumbers of the growth bands. In the CO stretching region, a strong band observed at 2138 cm^{-1} assignable to the CO stretching continued to grow during the prolonged irradiation period. A band at 2155 cm^{-1} showed growth and decay behavior accompanied with the bands at 1293 and 934 cm^{-1}, whose frequencies are consistent with those of CCl$_2$=C=O observed in the CCl$_3$COCl photolysis in Ar [16]. The bands at 2844 and 2836 cm^{-1} were assigned to the stretching vibration of HCl complexed with the CCl$_2$=C=O. With the different growth rate from those of CO and CCl$_2$=C=O, three bands at 2150, 1297 and 1113 cm^{-1} showed continuous growth which are assignable to the C=O stretching, C=C stretching, and C-H in-plane bending vibrations of CHCl=C=O, respectively [12]. The C-Cl stretching band observed in the photolysis of CH$_2$ClCOCl in Ar was difficult to be discerned due to the overlapping with the strong depletion band of *syn*-CHCl$_2$COCl. A band at 1878 cm^{-1} was assigned to the CO stretching vibration of COCl [22]. Photolysis counterpart of COCl, CHCl$_2$, showed the C-H bending and CCl$_2$ antisymmetric stretching vibrations at 1219 and 898 cm^{-1}, respectively [23]. Prolonged irradiation caused the depletion in intensities of the bands due to CCl$_2$=C=O as shown in **Figure 1(b)**. A band at 1969 cm^{-1} showing an induction period was assigned to the CO stretching vibration of CCO [24]. A band at 766 cm^{-1} grew continuously to be the strongest in the spectrum after 360 min irradiation, which was assigned to the C-Cl stretching vibration of CHCl$_3$. The C-H bending vibration of CHCl$_3$ was observed at 1223 cm^{-1}.

3.2. CHCl$_2$COCl/Kr, CHCl$_2$COCl/Xe

Figure 2 shows the infrared difference spectra obtained upon $\lambda > 253.7$ nm irradiation of the matrix CHCl$_2$COCl/Xe. In Kr, similar results were obtained. In addition to the photolysis products in Ar, the products of Kr$_2$H$^+$ and Xe$_2$H$^+$ were observed in Kr and Xe, respectively [25]. The growth bands at 1814, 1262, 987, and 740 cm^{-1} in Kr and 1809, 1259, 984, and 736 cm^{-1} in Xe were assigned to the C=O stretching, CH bending, C-C stretching, and CCl$_2$ symmetric stretching vibrations of *gauche*-CHCl$_2$COCl, respectively [9]. It is controversial

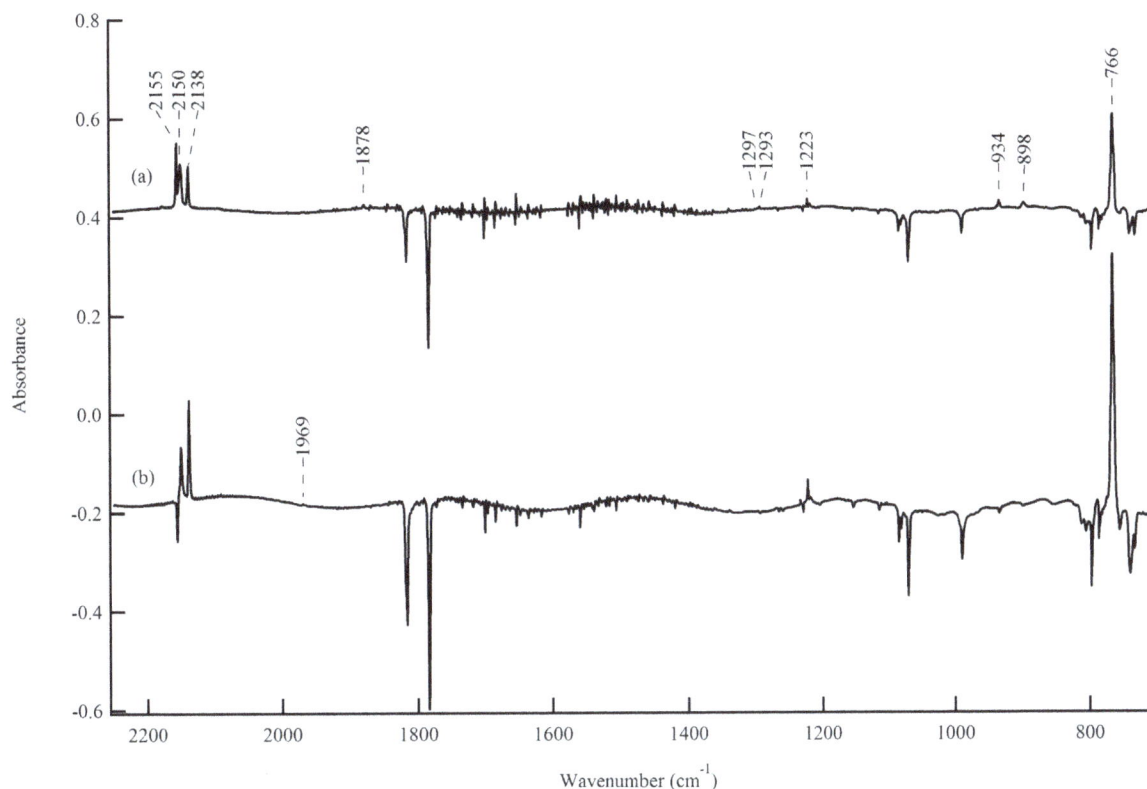

Figure 1. Infrared difference spectra upon $\lambda > 253.7$ nm irradiation of the matrix CHCl$_2$COCl/Ar = 1/1000. (a) 60 - 0 min and (b) 360 - 60 min.

which of the two conformers is more stable [11]. **Table 2** compares the relative electronic energies calculated at the several calculation levels. The barrier height for the conversion from the *syn* to *gauche* rotamer is calculated to be approximately 1200 cm^{-1} in the S$_0$ ground state indicating that the conversion between the *syn* and *gauche* rotamers is not expected to occur at 7 K in the absence of UV irradiation. UV irradiation yielded an increase of the population of the less stable rotamer.

3.3. CHCl$_2$COCl/O$_2$

In order to clarify the route of the ketenes and CHCl$_3$ formation *i.e.* the radical or concerted mechanism, the reactive O$_2$ matrix was used. **Figure 3** shows the infrared difference spectrum obtained after 480 min irradiation of CHCl$_2$COCl. The product bands were assigned by comparison with the spectrum observed in the photolysis of the matrix CCl$_3$COCl/O$_2$. Due to the photolysis in O$_2$ at 253.7 nm, ozone formation is prominent at 1038 cm^{-1} (v_3) [26]. Other O$_3$ absorption bands were observed at 2107 ($v_1 + v_3$) and 1101 cm^{-1} (v_1) [26] [27]. The 2342 and 2276 cm^{-1} bands are assigned to v_3 vibrations of ^{12}CO$_2$ and ^{13}CO$_2$, respectively. The 2037 cm^{-1} band is attributed to CO$_3$ complexed with Cl [16]. A broad band at 1436 cm^{-1} was assigned to ClOO v_1 [28]. In O$_2$, compared with the ratio of CHCl$_3$ or CO absorbance with CHCl$_2$COCl absorbance in **Figure 1**, the CO and CHCl$_3$ formation was depressed. Formation of CHCl$_2$ and ketenes was negligible. Instead major product was found to be CO$_2$ which would be produced via reactions of COCl and CHCl$_2$ with O$_2$. These indicate the reaction predominantly proceed by radical mechanism in the photolysis of CHCl$_2$COCl similar to that of CCl$_3$COCl.

3.4. Reaction Mechanism

Figure 4 shows the integrated absorbance changes of *syn*-CHCl$_2$COCl (1784 cm^{-1}), *gauche*-CHCl$_2$COCl (1816 cm^{-1}), CHCl$_3$ (766 cm^{-1}), CHCl=C=O (2150 cm^{-1}), CO (2138 cm^{-1}), CCl$_2$=C=O (2155 cm^{-1}), and CHCl$_2$ (898 cm^{-1}) observed in Ar, where the IR intensities of these absorption bands were calculated to be 283, 242, 320, 618, 80, 621, and 163 km mol^{-1}, respectively, at the B3LYP/aug-cc-pV(T+d)Z level. The *syn*- and *gauche*-

Table 1. FTIR spectra of the CHCl$_2$COCl photolysis products in the Ar, Kr, Xe, and O$_2$ matrices.

Wavenumber (cm^{-1})				Assignment
Ar	Kr	Xe	O$_2$	
3112				CHCl=C=O
3060				CHCl$_3$
3054				CHCl$_3$
2844/2836	2836/2827		2838	HCl···CCl$_2$=C=O
2809	2809		2807	HCl
2789	2788			HCl
			2342	CO$_2$
			2276	^{13}CO$_2$
2176				
2155	2154/2151	2148	2157	CCl$_2$=C=O
2150	2146	2143	2148	CHCl=C=O
2138	2136	2134	2137	CO
			2107	O$_3$v$_1$ + v$_3$ [a]
2094		2090		^{13}CO
			2037	CO$_3$
1969				CCO [b]
1878	1877	1877		COCl [c]
	1814	1809		gauche-CHCl$_2$COCl
			1436	ClOO [d]
1297	1296	1293		CHCl=C=O
1293	1292	1291		CCl$_2$=C=O
	1262	1259		gauche-CHCl$_2$COCl
1223	1220	1216		CHCl$_3$
1219	1214			CHCl$_2$ [e]
1113				CHCl=C=O
1107				CHCl=C=O
			1101	O$_3$v$_1$ [f]
		1055		
			1037	O$_3$v$_3$ [f]
	987	984		gauche-CHCl$_2$COCl
	965			
		954		Xe$_2$H$^+$ [g]
934	932	932		CCl$_2$=C=O
898	896	894		CHCl$_2$ [e]
864	861	859		^{13}CHCl$_2$
	852			Kr$_2$H$^+$ [g]
		843		Xe$_2$H$^+$ [g]
			839	COCl$_2$
766	764	762		CHCl$_3$
	740	736		gauche-CHCl$_2$COCl

[a]Ref. [27]. [b]Ref. [24]. [c]Ref. [22]. [d]Ref. [28]. [e]Ref. [23]. [f]Ref. [26]. [g]Ref. [25].

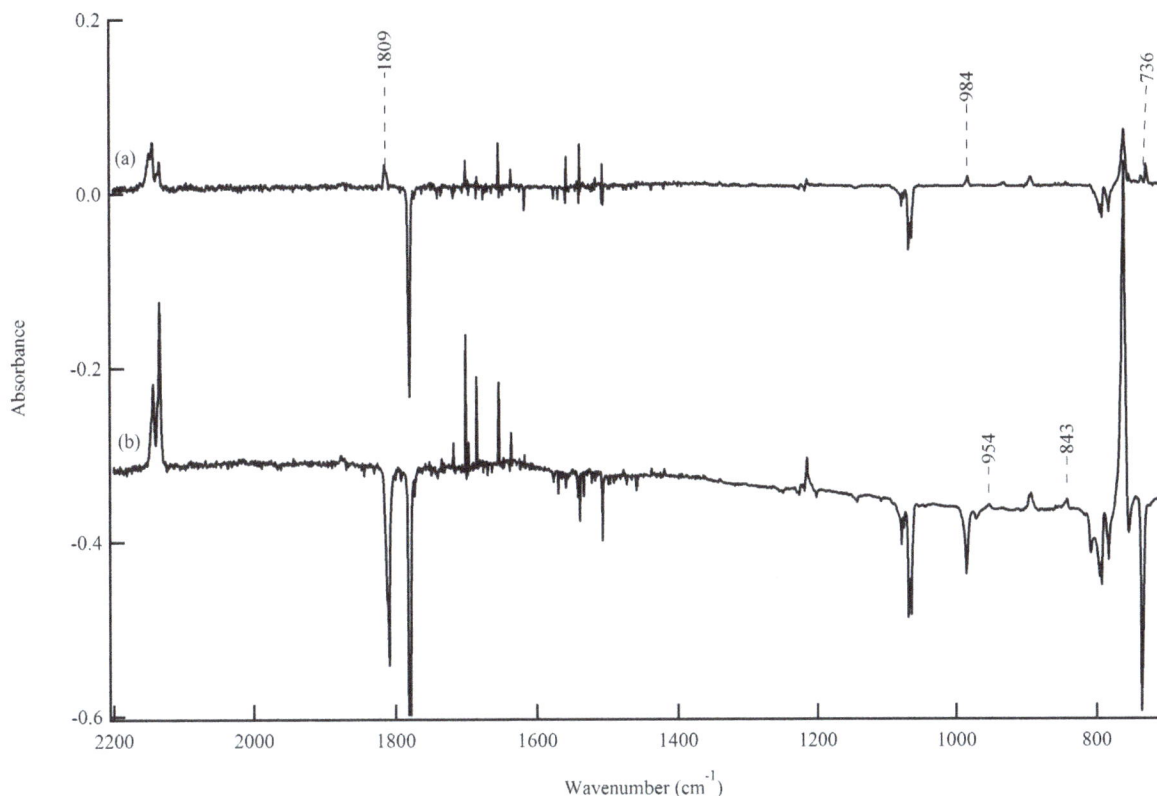

Figure 2. Infrared difference spectra upon $\lambda > 253.7$ nm irradiation of the matrix CHCl$_2$COCl/Xe = 1/1000. (a) 30 - 0 min and (b) 420 - 30 min.

Table 2. Calculated relative electronic energies in cm^{-1} including zero-point vibrational energy corrections.

Method	Difference (*gauche* to *syn*)	Barrier (*syn* to *gauche*)
B3LYP/6-311++G(3df,3pd)	157	1208
B3LYP/aug-cc-pV(T+d)Z	117	1161
CAM-B3LYP/6-311++G(3df,3pd)	180	1167
CAM-B3LYP/aug-cc-pV(T+d)Z	149	1123
M06-2X/6-311++G(3df,3pd)	211	1186
M06-2X/aug-cc-pV(T+d)Z	186	1164
MP2/6-311++G(3df,3pd)	225	1316
MP2/aug-cc-pV(T+d)Z	191	1234
CCSD/aug-cc-pVDZ	132	1239

CHCl$_2$COCl possess the different decay rates. The CCl$_2$=C=O and CHCl$_2$ showed the growth and decay profiles. The relative yield of CHCl$_3$:CHCl=C=O:CCl$_2$=C=O at the irradiation time of 360 min was found to be 1:0.09:0.008. There is an obvious contrast as compared with the relative yield obtained in the photolysis of the matrix CH$_2$COCl/Ar where the ratio of CH$_2$Cl$_2$:CHCl=C=O was found to be 1:7.5 [12].

Even in O$_2$, the ketene species were found to be produced, though the yields decreased greatly. It indicates the majority of the ketene species were formed in the triplet state by the radical mechanism. It seems plausible to explain the dominant radical mechanism in the triplet state by the enhanced intersystem crossing from S$_1$ caused by substitution of the chlorine atoms with methyl hydrogen atoms of acetyl chloride. Therefore, we focus on the triplet surface reaction after intersystem crossing and the ground state reaction after internal conversion. **Figure 5** shows the energy diagram for the CHCl$_2$COCl photolysis initiated by 253.7 nm irradiation. The photon energy at a wavelength of 253.7 nm corresponded to 113 kcal·mol^{-1}. The reaction enthalpies of three elementary reac-

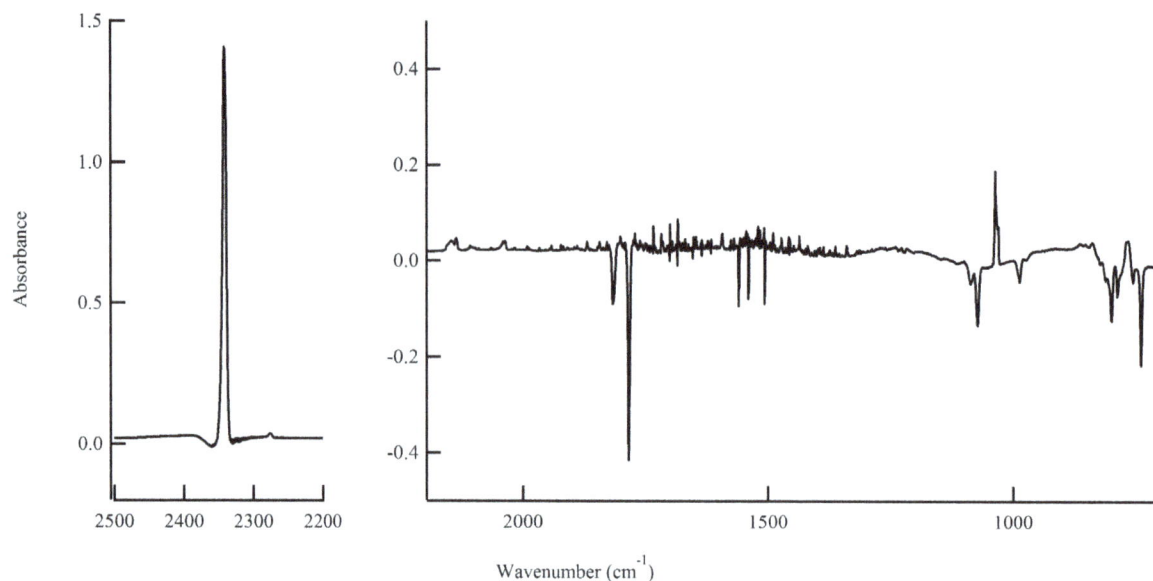

Figure 3. Infrared difference spectrum upon $\lambda > 253.7$ nm irradiation of the matrix $CHCl_2COCl/O_2 = 1/1000$ for 480 min.

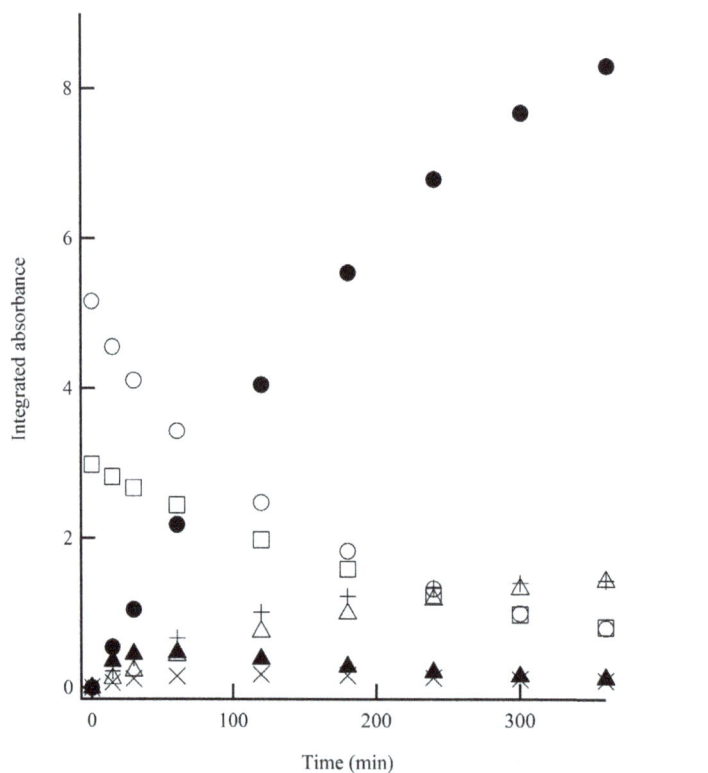

Figure 4. Integrated absorbance changes of ($\circ$) *syn*-$CHCl_2COCl$, ($\square$) *gauche*-$CHCl_2COCl$, ($\bullet$) $CHCl_3$, ($+$) $CHCl=C=O$, ($\triangle$) CO, ($\blacktriangle$) $CCl_2=C=O$, and ($\times$) $CHCl_2$ upon $\lambda > 253.7$ nm irradiation of the matrix $CHCl_2COCl/Ar = 1/1000$.

tions, $C(O)$-Cl, C-C, and $CHCl$-Cl bond cleavages from the T_1 equilibrium states are calculated to be -1.7, -14.8, and -19.5 kcal·mol^{-1} for *syn*-$CHCl_2COCl$ and -2.8, -15.1, and -20.9 kcal·mol^{-1} for *gauche*-$CHCl_2COCl$, respectively, where the reaction barriers are calculated to be 4.4, 5.8, and 2.6 kcal·mol^{-1} for *syn*-$CHCl_2COCl$ and3.7, 5.6, and 2.2 kcal·mol^{-1} for *gauche*-$CHCl_2COCl$, respectively. The C-C dissociation on the T_1 surface possesses the highest barrier, while CHCl-Cl dissociation the lowest barrier. Radical species $CHCl_2$ and $COCl$

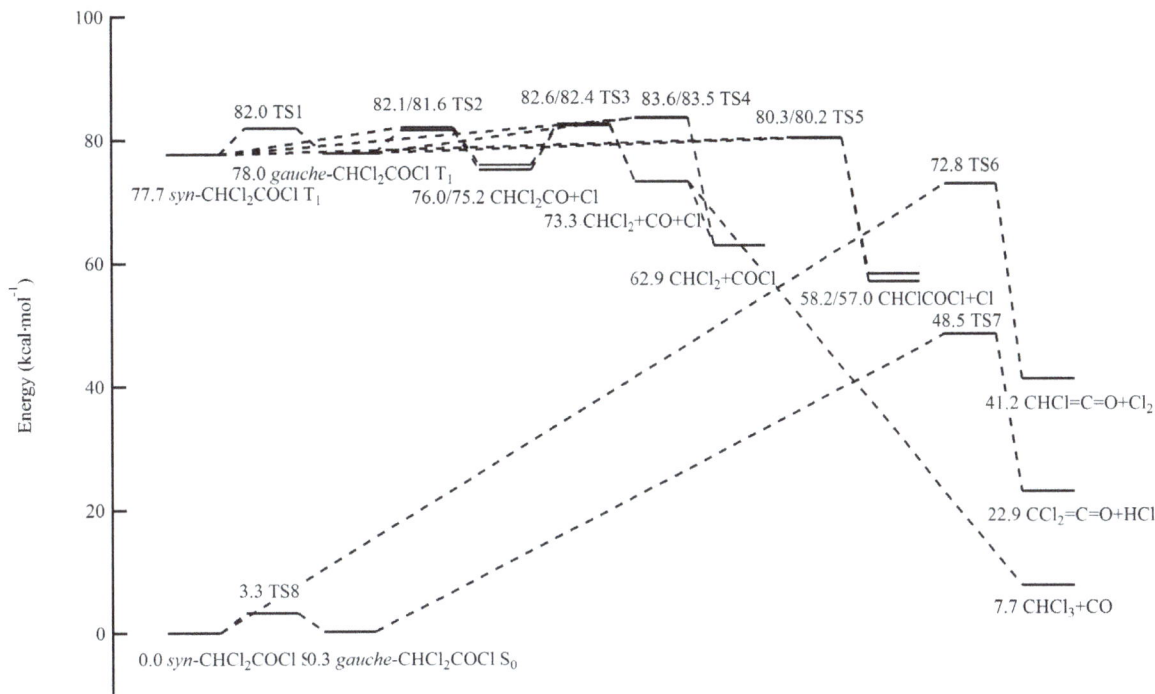

Figure 5. Energy diagram for the CHCl$_2$COCl photolysis.

can be also produced from the dissociation of CHCl$_2$CO into CHCl$_2$ and CO, followed by the recombination of CO with Cl. The CHClCOCl would be further photodissociated. The reaction barrier for the formation of CHCl=C=O + Cl$_2$ in the S$_0$ state was calculated to be higher compared with that for the formation of CCl$_2$=C=O + HCl. The SAC-CI calculation showed the S$_1$ and T$_1$ states of CCl$_2$=C=O possess the mixing characters of $\pi\sigma^*_{C-Cl}$ and πRydberg, −0.87 (HOMO → LUMO) + 0.30 (HOMO → LUMO+3). Upon UV irradiation the C-Cl bond dissociation would occur to form CCO.

For the CHCl$_2$COCl photolysis in the rare gas matrices, the C-C bond cleaved CHCl$_3$, CO, CHCl$_2$, and COCl were dominantly produced similar to the CCl$_3$COCl photolysis and contrary to the CH$_2$ClCOCl photolysis, where ketene formation was a major process. For the CHCl$_2$COCl photolysis in O$_2$, both ketene and CHCl$_3$ formations were greatly depressed, while for CH$_2$ClCOCl, the formation of ketene was slightly depressed. On the basis of these results it will be reasonable to consider that the reaction mechanism drastically changed between CH$_2$ClCOCl and CHCl$_2$COCl from the concerted mechanism in the S$_0$ state to the radical mechanism in the T$_1$ state.

4. Conclusion

UV light photolysis of CHCl$_2$COCl was investigated in cryogenic Ar, Kr, Xe, and O$_2$ matrices. In Ar, Kr, and Xe, the formation of CHCl$_3$ and CO became the dominant process over the ketene formation. The C-C bond cleaved products CHCl$_2$ and COCl were also observed. In Kr and Xe, photoisomerization from *syn*- to *gauche*-CHCl$_2$COCl was observed at the early stage of the irradiation. As the number of the chlorine atom substitution to methyl group of acetyl chloride increased, the C-C bond cleaved product yield in the triplet state increased, which can be attributed to an internal heavy-atom effect where the intersystem crossing rate was enhanced.

Acknowledgements

The author thanks Prof. Tsuneo Fujii and Prof. Hiromasa Nishikiori (Shinshu University) for their helpful discussions.

References

[1]   Haag, W.R., Johnson, M.D. and Scofield, R. (1996) Direct Photolysis of Trichloroethene in Air: Effect of Cocontami-

nants, Toxicity of Products, and Hydrothermal Treatment of Products. *Environmental Science & Technology*, **30**, 414-421. http://dx.doi.org/10.1021/es950047y

[2]   Oki, K., Tsuchida, S., Nishikiori, H., Tanaka, N. and Fujii, T. (2003) Photocatalytic Degradation of Chlorinated Ethenes. *International Journal of Photoenergy*, **5**, 11-15. http://dx.doi.org/10.1155/S1110662X03000059

[3]   Zuo, G.M., Cheng, Z.X., Xu, M. and Qiu, X.Q. (2003) Study on the Gas-Phase Photolytic and Photocatalytic Oxidation of Trichloroethylene. *Journal of Photochemistry and Photobiology A—Chemistry*, **161**, 51-56. http://dx.doi.org/10.1016/S1010-6030(03)00271-5

[4]   Wiltshire, K.S., Almond, M.J. and Mitchell, P.C.H. (2004) Reactions of Hydroxyl Radicals with Trichloroethene and Tetrachloroethene in Argon Matrices at 12 K. *Physical Chemistry Chemical Physics*, **6**, 58-63. http://dx.doi.org/10.1039/b310495h

[5]   Hasson, A.S. and Smith, I.W.M. (1999) Chlorine Atom Initiated Oxidation of Chlorinated Ethenes: Results for 1,1-Dichloroethene ($H_2C=CCl_2$), 1,2-Dichloroethene ($HClC=CClH$), Trichloroethene ($HClC=CCl_2$), and Tetrachloroethene ($Cl_2C=CCl_2$). *Journal of Physical Chemistry A*, **103**, 2031-2043. http://dx.doi.org/10.1021/jp983583w

[6]   Miyake, A., Nakagawa, I., Miyazawa, T., Ichishima, I., Shimanouchi, T. and Mizushima, S. (1958) Infra-Red and Raman Spectra of Dichloroacetyl Chloride in Relation to Rotational Isomerism. *Spectrochimica Acta*, **13**, 161-167. http://dx.doi.org/10.1016/0371-1951(58)80073-9

[7]   Woodward, A.J. and Jonathan, N. (1970) Rotational Isomerism in Dichloroacetyl Halides. *Journal of Physical Chemistry*, **74**, 798-805. http://dx.doi.org/10.1021/j100699a022

[8]   Fausto, R. and Teixeira-Dias, J.J.C. (1986) Conformational and Vibrational Spectroscopic Analusis of $CHCl_2COX$ and CCl_3COX (X=Cl, OH, OCH_3). *Journal of Molecular Structure*, **144**, 241-263. http://dx.doi.org/10.1016/0022-2860(86)85004-9

[9]   Durig, J.R., Bergana, M.M. and Phan, H.V. (1991) Conformational Stability, Barriers to Internal Rotation, Abinitio Calculations and Vibrational Assignment of Dichloroacetyl Chloride. *Journal of Molecular Structure*, **242**, 179-205. http://dx.doi.org/10.1016/0022-2860(91)87135-5

[10] Shen, Q., Hilderbrandt, R.L. and Hagen, K. (1980) The Structure and Conformation of Dichloroacetyl Chloride. *Journal of Molecular Structure*, **71**, 161-169. http://dx.doi.org/10.1016/0022-2860(81)85113-7

[11] Soifer, G.B. and Feshin, V.P. (2006) Molecular Structure and Conformational Transitions of Dichloroacetylchloride. *Journal of Structural Chemistry*, **47**, 371-374. http://dx.doi.org/10.1007/s10947-006-0309-5

[12] Tanaka, N. and Nakata, M. (2014) Matrix Isolation and Theoretical Study on the Photolysis of $CH_2ClCOCl$. *International Research Journal of Pure and Applied Chemistry*, **4**, 762-772. http://dx.doi.org/10.9734/IRJPAC/2014/12002

[13] Davidovics, G., Monnier, M. and Allouche, A. (1991) FT-IR Spectral Data and *ab Initio* Calculations for Haloketenes. *Chemical Physics*, **150**, 395-403. http://dx.doi.org/10.1016/0301-0104(91)87112-9

[14] Kogure, N., Ono, T., Suzuki, E. and Watari, F. (1993) Photolysis of Matrix-Isolated Acetyl Chloride and Infrared Spectrum of the 1:1 Molecular Complex of Hydrogen Chloride with Ketene in Solid Argon. *Journal of Molecular Structure*, **296**, 1-4. http://dx.doi.org/10.1016/0022-2860(93)80111-8

[15] Rowland, B. and Hess, W.P. (1997) UV Photochemistry of Thin Film and Matrix-Isolated Acetyl Chloride by Polarized FTIR. *Journal of Physical Chemistry A*, **101**, 8049-8056. http://dx.doi.org/10.1021/jp971980l

[16] Tamezane, T., Tanaka, N., Nishikiori, H. and Fujii, T. (2006) Matrix Isolation and Theoretical Study on the Photolysis of Trichloroacetyl Chloride. *Chemical Physics Letters*, **423**, 434-438. http://dx.doi.org/10.1016/j.cplett.2006.04.031

[17] Becke, A.D. (1993) Density-Functional Thermochemistry. III. The Role of Exact Exchange. *Journal of Chemical Physics*, **98**, 5648-5652. http://dx.doi.org/10.1063/1.464913

[18] Lee, C., Yang, W. and Parr, R.G. (1988) Development of the Colle-Salvetti Correlation-Energy Formula into a Functional of the Electron Density. *Physical Review B*, **37**, 785-789. http://dx.doi.org/10.1103/PhysRevB.37.785

[19] Yanai, T., Tew, D.P. and Handy, N.C. (2004) A New Hybrid Exchange-Correlation Functional Using the Coulomb-Attenuating Method (CAM-B3LYP). *Chemical Physics Letters*, **393**, 51-57. http://dx.doi.org/10.1016/j.cplett.2004.06.011

[20] Truhlar, D.G. and Zhao, Y. (2008) The M06 Suite of Density Functionals for Main Group Thermochemistry, Thermochemical Kinetics, Noncovalent Interactions, Excited States, and Transition Elements: Two New Functionals and Systematic Testing of Four M06-Class Functionals and 12 Other Functionals. *Theoretical Chemistry Accounts*, **120**, 215-241. http://dx.doi.org/10.1007/s00214-007-0310-x

[21] Frisch, M.J., Trucks, G.W., Schlegel, H.B., Scuseria, G.E., Robb, M.A., Cheeseman, J.R., Scalmani, G., Barone, V., Mennucci, B., Petersson, G.A., Nakatsuji, H., Caricato, M., Li, X., Hratchian, H.P., Izmaylov, A.F., Bloino, J., Zheng, G., Sonnenberg, J.L., Hada, M., Ehara, M., Toyota, K., Fukuda, R., Hasegawa, J., Ishida, M., Nakajima, T., Honda, Y., Kitao, O., Nakai, H., Vreven, T., Montgomery, J.A., Peralta, J.E., Ogliaro, F., Bearpark, M., Heyd, J.J., Brothers, E.,

Kudin, K.N., Staroverov, V.N., Kobayashi, R., Normand, J., Raghavachari, K., Rendell, A., Burant, J.C., Iyengar, S.S., Tomasi, J., Cossi, M., Rega, N., Millam, N.J., Klene, M., Knox, J.E., Cross, J.B., Bakken, V., Adamo, C., Jaramillo, J., Gomperts, R., Stratmann, R.E., Yazyev, O., Austin, A.J., Cammi, R., Pomelli, C., Ochterski, J.W., Martin, R.L., Morokuma, K., Zakrzewski, V.G., Voth, G.A., Salvador, P., Dannenberg, J.J., Dapprich, S., Daniels, A.D., Farkas, Ö., Foresman, J.B., Ortiz, J.V., Cioslowski, J. and Fox, D.J. (2010) Gaussian 09, Revision B.01. Gaussian, Inc., Wallingford.

[22] Jacox, M.E. and Milligan, D.E. (1965) Matrix Isolation Study of the Reaction of Cl Atoms with CO. The Infrared Spectrum of the Free Radical ClCO. *Journal of Chemical Physics*, **43**, 866-870. http://dx.doi.org/10.1063/1.1696861

[23] Granville, T. and Andrews, L. (1969) Matrix Infrared Spectrum and Bonding in the Dichloromethyl Radical. *Journal of Chemical Physics*, **50**, 4235-4245. http://dx.doi.org/10.1063/1.1670888

[24] Jacox, M.E., Milligan, D.E., Moll, N.G. and Thompson, W.E. (1965) Matrix-Isolation Infrared Spectrum of the Free Radical CCO. *Journal of Chemical Physics*, **43**, 3734-3746. http://dx.doi.org/10.1063/1.1696543

[25] Kunttu, H.M. and Seetula, J.A. (1994) Photogeneration of Ionic Species in Ar, Kr and Xe Matrices Doped with HCl, HBr and HI. *Chemical Physics*, **189**, 273-292. http://dx.doi.org/10.1016/0301-0104(94)00273-8

[26] Schriver-Mazzuoli, L., de Saxcé, A., Lugez, C., Camy-Peyret, C. and Schriver, A. (1995) Ozone Generation through Photolysis of an Oxygen Matrix at 11 K: Fourier Transform Infrared Spectroscopy Identification of the O$\cdots$O$_3$ Complex and Isotopic Studies. *Journal of Chemical Physics*, **102**, 690-701. http://dx.doi.org/10.1063/1.469181

[27] Schriver-Mazzuoli, L., Schriver, A., Lugez, C., Perrin, A., Camy-Peyret, C. and Flaud, J.M. (1996) Vibrational Spectra of the ^{16}O/^{17}O/^{18}O Substituted Ozone Molecule Isolated in Matrices. *Journal of Molecular Spectroscopy*, **176**, 85-94. http://dx.doi.org/10.1006/jmsp.1996.0064

[28] Johnsson, K., Engdahl, A. and Nelander, B. (1993) A Matrix-Isolation Study of the ClOO Radical. *Journal of Physical Chemistry*, **97**, 9603-9606. http://dx.doi.org/10.1021/j100140a013

18

Comparative Isotherms Studies on Adsorptive Removal of Congo Red from Wastewater by Watermelon Rinds and Neem-Tree Leaves

M. B. Ibrahim[1], S. Sani[2]

[1]Department of Pure and Industrial Chemistry, Bayero University, Kano, Nigeria
[2]Department of Applied Chemistry, Federal University, Dutsin-Ma, Nigeria
Email: mbibrahim.chm@buk.edu.ng, sadiqsani123@gmail.com

Abstract

Equilibrium adsorption studies for detoxification of Congo Red (CR) dye from single component model wastewater by powdered watermelon rinds and neem leaves adsorbents were carried out with the view to test the applicability of the adsorption process to Langmuir, Freundlich, Temkin, Dubinin-Radushkevich and Harkins-Jura isotherm models. The values of correlation coefficient, R^2 (0.9359 - 0.9998), showed that all the experimental data fitted the linear plots of the tested isotherm models. Dubinin-Radushkevich's monolayer maximum adsorption capacity q_D (20.72 - 26.06 mg/g) is better than Langmuir's q_m (18.62 - 24.75 mg/g) for both adsorbents with the capacities higher for adsorption on watermelon rind than on neem leaves. Values of Langmuir separation factor (R_L) suggest unfavourable adsorption processes (*i.e.* chemisorption) of the dye on both the adsorbents, while Freundlich constant (n_F) indicates unfavourable process only for CR adsorption onto neem leaves. The Dubinin-Radushkevich's mean free energy of adsorption, E (0.29 - 0.32 kJ/mol), suggests physical adsorption processes. Values for Temkin's heat of adsorption, b_T (-0.95 to 0.74 kJ/mol), also show physical adsorption process.

Keywords

Adsorption Isotherms, Congo Red, Neem Leaves, Watermelon Rinds

1. Introduction

Water is an essential necessity for human survival whose global demand doubles every 21 years and its scarcity affects 40% of the world population (about 1.2 billion) projected to reach 2.7 billion by 2025 with water borne diseases claiming annual death rate of 5 to 10 million [1]. About 71% of the earth surface is occupied by water of which only about 0.05% is accessible for human consumption while the bulk of the remaining comprises of the inaccessible seawater, groundwater, swamps and frozen polar ice caps [2]. The scarcity of water is due to rapid population growth, increased industrialization and decreased amounts of rainfall in the previous decades [3]. More so, water pollution by untreated synthetic dye effluents released from industries has been identified as one of the consequences of worsening situation of water scarcity in the society.

Dyes are complex chemical substances that bear stable aromatic rings synthesized to impart strong and persistent colour that does not degrade on exposure to light [4] [5]. Although natural dyes are still in rare use, almost all dyes in use today are synthetic with annual production of over 7×10^5 tonnes of which azo dyes account for 60% - 70% [6]. About 10% - 15% of these dyes are discharged as untreated effluents during the dyeing process [7] [8]. The untreated effluent discharged from textile, cosmetics, pulp and paper, paint, pharmaceutical, food, carpet and printing industries is highly coloured due to large amounts of unfixed dyes that remained during colouring and washing [9].

Untreated dye effluents are toxic and non-biodegradable environmental pollutants that prevent re-establishment of microbial populations, degrade water quality permanently, cause allergy, dermatitis, cancer, skin irritation, dysfunction of kidneys, liver and reproductive system in humans [10] [11]. In other words, it could leach into and pollute surface and ground waters used for drinking; affect the photosynthesis of aquatic plants by hindering penetration of light into the water; and may cause suffocation of aquatic flora and fauna due to anaerobic degradation of azo dyes into highly lethal substances [12]-[14].

Thus, to overcome the challenges of water scarcity and safe exploitation that attracted much attention from government organizations and water industries globally, it has become necessary to develop cost-effective technologies for water/wastewater treatment, reclamation, recycling and reuse for sustainable industrial and agricultural development.

Traditional and conventional techniques usually employed for the treatment of dye wastewater consist of biological, physical and chemical methods most of which are becoming inadequate due to large variability of the composition of dye wastewaters. In other words, most of these techniques are often ineffective, expensive, complicated, time-consuming and require highly-skilled personnel especially when the levels of dissolved dye adsorbates are in the range of 1 - 100 mg/L [15]. Similarly, adsorption methods using conventional adsorbents (e.g. activated carbons) poses the disadvantages of sludge disposal problems and high costs of operation, maintenance, adsorbent purchase and sludge regeneration [16] [17].

However, proposed adsorption techniques using living and dead biomass as adsorbents are relatively cheaper, environmentally friendlier and more efficient than conventional adsorbents for the removal of dyes from wastewater even at trace level. Non-conventional adsorption method utilizes the ability of agricultural waste materials to accumulate dye pollutants from waste streams by purely physico-chemicals pathways of uptake. Their adsorption capacities are studied using adsorption equilibrium isotherms under such optimized conditions as agitation time, adsorbent dose, adsorbents particle size, initial dye concentration and initial pH of dye [17]-[19].

Agricultural solid wastes and by-products are renewable resources available in large quantities with little or no value in most countries. Their utilization as good source of raw materials for dye removal poses the dual advantages of effective wastewater treatment and waste management. They usually have high molecular weight due to the presence of lignin, cellulose and hemicelluloses components [20]. Many agricultural waste adsorbents (rice husks, corncob, coir-pith, plum kernels, bagasse pith, nut shells, fruit peels, leaf powders, spent tea leaves, fruit shells, seed husk, sawdust, hyacinth root, etc.) were reported as cost-effective alternative low cost adsorbents removal of dyes from wastewater in the recent past decades [21].

2. Materials and Methods

2.1. Adsorbents Preparation

Neem tree (*Azadirachta indica*) leaves were collected from twigs of a number of matured tall neem trees within and near the main campus of Umaru Musa Yar'adua University, Katsina. The samples were thoroughly washed with tap-water, rinsed copiously with distilled water to remove dust and any other soluble substances. The leaves were allowed to air dry under shade at room temperature until they become crisp. The dried leaves samples were then pulverized with a mechanical grinder into a powdered; and then dried overnight for 16 hours in an oven at a temperature of 65°C. The oven-dried neem-tree leaves powder (NLP) samples were sieved to the working sizes of 75 - 300 μm using electronic sieve shaker and the fractions preserved in separately labelled air-tight plastic containers according to their particle sizes. Similar procedure was carried out on sliced pieces of fresh watermelon (*Citrullus lanatus*) rinds samples, collected from local fruit vendors at Kofar Kaura and Central Market in Katsina Metropolis, with the powdered fractions (WRP) separately preserved in plastic containers [21] [22].

The analytical grade Congo red dye supplied by BDH Laboratory was used as received. Stock solution of the dye was prepared by dissolving 1 g solute in 1000 cm^3 volumetric flasks to make 1000 mg/L of the dye solution [23]. Model and working calibration standards were prepared by serial dilution of the stock solution.

2.2. Batch Adsorption Technique

Experiments on the adsorption of Congo red by the adsorbents (WRP and NLP) were carried out by batch method and the influence of various parameters such as contact time (5 - 240 min), adsorbent dosage (100 - 500 mg), particle size (≤ 75 μm, ≤ 150 μm, ≤ 250 μm, ≤ 300 μm and > 300 μm), initial dye concentration (5 - 300 mg/L) and initial dye pH (2 - 12) were studied at constant agitation speed of 300 rpm and room temperature (25°C) in triplicates. The adsorption measurements were conducted by mixing various amounts of adsorbent in 150 cm^3 Erlenmeyer glass flasks containing 50 cm^3 of dye solution of known concentration. The initial pH of the dye solutions were adjusted to the desired values by adding few drops of 0.1 M HCl or 0.1 M NaOH aqueous solutions. The solutions were agitated using orbital shaker for a predetermined time to attain equilibrium after which, the samples were removed and the supernatant solution was separated from the adsorbent by filtration using Whatman No. 41 filter paper, discarding the first few volume (3 - 4 drops) of the filtrate [24]. The filtrates were used for analyses using UV-visible spectrophotometer, reporting each data point as an average value of the triplicates recorded. In each case, the percentage adsorption and substrate's equilibrium adsorption capacity, q_e (mg/g) were evaluated using Equations (1) and (2) respectively.

$$\% \text{ Adsorption} = \left[\frac{C_o - C_e}{C_o} \right] \times 100 \tag{1}$$

$$q_e = \frac{V(C_o - C_e)}{W} \tag{2}$$

where C_o (mg/L) is the initial dye concentration, C_e is the concentration at equilibrium or predetermined time t, V (L) is the volume of dye solution used and W (g) is the weight of the adsorbent. The data obtained were tested against the linear forms of Langmuir, Freundlich, Temkin, Dubinin-Radushkevich (D-R) and Harkins-Jura isotherms, respectively represented as;

$$\frac{C_e}{q_e} = \frac{1}{K_L q_m} + \frac{C_e}{q_m} \tag{3}$$

$$\ln q_e = \ln K_F + n \ln C_e \tag{4}$$

$$q_e = B_T \ln A_T + B_T \ln C_e \tag{5}$$

$$\ln q_e = \ln q_D - B_D \varepsilon^2 \tag{6}$$

$$\frac{1}{q_e^2} = \left(\frac{B_{HJ}}{A_{HJ}} \right) - \left(\frac{1}{A_{HJ}} \right) \log_{10} C_e \tag{7}$$

where C_e is any liquid phase concentration of the dye in equilibrium with the adsorbent, q_e is equilibrium adsorption capacity of the adsorbent, q_m is monolayer capacity, q_D (mg/g) is the theoretical monolayer saturation capacity of adsorbent, K_L is Langmuir adsorption constant, K_F is Freundlich constant for relative adsorption capacity of adsorbent, A_T is the Temkin isotherm equilibrium binding constant (L/g), B_T is the Temkins heat of adsorption, A_{HJ} is Harkins-Jura isotherm parameter which accounts for multilayer adsorption and explains the existence of heterogeneous pore distribution, while B_{HJ} is the isotherm constants [24].

3. Results and Discussion

To have an insight into the adsorption behaviours of CR dye onto watermelon rinds and neem leaves samples and to gain the optimal fitting of theoretical model, the experimental data from batch experiment were analyzed using five two-parameter isotherm equations (Langmuir, Freundlich, Dubinin-Radushkevich (D-R), Temkin and Harkins-Jura), in which linear regression analysis was used to evaluate whether the theoretical models have better or worse fit for the experimental data. The respective parameters of these isotherm models have been enumerated in **Table 1**.

3.1. Langmuir Isotherm

Based on the relationship of adsorption capacity for CR dye adsorption onto the adsorbents and the equilibrium concentrations, the Langmuir adsorption isotherms are modeled and presented in **Figure 1**. According to these

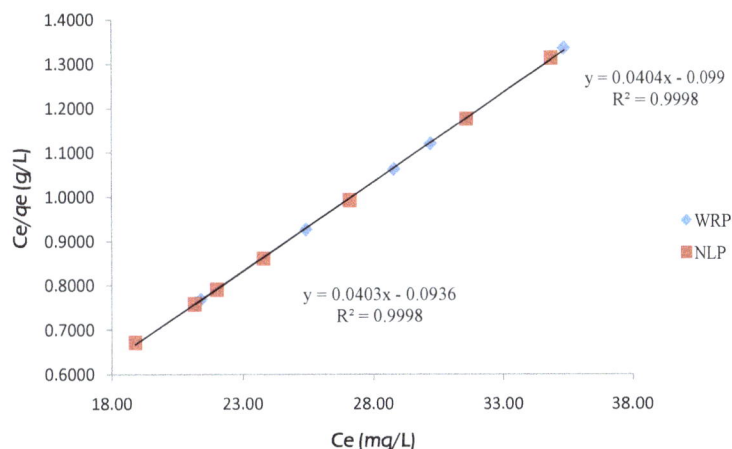

Figure 1. Langmuir isotherm plot for CR adsorption onto WRP and NLP.

Table 1. Isotherm parameters for CR adsorption onto WRP and NLP.

Adsorption Isotherm Models and Parameters		Watermelon Rinds	Neem-Tree Leaves
Langmuir	q_m (mg/g)	24.75	24.81
	K_L	−0.4081	−0.4306
	R_L	−0.0082	−0.0078
	R^2	0.9998	0.9998
Freundlich	$|n_F|$	10.13	3.78
	K_F	37.7204	37.3488
	R^2	0.9935	0.9919
Temkin	b_T (kJ/mol)	−0.9212	0.7374
	B_T	−2.6896	−2.6164
	A_T	1.47×10^{-6}	1.11×10^{-6}
	R^2	0.9947	0.9932
Dubinin-Redushkevich	B_D (mol²/J²)	6×10^{-6}	5×10^{-6}
	E (kJ/mol)	0.2887	0.3162
	q_D (mg/g)	25.94	26.06
	R^2	0.9574	0.9359
Harkins-Jura	A_{HJ}	−1667	−1667
	B_{HJ}	−0.8333	−0.8333
	R^2	0.9908	0.9889

isotherm curves, the Langmuir isotherm parameters are calculated and listed in **Table 1**.

As shown in **Table 1**, CR adsorption on both adsorbents have the same value of linear regression coefficient, R^2 (0.9998), suggesting that the experimental data agreed closely with each other. However, the negative values of R_L and K_L indicates unfavourable adsorption of the dye onto the adsorbents [25].

3.2. Freundlich Isotherm

Based on the relationship of adsorption capacity for CR dye adsorption onto the adsorbents and the equilibrium concentrations, the Freundlich adsorption isotherms are correlated and given in **Figure 2**, while the isotherm parameters are as presented in **Table 1**.

From the **Table 1**, CR adsorption on the adsorbents have a range of values of linear regression coefficient, R^2 (0.9913 - 0.9935), demonstrating that the experimental data fitted well with the Freundlich isotherm equation, third to the Langmuir isotherm. Moreover, it was reported that the Freundlich isotherm constant can be used to explore the favourability of adsorption process. The adsorption process is said to be favourable when the value of n_F satisfies the condition $|1| < n_F < |10|$, otherwise it is unfavourable. While the n_F values in **Table 1** for adsorption of CR on watermelon are situated outside the range of 1 - 10 indicating unfavourable adsorption process, the values for CR adsorption on neem leaves are within the range of 1 - 10, demonstrating favourable adsorption process [26] [27].

3.3. Temkin Isotherm

Figure 3 illustrates the Temkin isotherm model for the dye adsorption onto the adsorbents from which the relevant isotherm parameters are listed in **Table 1**. It can be discovered in **Table 1** that the values of R^2 are positioned within 0.9932 - 0.9947, which gave a close fit to the CR adsorption on watermelon rind and neem leaves samples, values next only to Langmuir's model. This outcome suggests that the experimental data fitted better with the Temkin isotherm model [28]. Furthermore, it can also be discovered in **Table 1** that the adsorption heat of CR adsorption on watermelon rind and neem leaves samples was restricted within −0.92 to 0.74 kJ/mol.

3.4. Dubinin-Redushkevich Isotherm

Making the linear plot according to adsorption capacity for CR dye adsorption onto the adsorbents and the equilibrium concentrations, the Dubinin-Radushkevich (D-R) adsorption isotherms (**Figure 4**) was obtained. Corresponding to which, the isotherm parameters were calculated as in **Table 1**.

The values of linear regression coefficient (R^2) are in the range of 0.9359 - 0.9574, revealing that the experimental data fitted well with the Dubinin-Radushkevich (D-R) isotherm model. Moreover, it is reported that when the value of E is below 8 kJ/mol, the adsorption process can be considered as the physical adsorption. In contrast, if the value of E is located in the range of 8 - 16 kJ/mol, it is the chemical adsorption. From **Table 1**, it can be observed that the obtained values of mean free energy, E, are limited within the range of 0.29 - 0.32

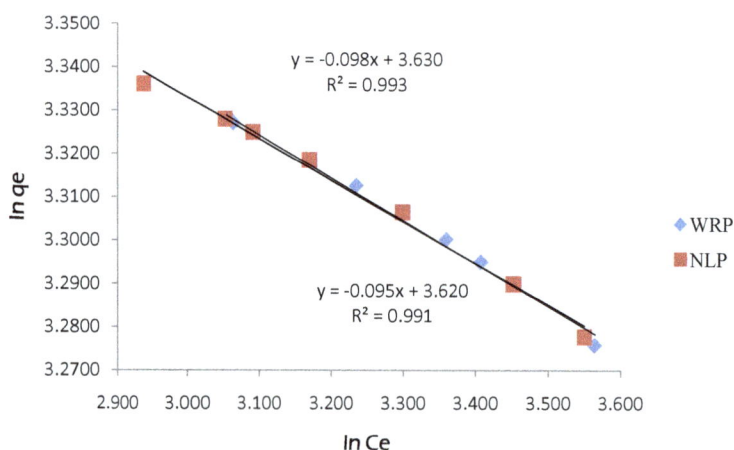

Figure 2. Freundlich isotherm plot for CR Adsorption onto WRP and NLP.

kJ/mol. Based on these data, it can thus be concluded that the effect of physical adsorption will play a dominating role in the adsorption process of CR dye adsorption onto the adsorbents [25] [29].

3.5. Harkins-Jura Isotherm

The Harkins-Jura isotherm models for CR adsorption onto watermelon rinds and neem-tree leaves samples are presented in **Figure 5** and the relevant isotherm parameters (**Table 1**) shows that the values of R^2 are located in the range of 0.9889 - 0.9908, which indicate a better fits to the CR adsorption onto watermelon rinds and neem-

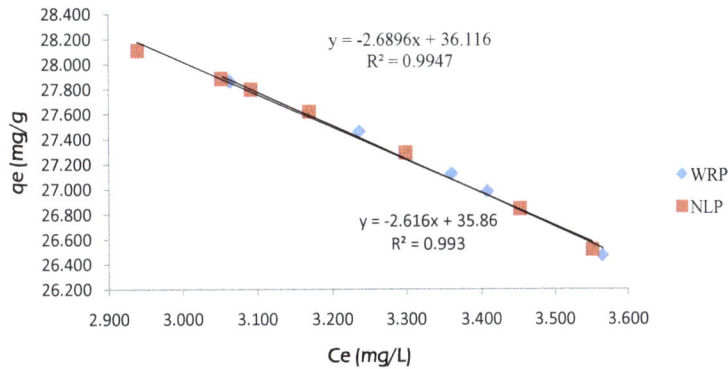

Figure 3. Temkin isotherm plot for CR adsorption onto WRP and NLP.

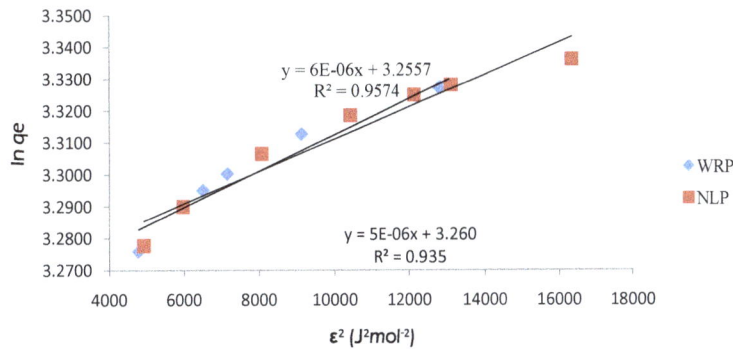

Figure 4. Dubinin-Redushkevich isotherm plot for CR adsorption onto WRP and NLP.

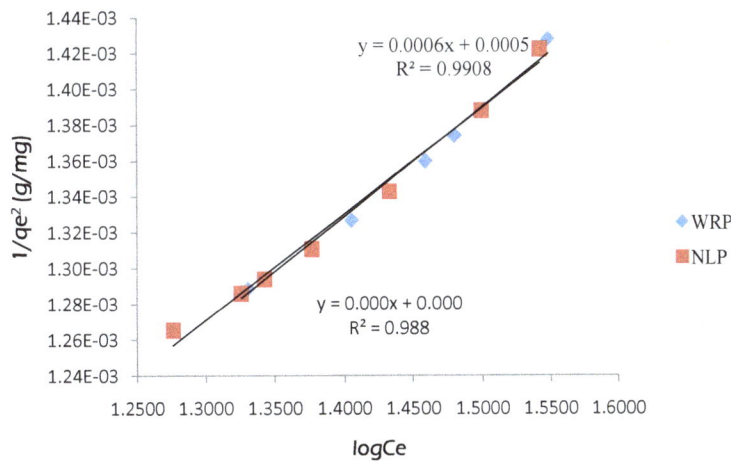

Figure 5. Harkins-Jura isotherm plot for CR adsorption onto WRP and NLP.

tree leaves samples. This result reveals that CR adsorption onto watermelon rinds and neem-tree leaves samples is in support of the multilayer adsorption rule [30] [31].

4. Conclusion

Adsorption of CR dye onto watermelon rinds- and neem leaves-derived adsorbents has been modeled using five two-parameter isotherm equations. The results achieved suggested that all the experimental data followed the tested isotherm models and D-R, Temkin and Harkins-Jura model suggest that the dye is removed from aqueous medium by a multilayer adsorption process.

References

[1]   Ahuja, S. (2009) Handbook of Water Purity and Quality. IWA Publishing, Great Britain, 1-7. http://dx.doi.org/10.1016/B978-0-12-374192-9.00001-7

[2]   Rijberman, F.R. (2006) Water Scarcity: Fact or Fiction. *Agricultural Water Management*, **80**, 5-22. http://dx.doi.org/10.1016/j.agwat.2005.07.001

[3]   Lee, B., Preston, F., Kooroshy, J., Bailey, R. and Lahn, G. (2012) Resources Futures—A Chatham House Report. The Royal Institute of International Affairs, London.

[4]   Damarji, B., Khalaf, H., Duclaux, L. and David, B. (2009) Preparation of TiO_2-Pillard Montmorillonite as Photocatalyst Part II: Photocatalytic Degradation of a Textile Azo Dye. *Applied Clay Science*.

[5]   Unuabonah, E.I., Adebowale, K.O. and Dwodu, F.A. (2008) Equilibrium, Kinetics and Sober Design Studies on the Adsorption of Aniline Blue Dye by Sodium Tetraborate-Modified Kaolinite Clay Adsorbent. *Journal of Hazardous Materials*, **157**, 397-409. http://dx.doi.org/10.1016/j.jhazmat.2008.01.047

[6]   Zohra, B., Aicha, K., Fatima, S., Nourredin, B. and Zoubir, D. (2008) Adsorption of Direct Red 2 on Bentonite Modified by Cetyltrimethylammonium Bromide. *Chemical Engineering Journal*, **136**, 295-305. http://dx.doi.org/10.1016/j.cej.2007.03.086

[7]   Gómez, V., Larrechi, M.S. and Callao, M.P. (2007) Kinetic and Adsorption Study of Acid Dye Removal Using Activated Carbon. *Chemosphere*, **69**, 1151-1158. http://dx.doi.org/10.1016/j.chemosphere.2007.03.076

[8]   Gupta, V.K., Kumar, R., Nayak, A., Saleh, T.A. and Barakat, M.A. (2013) Adsorptive Removal of Dyes from Aqueous Solution onto Carbon Nanotubes: A Review. *Advances in Colloid and Interface Science*, **193-194**, 24-34. http://dx.doi.org/10.1016/j.cis.2013.03.003

[9]   Santos, S.C.R. and Boaventura, R.A.R. (2008) Adsorption Modelling of Textile Dyes by Sepiolite. *Applied Clay Science*, **42**, 137-145. http://dx.doi.org/10.1016/j.clay.2008.01.002

[10]   Bulut, E., Ozcar, M. and Sengil, I.A. (2008) Equilibrium and Kinetic Data and Process Design for Absorption of Congo Red on Bentonite. *Journal of Hazardous Materials*, **154**, 613-622. http://dx.doi.org/10.1016/j.jhazmat.2007.10.071

[11]   Özcan, A.S. and Özcan, A. (2004) Adsorption of Acid Dyes from Aqueous Solution onto Acid-Activated Bentonite. *Journal of Colloidal and Interface Sciences*, **276**, 39-46. http://dx.doi.org/10.1016/j.jcis.2004.03.043

[12]   Acemioğlu, B. (2004) Adsorption of Congo Red from Aqueous Solution onto Calcium-Rich Fly Ash. *Journal of Colloid and Interface Science*, **274**, 371-379. http://dx.doi.org/10.1016/j.jcis.2004.03.019

[13]   Purkait, M.K., Maiti, A., DasGupta, S. and De, S. (2007) Removal of Congo Red Using Activated Carbon and Its Regeneration. *Journal of Hazardous Materials*, **145**, 289-295. http://dx.doi.org/10.1016/j.jhazmat.2006.11.021

[14]   Song, Y.L., Li, J.T. and Chen H. (2009) Degradation of C.I. Acid Red 88 Aqueous Solution by Combination of Fenton's Reagent and Ultrasound Irradiation. *Journal of Chemical Technology and Biotechnology*, **84**, 578-583. http://dx.doi.org/10.1002/jctb.2083

[15]   Liu, C., Ngo, H.H., Guo, W. and Tung, K. (2012) Optimal Conditions for Preparation of Banana Peels, Sugarcane Bagasse and Watermelon Rind in Removing Copper from Water. *Bioresource Technology*, **119**, 349-354. http://dx.doi.org/10.1016/j.biortech.2012.06.004

[16]   Yadla, S.V., Sridevi, V. and Lakshmi, M.V.V.C. (2012) A Review on Adsorption of Heavy Metals from Aqueous Solution. *Journal of Chemical, Biological and Physical Sciences*, **2**, 1585-1593.

[17]   Gönen, F. and Serin, D.S. (2012) Adsorption Study on Orange Peel: Removal of Ni (II) Ions from Aqueous Solution. *African Journal of Biotechnology*, **11**, 1250-1258.

[18]   Neomagus, H.W.J.P. (2005) Bio-Polymeric Heavy Metal Adsorbing Materials for Industrial Wastewater Treatment (WRC Report No. 1072/1/05).

[19]   Nită, I., Iorgulescu, M., Spiroiu, M.F., Ghiurea, M., Petcu, C. and Cinteză, O. (2007) The Adsorption of Heavy Metal

Ions on Porous Calcium Alginate Microparticles. *Analele Universităt din Bucureşti-Chimie, Anul XVI (serie nouă)*, **1**, 59-67.

[20] Salleh, M.A.M., Mahmoud, D.K., Karim, W.A. and Idris, A. (2011) Cationic and Anionic Dye Adsorption by Agricultural Solid Wastes: A Comprehensive Review. *Desalination*, **280**, 1-13. http://dx.doi.org/10.1016/j.desal.2011.07.019

[21] Sharma, N., Tiwari, D.P. and Singh, S.K. (2012) Decolourization of Synthetic Dyes by Agricultural Waste: A Review. *International Journal of Scientific & Engineering Research*, **3**, 1-10. http://dx.doi.org/10.15373/22778179/MARCH2014/139

[22] Gopalakrishnan, K., Manivannan, V. and Jeyadoss, T. (2010) Comparative Study of Zn(II), Cu(II) and Cr(VI) from Textile Dye Effluent Using Sawdust and Neem Leaves Powder. *E-Journal of Chemistry*, **7**, S504-S510. http://dx.doi.org/10.1155/2010/506424

[23] Haddadian, Z., Shavandi, M.A., Abidin, Z.Z., Fakhru'l-Razi, A. and Ismail, M.H.S. (2013) Removal of Methyl Orange from Aqueous Solutions Using Dragon Fruit (*Hylacereusundatus*) Foliage. *Chemical Science Transactions*, **2**, 900-910.

[24] Suyamboo, B.K. and Perumal R.S. (2012) Equilibrium, Thermodynamic and Kinetic Studies on Adsorption of a Basic Dye by *Citrullus lanatus* Rind. *Iranica Journal of Energy & Environment*, **3**, 23-34. http://dx.doi.org/10.5829/idosi.ijee.2012.03.01.0130

[25] Dada, A.O., Olalekan, A.P., Olatunya, A.M. and Dada, O. (2012) Langmuir, Freundlich, Temkin and Dubinin-Radushkevich Isotherms Studies of Equilibrium Sorption of Zn^{2+} unto Phosphoric Acid Modified Rice Husk. *IOSR Journal of Applied Chemistry*, **3**, 38-45. http://dx.doi.org/10.9790/5736-0313845

[26] Bhattacharyya, K.G. and Sharma, A. (2004) *Azadirachta indica* Leaf Powder as an Effective Biosorbent for Dyes: A Case Study with Aqueous Congo Red Solutions. *Journal of Environmental Management*, **71**, 217-229. http://dx.doi.org/10.1016/j.jenvman.2004.03.002

[27] Vimonses, V., Lei, S., Jin, B., Chow, C.W.K. and Saint, C. (2009) Kinetic Study and Equilibrium Isotherm Analysis of Congo Red Adsorption by Clay Materials. *Chemical Engineering Journal*, **148**, 354-364. http://dx.doi.org/10.1016/j.cej.2008.09.009

[28] Yaneva, Z.L. and Georgieva, N.V. (2012) Insights into Congo Red Adsorption on Agro-Industrial Materials-Spectral, Equilibrium, Kinetic, Thermodynamic, Dynamic and Desorption Studies. A Review. *International Review of Chemical Engineering*, **4**, 127-146.

[29] Samarghandi, M.R., Hadi, M., Moayedi, S. and Askari, F.B. (2009) Two-Parameter Isotherms of Methyl Orange Sorption by Pinecone Derived Activated Carbon. *Iranica Journal of Environmental Health, Science and Engineering*, **6**, 285-294.

[30] Abdullah, N.M., Othaman, R., Abdullah, I., Jon, N. and Baharum, A. (2012) Studies on the Adsorption of Phenol Red Dye Using Silica-Filled enr/pvc Beads. *Journal of Emerging Trends in Engineering and Applied Sciences (JETEAS)*, **3**, 845-850.

[31] Liu, J. and Wang, X. (2013) Novel Silica-Based Hybrid Adsorbents: Lead(II) Adsorption Isotherms. *The Scientific World Journal*, **2013**, Article ID: 897159. http://dx.doi.org/10.1155/2013/897159

19

Thermochemical Parameters of Tetramethylthiourea Adducts of Certain Metal(II) Bromides

Pedro Oliver Dunstan

Instituto de Química, Universidade Estadual de Campinas, Campinas, Brazil
Email: dunstan@iqm.unicamp.br

Abstract

Complexes of the general formula $[MBr_2(TMTU)_n]$ (where M is Mn, Fe, Co, Ni, Cu, Zn or Cd; TMTU is Tetramethylthiourea; n is 0.75, 2 or 3) were obtained by the reaction of salts and ligand in solution. The bromides were selected among several other salts because they had thermochemical data in the literature. Properties as capillary melting points; C, H, N, Br and metal contents; TG/DTG and DSC curves; and IR and electronic spectra were determined. The values of several thermodynamic parameters for the complexes were found by solution calorimetry. From them, the standard enthalpies of the metal-sulphur coordinated bonds were calculated. The standard enthalpies of the formation of the gaseous phase adducts also were estimated.

Keywords

Enthalpies of Formation, Thermodynamic, Coordinated Bond Enthalpies, Solution Calorimetry, Dissolution Enthalpies, Tetramethylthiourea Adducts

1. Introduction

Complexes formed by salts of transition and representative elements with thioamides are mentioned in the literature. M. V. Raja *et al.* [1] studied thioamide complexes of aryl bromides. R. Sah *et al.* [2] synthesized and characterized transition Metal(II) complexes of heterocyclic thiomides. H. Ajaz *et al.* [3] got the crystal structure of complexes of Antimony(II) with thioamides. T. Singh *et al.* [4] studied thioamide complexes of ruthenium. S. Nadeem *et al.* [5] studied thioamide complexes of Palladium(II) bromides. I. I. Ozturki *et al.* [6] characterized thioamide complexes of Antimony(III) bromide. J. Sola *et al.* [7] studied thioamide complexes of Silver(I) ion. L. S. Sbirna *et al.* [8] studied several complexes of transition metals with bidentated heterocyclic (N, S) ligands.

S. K. Misra *et al.* [9] studied the complexing behavior of thioamide in several transition metal complexes. V. Muresan [10] prepared new thioamide complexes of transition metals. C. Neagoe [11] *et al.* studied tertiary thioamide complexes of transition metals. F. Zalaru *et al.* [12] synthesized thioamide complexes of Cu(II). P. O. Dunstan *et al.* [13] characterized thioamide complexes of arsenic trihalides. H. O. Desseyn [14] studied chelates of thioamides. S. Neagoe *et al.* [15] studied thioamide complexes of transition metals. C. E. Carraher Jr. *et al.* [16] studied complexes of polythioamides with Palladium(II). B. Singh [17] studied thioamide complexes of Oxovanadium(IV). G. R. Burns *et al.* [18] studied thioamide complexes of Chromium(III). J. M. Bret *et al.* [19] studied thioamide complexes of platinum. V. Muresan *et al.* [20] prepared several thioamide complexes of transition metals. Yu. N. Kukushkin *et al.* [21] studied thioamide complexes of Platinum(I). A. J. Aarts *et al.* [22] studied thioamide complexes of Palladium(II). V. V. Sibirskaya *et al.* [23] prepared thioamide complexes of platinum metals. M. Molina *et al.* [24] studied the interaction of thioamides with metals. J. G. H. Du Preez *et al.* [25] studied thioamide complexes of oxovanadium dichloride.

In this work, complexes formed by tetramethylthiourea with bromides of Manganese(II), Iron(II), Cobalt(II), Nickel(II), Cooper(II), Zinc(II) and Cadmium(II) were studied. Thermodynamic data found in the literature concerning the standard enthalpies of formation of coordinated bonds in this kind of complexes are limited. The knowledge of the thermodynamic properties of these compounds is important to find their applications in catalysis and in the chromatographic separation of metal ions. It could be inferred the affinity order of the metallic ions for stationary-movable chromatographic phases by the knowledge of the formation enthalpies of complexes. Also, the thermochemical parameters can be used in catalysis for finding the more adequate complex to accelerate a given reaction. In this article, calorimetric measurements were made to measure the strength of the metal-sulphur coordinated bonds. Correlations of the thermodynamic properties of the complexes were got. The enthalpies of formation of the complexes derived from the gaseous-phase metal ions, bromide ions and tetramethylthiourea were determined.

2. Materials and Methods

2.1. Reagents

Tetramethylthiourea (RP, Fluka AG Buchs SG) was purified by recrystallization from methanol (MP 75°C - 76°C). The anhydrous Metal(II) bromides used in the synthesis of the complexes were of analytical grade. Solvents were distilled and stocking over Linde 4Å molecular sieves before using.

2.2. Experimental Procedure

The complexes were prepared by the reaction between Metal(II) bromides and tetramethylthiourea in a molar ratio salt/ligand of 1/4 in hot tert-buthyl alcohol solution or in a hot mixture ethanol/chloroform solution. Following, an example of one preparation: 1.00 g of $NiBr_2$ (4.58 mmol) was dissolved in 25 mL of hot tert-butyl alcohol and 2.41 g (18 mmol) of tetramethylthiourea was dissolved in 20 mL of hot tert-butyl alcohol. The later solution was poured into the solution of the salt, slowly and dropwise with stirring. The mixture was refluxed by five hours after which, the solid that formed was filtered and washed with 60 mL of petroleum ether divided in three portions. The compound obtained was maintained in a vacuum over twelve hours. It was stocked in desiccator over $CaCl_2$. The chemical analysis confirmed the contains proposed by the assumed stoichiometries. Microanalytical procedures [26] were used for the determination of C, H and N contents. Gravimetric analysis [27] was used to determine the bromine contents. The metal contents were complexometrically determined by using 0.01 M ethylenediaminetetraacetic acid solution [28]. Samples of the compounds in a KBr matrix were used to get the IR spectra. The region of spectra was from 4000 to 400 cm^{-1} and a Perkin Elmer 1600 series FTIR spectrophotometer was used. A UV-Vis-NIR spectrophotometer was used to record the spectra of the solid compounds in the region 350 - 2000 nm using a standard reflectance attachment to get the spectra. TG/DTG and DSC curves were recorded in an argon atmosphere in a Du Pont 951 analyzer. The mass of the compounds was initially between 6.37 and 8.79 mg (TG/DTG) and from 4.80 to 16.14 mg (DSC). A heating rate of 10 K·min^{-1} was used from 298 to 678 K (DSC) and from 298 to 1248 K (TG/DTG). The calibration for temperatures was conducted with metallic aluminum as a standard (MP = 933.49 K). The equipment performed the calibration for mass automatically. The DSC calibration was made with metallic indium as a standard (MP = 438.85 K, $\Delta_s^l H°$ = 28.4 J·g^{-1}). For the calorimetric study of the complexes, an LKB 8700-1 precision calorimeter was used at the measurements temperature of 298.15 ± 0.02 K. A thin-walled ampoule that contained reactant was broken in a glass reaction vessel filled with (100.00 mL) of calorimetric solvent [29]. The accuracy of the equipment was determined as previously reported [29] [30]. Three to eight replicate measurements were made on each compound and the uncertainty intervals are twice the standard deviations. The experimental deviations of the dissolution measurements stated between (1% - 3%).

3. Results

3.1. Complex Characterization

The interaction of MBr_2 with TMTU in solution leads to compounds of definite stoichiometry. Only in the case of $FeBr_2$ the compound formed had a fractional stoichiometry. The yields (Y), capillary melting points (MP), colors, appearance (A) and analytical data are reported in **Table 1**.

3.2. Infrared Studies

Table 2 presents the main IR bands of the complexes. A strong band is observed in the region 1097 - 1148 cm^{-1}. This band is attributed to the C=S stretching frequency (v_{CS}) [31]. It is observed in the complexes relative to the free ligand, negative shifts of this frequency and positive shifts of the v_{CN} frequency after coordination, indicating coordination of the ligand through the sulphur atom to the Metal(II) bromide [31]. **Figure 1** presents the IR spectra of the Zn(II) complex.

Table 1. Melting points, yields, appearance and analytical data of the complexes.

Compound*	Y %	MP[a] °C	A[b]	%C		%H		%N		%Br		%M	
				Cal.	Obs.	Cal.	Obs.	Cal.	Obs.	Cal.	Obs.	Cal.	Obs.
$[MnBr_2(L)_2]$	76	150 - 52	y. p.	25.06	25.04	5.05	4.64	11.69	11.54	33.35	33.00	11.46	11.66
$[FeBr_2(L)_{0.75}]$	49	165 - 68	b. re. pa.	14.31	14.17	2.88	3.03	6.67	6.53	50.76	50.55	17.74	17.66
$[CoBr_2(L)_3]$	86	82 - 84	gr. p.	29.27	29.57	5.90	5.95	13.66	13.98	25.97	25.38	9.58	9.88
$[NiBr_2(L)_2]$	62	160 - 62	g. p.	24.87	24.54	5.01	4.96	11.60	11.21	33.09	33.45	12.16	12.04
$[CuBr_2(L)_2]$	43	80 - 82	g. p.	24.62	24.87	4.96	5.35	11.49	11.85	32.76	33.01	13.03	13.25
$[ZnBr_2(L)_2]$	74	158 - 60	w. p.	24.53	24.75	4.94	5.04	11.44	11.45	32.64	32.40	11.35	11.62
$[CdBr_2(L)_2]$	86	169 - 71	w. p.	22.38	22.52	4.51	4.74	10.44	10.37	29.78	29.51	20.94	21.07

[a]Melting with decomposition. [b]Key: y., yellow; b., brown; g., green; w., white; re., redish; gr., greenish; p., powder; pa., paste; *L = TMTU.

Table 2. IR absorption bands (cm^{-1}) of the compounds.

Compound	Assigment[a] $v_{(CS)}$	$v_{(CN)}$
TMTU	1120s	1511s, 1372s
$[MnBr_2(TMTU)_2]$	1098s	1513m, 1367s
$[FeBr_2(TMTU)_{0.75}]$	1067s	1532w, 1393m
$[CoBr_2(TMTU)_3]$	1109s	1558s, 1377s
$[NiBr_2(TMTU)_2]$	1098s	1512m, 1361s
$[CuBr_2(TMTU)_2]$	1114s	1537s, 1380s
$[ZnBr_2(TMTU)_2]$	1110s	1557s, 1382s
$[CdBr_2(TMTU)_2]$	1108s	1556s, 1390s

[a]v, stretching. Intensity of bands: s, strong; m, medium; w, weak.

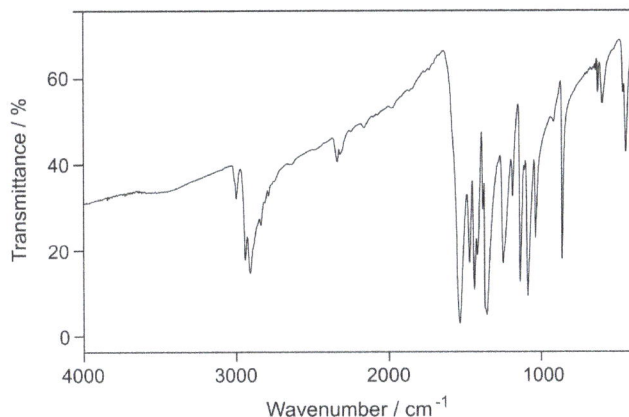

Figure 1. Infrared spectrum of the complex [ZnBr$_2$(TMTU)$_2$].

3.3. Thermal Studies

The thermogravimetry of the complex of Mn(II) shows the loss of the ligand in three steps of mass loss. In a fourth step it loses part of the bromine content leaving a residue that is part of the bromine content plus the metal content. The complex of Fe(II) shows the loss of the ligand in three steps. Part of the bromine content is lost together with part of the ligand in the third step of mass loss. Part of the bromine content is lost in the fourth and fifth steps of mass loss leaving a residue that is part of the bromine content plus the metal content. The complex of Co(II) shows the loss of ligand in three steps of mass loss. Part of the bromine content is lost together with part of the ligand in the third step. Part of the bromine content is lost in a fourth step of mass loss leaving a residue that is part of the bromine content plus the metal content. The complex of Ni(II) shows the loss of the ligand in the first step of mass loss follow by the loss of part of the bromine content in a second step of mass loss leaving a residue that is part of the bromine content plus the metal content. The complex of Cu(II) shows the loss of the ligand in the first step of mass loss together with part of the bromine content. The rest of the bromine content together with part of the metal content is lost in the second step of mass loss leaving a residue that is part of the metal content. The complex of Zn(II) shows the loss of the ligand in three steps of mass loss. The bromine content is lost together with part of the metal in the third step of mass loss. Part of the metal content is lost in the fourth step of mass loss leaving a residue that is part of the metal content. The complex of Cd(II) shows the loss of the ligand in two steps of mass loss follow by the loss of the bromine content and part of the metal content in the third step of mass loss leaving a residue that is part of the metal content. **Figure 2** presents the TG/DTG curve of the Co(II) complex. The DSC curves of the complexes are consistent with the TG data. They present endothermic peaks due to the elimination of part of the ligand or part of the bromine content alone or together with part of the ligand. An exothermic peak is observed in the DSC curve of the Fe(II) complex due to the decomposition of the complex. **Figure 3** presents the DSC curve of the Co(II) complex. **Table 3** presents the thermoanalytical data for the complexes.

3.4. Electronic Spectra

The ligand field parameters for the Ni(II) complex were calculated according to Reedijk *et al.* [32] and Lever [33]. The number and position of the observed bands and the magnitude of the crystal field parameters as compared with that of Bolster [34] indicates that the Ni(II) complex is pseudo-tetrahedral with the Ni(II) ion surrounded by two bromide ions and two sulphur atoms from two ligand molecules. The Co(II) complex shows bands attributed to pseudo-octahedral species with Co(II) ion surrounded by three bromide ions and three sulphur atoms from three ligand molecules in a dimeric structure of bridging bromide ions. The ligand field parameters were calculated according to Lever [33]. The electronic spectra of Cu(II) complex shows a rather broad and symmetrical band which position according to Bolster [34] corresponds to pseudo-tetrahedral species with the Cu(II) ion surrounded by two bromide ions and two sulphur atoms from two ligand molecules. The ligand field parameters of the Fe(II) complex were calculated according to Bolster [34]. It is concluded that Fe(II) ion is pseudo-octahedral with units [FeBr$_6$]$^{4-}$ and [FeBr$_5$L]$^{3-}$ in a molar relation of 1:3 in a polymeric chain of

Table 3. Thermal analysis of the compounds.

Compound[*]	Apparent MP/K	Mass loss/%		TG temperature range/K	Species lost	DSC peak temperature	$\Delta H^{\circ}/kJ\cdot mol^{-1}$
		Calcd.	Obs.				
[MnBr$_2$(L$_2$)]	423 - 25	1.93	1.84	325 - 353	−0.07L	346	1.53
		38.65	38.21	353 - 487	−1.41L	421	35.99
		14.63	16.19	487 - 564	−0.53L		
		28.34	28.39	884 - 962	−1.7Br		
			15.00[a]				
[FeBr$_2$(L)$_{0.75}$]	438 - 41	9.00	8.67	375 - 494	−0.21L	349	−5.66
		13.50	12.72	494 - 531	−0.32L	413	2.33
		26.26	26.08	531 - 700	−0.22L - 0.68Br	469	1.33
		19.29	19.69	700 - 858	−0.76Br	510	0.32
		4.06	4.07	858 - 915	−0.16Br		
			28.77[a]				
[CoBr$_2$(L$_3$)]	355 - 57	25.79	25.53	353 - 389	−1.20L	340	30.36
		26.86	26.61	389 - 538	−1.25L	385	9.77
		15.71	16.03	538 - 564	−0.55L - 0.3Br		
		18.17	18.67	564 - 861	−1.4Br		
			13.16[a]				
[NiBr$_2$(L$_2$)]	433 - 35	54.76	55.24	428 - 505	−2L	393	1.09
		28.12	27.44	505 - 890	−1.7Br	427	27.48
			17.32[a]			442	21.49
[CuBr$_2$(L)$_2$]	353 - 55	70.60	68.07	398 - 484	−2L - Br	352	38.19
		26.80	26.76	484 - 838	−Br - 0.8Cu	441	2.62
			5.17[a]				
[ZnBr$_2$(L)$_2$]	433 - 35	35.10	35.09	434 - 591	−1.3L	430	38.51
		13.54	13.24	591 - 615	−0.5L		
		46.05	46.26	615 - 760	−0.2L - 2Br - 0.6Zn		
		1.34	0.83	760 - 942	−0.1Zn		
			4.58[a]				
[CdBr$_2$(L)$_2$]	442 - 44	36.96	36.62	412 - 570	−1.5L	434	42.08
		12.32	13.15	570 - 580	−0.5L		
		49.67	49.12	784 - 838	−2Br - 0.95Cd		
			1.11[a]				

[a]Residue at 1243 K; [*]L = TMTU.

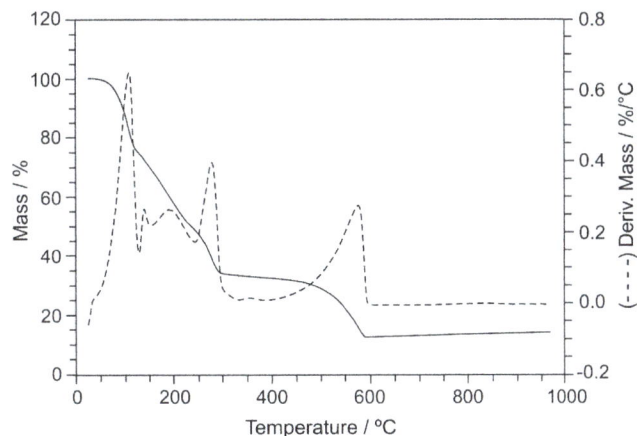

Figure 2. TG/DTG curve of the complex [CoBr$_2$(TMTU)$_3$].

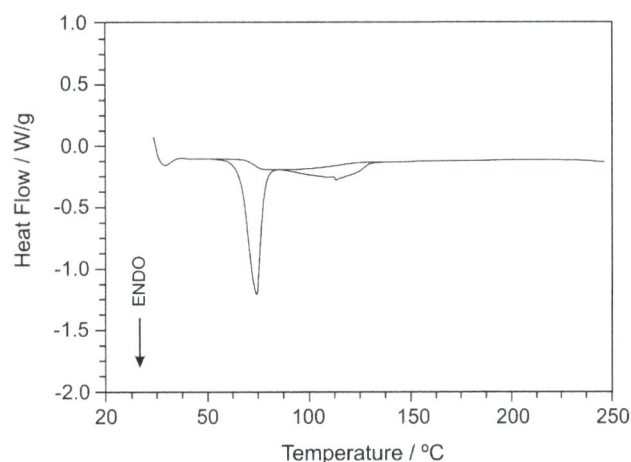

Figure 3. DSC curve of the complex [CoBr$_2$(TMTU)$_3$].

bridging bromide ions. The complex of Mn(II) according with the position of the absorption band [34] is pseudo-tetrahedral with the manganese ion surrounded by two bromide ions and two sulphur atoms from two ligand molecules. **Table 4** contains the band maxima assignments and calculated ligand field parameters of the complexes.

3.5. Calorimetric Measurements

The standard enthalpies of dissolution of Metal(II) bromides, TMTU and complexes were measured [35]. The standard enthalpies of the following reactions were obtained:

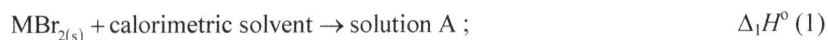

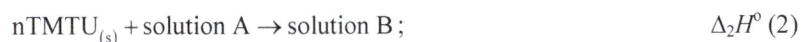

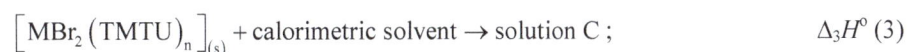

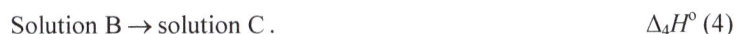

$$MBr_{2(s)} + \text{calorimetric solvent} \rightarrow \text{solution A} ; \qquad \Delta_1 H^\circ \text{ (1)}$$

$$nTMTU_{(s)} + \text{solution A} \rightarrow \text{solution B} ; \qquad \Delta_2 H^\circ \text{ (2)}$$

$$\left[MBr_2 \left(TMTU \right)_n \right]_{(s)} + \text{calorimetric solvent} \rightarrow \text{solution C} ; \qquad \Delta_3 H^\circ \text{ (3)}$$

$$\text{Solution B} \rightarrow \text{solution C} . \qquad \Delta_4 H^\circ \text{ (4)}$$

The application of the Hess' law to the reactions 1 to 4 gives the standard enthalpies of reaction ($\Delta_r H^\circ$):

$$MBr_2 + nTMTU_{(s)} \rightarrow \left[MBr_2 \left(TMTU \right)_n \right]_{(s)} ; \qquad \Delta_r H^\circ \text{ (5)}$$

being $\Delta_r H^\circ = \Delta_1 H^\circ + \Delta_2 H^\circ - \Delta_3 H^\circ$, because $\Delta_4 H^\circ = 0$ for the dissolution of solution B into solution C. **Table 5**

Table 4. Band maxima and calculated ligand field parameters for the complexes.

Complex[*]	Band maxima ($\times 10^3$ cm^{-1})							Interligand + charge transfer ($\times 10^3$ cm^{-1})
	d-d							
[MnBr$_2$(L$_2$)]	21.37, 26.32							31.95
[CuBr$_2$(L)$_2$]	12.50							26.81
	ν_1	Dq (cm^{-1})						
[FeBr$_2$(L)$_{0.75}$]	7.26[a]	726						
	ν_1	ν_2	ν_3	Dq (cm^{-1})	B	Dq/B	β^+	
[CoBr$_2$(L)$_3$]	7.66[b]	13.46[c]	19.08[d]	580	1002	0.579	1.032	19.08, 20.75, 26.95
[NiBr$_2$(L)$_2$]	9.07[e]	15.20[f]		236	275	0.858	0.267	21.93

[a] $\nu_1 = {}^5E_g \leftarrow {}^5T_{2g}$; [b] $\nu_1 = {}^4T_{2g} \leftarrow {}^4T_{1g}(F)$; [c] $\nu_2 = {}^4A_{2g} \leftarrow {}^4T_{1g}(F)$; [d] $\nu_3 = {}^2P, {}^2G, {}^4T_{1g}(P) \leftarrow {}^4T_{1g}(F)$; [e] $\nu_2 = {}^3A_2 \leftarrow {}^3T_1(F)$; [f] $\nu_3 = {}^3T_1(P) \leftarrow {}^3T_1(F)$; [+] $\beta = B/B_0$; $B_0 = 971$ cm^{-1} (Co^{2+}); $B_0 = 1030$ cm^{-1} (Ni^{2+}); [*] L = TMTU.

Table 5. Enthalpies of dissolution at 298.15 K.

Compound	Calorimetric solvent	Number of experiments	i	$\Delta_i H^o$ (kJ·mol^{-1})
MnBr$_{2(s)}$	1.2 M HCl	5	1	-43.96 ± 1.73
TMTU$_{(s)}$	2:1 MnBr$_2$ - 1.2 M HCl	4	2	21.99 ± 1.41
[MnBr$_2$(TMTU)$_2$]$_{(s)}$	1.2 M HCl	3	3	-20.06 ± 0.74
CoBr$_{2(s)}$	1.2 M HCl	6	1	-69.83 ± 0.80
TMTU$_{(s)}$	3:1 CoBr$_2$ - 1.2 M HCl	4	2	30.10 ± 1.80
[CoBr$_2$(TMTU)$_3$]$_{(s)}$	1.2 M HCl	4	3	21.97 ± 0.62
NiBr$_{2(s)}$	1.2 M HCl	5	1	-55.59 ± 2.11
TMTU$_{(s)}$	2:1 NiBr$_2$ - 1.2 M HCl	4	2	24.05 ± 1.73
[NiBr$_2$(TMTU)$_2$]$_{(s)}$	1.2 M HCl	4	3	-19.86 ± 0.88
CuBr$_{2(s)}$	1.2 M HCl	6	1	-24.63 ± 1.54
TMTU$_{(s)}$	2:1 CuBr$_2$ - 1.2 M HCl	3	2	14.11 ± 0.63
[CuBr$_2$(TMTU)$_2$]$_{(s)}$	1.2 M HCl	4	3	30.38 ± 0.80
ZnBr$_{2(s)}$	1.2 M HCl	8	1	-42.23 ± 0.88
TMTU$_{(s)}$	2:1 ZnBr$_2$ - 1.2 M HCl	4	2	19.80 ± 0.81
[ZnBr$_2$(TMTU)$_2$]$_{(s)}$	1.2 M HCl	3	3	19.40 ± 0.46
CdBr$_{2(s)}$	1.2 M HCl	5	1	23.83 ± 2.79
TMTU$_{(s)}$	2:1 CdBr$_2$ - 1.2 M HCl	4	2	19.95 ± 0.23
[CdBr$_2$(TMTU)$_2$]$_{(s)}$	1.2 M HCl	3	3	62.78 ± 1.32

presents the values observed for the enthalpies of dissolution of MBr$_2$($\Delta_1 H^o$), for the enthalpies of dissolution of TMTU into the solution of MBr$_2$($\Delta_2 H^o$) and that of the complexes ($\Delta_3 H^o$). It was not possible to measure the enthalpies of dissolution of the complex of Fe(II) due to its paste consistency that made it difficult its manipulation. Uncertainty intervals given in this table are twice the standard deviations of the mean of 3 to 8 replicate measurements.

4. Discussions

Using the standard enthalpies of reaction ($\Delta_r H^o$) and appropriate thermochemical cycles [35], the following thermochemical parameters were got: the standard enthalpies of formation ($\Delta_f H^o$) from Equation (5),

$$\Delta_f H^\circ (\text{complex}) = \Delta_r H^\circ + \Delta_r H^\circ \left(\text{MBr}_{2(s)}\right) + n\Delta_f H^\circ \left(\text{TMTU}_{(s)}\right)$$

the standard enthalpies of decomposition ($\Delta_D H^o$) from Equation (6),

$$\left[\text{MBr}_2\left(\text{TMTU}\right)_n\right]_{(s)} \rightarrow \text{MBr}_{2(s)} + n\text{TMTU}_{(g)} ; \qquad \Delta_D H^\circ \ (6)$$

being $\Delta_D H^\circ = \Delta_r H^\circ + n\Delta_s^g H^\circ \left(\text{TMTU}_{(s)}\right)$. The standard lattice enthalpy ($\Delta_M H^o$), from Equation (7),

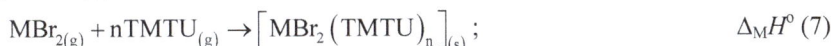

$$\text{MBr}_{2(g)} + n\text{TMTU}_{(g)} \rightarrow \left[\text{MBr}_2\left(\text{TMTU}\right)_n\right]_{(s)} ; \qquad \Delta_M H^\circ \ (7)$$

being $\Delta_M H^\circ = -\Delta_D H^\circ - \Delta_s^g H^\circ \left(\text{MBr}_{2(s)}\right)$. The enthalpy of reaction in the gaseous phase, from Equation (8)

$$\text{MBr}_{2(g)} + n\text{TMTU}_{(g)} \rightarrow \left[\text{MBr}_2\left(\text{TMTU}\right)_n\right]_{(s)} ; \qquad \Delta_r H^\circ(g) \ (8)$$

being $\Delta_r H^\circ (g) = -\Delta_s^g H^\circ \left(\text{MBr}_{2(s)}\right) - n\Delta_s^g H^\circ \left(\text{TMTU}_{(s)}\right) + \Delta_r H^\circ + \Delta_s^g H^\circ \left(\text{complex}_{(s)}\right)$.

As the complexes decomposed on heating, the enthalpies of sublimation of the complexes were estimated [36]. From Equation (8) it is got the standard enthalpies of the metal-sulphur bonds: $D_{(M-S)} = -\left(\Delta_r H^\circ (g)\right)/n$. **Table 6** presents the values obtained for all these enthalpies. The formation enthalpies of the complexes in the gaseous phase, according to the Equation (9)

$$\text{M}^{2+}_{(g)} + 2\text{Br}^-_{(g)} + n\text{TMTU}_{(g)} \rightarrow \left[\text{MBr}_2\left(\text{TMTU}\right)_n\right]_{(g)} ; \qquad \Delta_{fl} H^\circ \ (9)$$

are equal to $\Delta_{fl} H^\circ = \Delta_f H^\circ (\text{complex}) - \Delta_f H^\circ \left(\text{M}^{2+}_{(g)}\right) - 2\Delta_f H^\circ \left(\text{Br}^-_{(g)}\right) - n\Delta_f H^\circ \left(\text{TMTU}_{(g)}\right)$. **Table 7** shows the values obtained for these enthalpies values.

The acidity order obtained based on $\Delta_r H^o$ values for the complexes of the same stoichiometry is: Zn(II) > Cu(II) > Cd(II) > Ni(II) > Mn(II). Using the $D_{(M-S)}$ values the, order is: Cu(II) > Zn(II) > Mn(II) > Ni(II) > Cd(II).

Table 6. Summary of the thermochemical results (kJ·mol^{-1}) for the compounds.

Compound	$\Delta_r H^o$	$\Delta_f H^o$	$\Delta_s^g H^o$	$\Delta_M H^o$	$\Delta_D H^o$	$\Delta_r H^o(g)$	$D_{(M-S)}$
MnBr$_{2(s)}$		−384.9[a]	205.9[a]				
CoBr$_{2(s)}$		−220.9[a]	183[a]				
NiBr$_{2(s)}$		−212.1[a]	170[a]				
CuBr$_{2(s)}$		−141.8[a]	182.4[a]				
ZnBr$_{2(s)}$		−328.65[a]	159.7[a]				
CdBr$_{2(s)}$		−316.2[b]	151.2[c]				
TMTU$_{(S)}$		−38.3 ± 2.3[d]	82.36 ± 0.20[d]				
[MnBr$_2$(TMTU)$_2$]$_{(s)}$	−1.91 ± 2.35	−463.4 ± 5.5	−228.4 ± 17.6	−372.5 ± 3.1	166.63 ± 2.38	−228.4 ± 17.6	114.2 ± 8.8
[CoBr$_2$(TMTU)$_3$]$_{(s)}$	−61.70 ± 2.07	−396.6 ± 7.5	133 ± 16	−492 ± 2	308.78 ± 2.16	−359 ± 16	120 ± 5
[NiBr$_2$(TMTU)$_2$]$_{(s)}$	−11.68 ± 2.87	−300.4 ± 5.6	126.2 ± 15.1	−346.5 ± 3.1	176.40 ± 2.90	−220.3 ± 15.4	110.2 ± 7.7
[CuBr$_2$(TMTU)$_2$]$_{(s)}$	−40.90 ± 1.85	−259.3 ± 3.6	132.4 ± 15.9	−388 ± 2	205.62 ± 1.89	−256 ± 16	128 ± 8
[ZnBr$_2$(TMTU)$_2$]$_{(s)}$	−41.83 ± 1.28	−447.1 ± 5.2	121.0 ± 14.5	−366.3 ± 2.4	206.55 ± 1.34	−245.3 ± 14.7	122.7 ± 7.4
[CdBr$_2$(TMTU)$_2$]$_{(s)}$	−19.00 ± 3.10	−131.8 ± 5.6	116.8 ± 14.0	−334.9 ± 3.3	187.72 ± 3.13	−218.1 ± 14.4	109.1 ± 7.2

[a][37]; [b][38]; [c][39]; [d][40].

Table 7. Auxiliary data and enthalpy changes of the complex formation process in the gaseous phase (kJ·mol^{-1}).

Compound	$\Delta_f H^\circ$	$\Delta_r H^\circ(g)$	$\Delta_{fl} H^\circ$
$Br^-_{(g)}$	-219.07^a		
$Mn^{2+}_{(g)}$	2522.0 ± 0.1^b		
$Co^{2+}_{(g)}$	2841.7 ± 3.4^b		
$Ni^{2+}_{(g)}$	2930.5 ± 1.5^b		
$Cu^{2+}_{(g)}$	3054.5 ± 2.1^b		
$Zn^{2+}_{(g)}$	2781.0 ± 0.4^b		
$Cd^{2+}_{(g)}$	2623.54^a		
$[MnBr_2(TMTU)_2]_{(g)}$	-319.2 ± 18.2	-228.4 ± 17.6	-2491.3 ± 18.3
$[CoBr_2(TMTU)_3]_{(g)}$	-265 ± 17	-359 ± 16	-2801 ± 19
$[NiBr_2(TMTU)_2]_{(g)}$	-174 ± 16	-220.3 ± 15.4	-2755 ± 17
$[CuBr_2(TMTU)_2]_{(g)}$	-127 ± 17	-256 ± 16	-2832 ± 18
$[ZnBr_2(TMTU)_2]_{(g)}$	-326.1 ± 15.4	-245.3 ± 14.7	-2757.2 ± 16.1
$[CdBr_2(TMTU)_2]_{(g)}$	-295 ± 15	-218.1 ± 14.4	-3445 ± 16

^{a}[38]; ^{b}[41].

5. Conclusion

Solid state complexes were obtained from the interaction in hot tert-buthyl alcohol solution of tetramethylurea with certain divalent transition metal bromides. The complexes decomposed on heating. The dissolution enthalpies were determined for complexes, salts and ligand. By using thermochemical cycles, the energies of the Metal(II)-sulphur coordinated bonds as well as the values of other thermochemical parameters were estimated. The energies of the coordinated bonds have values between 109 and 128 kJ·mol^{-1}.

References

[1]   Raja, M.U., Ramesh, R. and Liu, Y. (2011) New Binuclear Pd(II) Thioamide Complexes for the Heck Reaction of Aryl Bromides. *Tetrahedron Letters*, **52**, 5427-5430. http://dx.doi.org/10.1016/j.tetlet.2011.07.080

[2]   Sah, R., Kumari, S. and Kumar, A. (2011) Synthesis and Characterization of Transition Metal(II) Complexes with HeterocyclicThioamides. *Journal of Chemistry*, **23**, 3563-3565.

[3]   Ajaz, H., Hussain, S., Altaf, M., Stoeckli-Evans, H., Isab, A.A., Mahmood, R., Altaf, S. and Ahmad, S. (2011) Synthesis and Characterization of Antimony(III) Complexes of Thioamides, and Crystal Structure of Sb(Imt)$_2$Cl$_2$]$_2$(μ_2-Imt)}Cl$_2$ (Imt = Imidazolidine-2-Thione). *Chinese Journal of Chemistry*, **29**, 254-258. http://dx.doi.org/10.1002/cjoc.201190074

[4]   Singh, T. and Singh, K.K. (2010) Synthesis and Characterization of New Ruthenium-Aromatic Thioamidenitrosyls. *Oriental Journal of Chemistry*, **26**, 1171-1174.

[5]   Nadeem, S., Rauf, M.K., Bolte, M., Ahmad, S., Tirmizi, S.A., Asna, M. and Hameed, A. (2010) Synthesis, Characterization and Antibacterial Activity of Palladium(II) Bromide Complexes of Thioamides; X-Ray Structure of [Pdf(Tetrametylthiourea)$_4$]Br$_2$. *Transition Metal Chemistry*, **35**, 555-561. http://dx.doi.org/10.1007/s11243-010-9363-0

[6]   Ozturki, I.I., Hadjikakou, S.K., Hadjiliardis, N., Kourkoumelis, N., Kubicki, M., Tasiopoulos, A.J., Scleiman, H., Barsan, M.M., Butler, I.S. and Balzarini, J. (2009) New Antimony(III) Bromide Complexes with Thioamides: Synthesis, Characterization and Cytotatic Properties. *Inorganic Chemistry*, **48**, 2233-2245. http://dx.doi.org/10.1021/ic8019205

[7]   Sola, J., Lopez, A., Coxall, R.A. and Clegy, W. (2004) Hidrogen-Bonding Network and Layerd Supra-Molecular Structures Assembled from ClO$_4^-$ Counterions with Unprecedented Monomeric [AgL$_2$]$^+$ and Chain Polymeric [AgL$_2$]$^{nm+}$ Complex Cations (L = Thioamide or Thiourea-Like Ligands). *Journal of Inorganic Chemistry*, **24**, 4871-4881.

[8]   Sbirna, L.S., Muresan, V., Sbirna, S., Muresan, N. and Lepadatu, C.I. (2004) Complex Compounds of Ions with d6-10 Configuration with Bidentate Heterocyclic Ligands (N, S). *Journal of the Indian Chemical Society*, **81**, 150-152.

[9]   Misra, S.K. and Tewari, U.C. (2002) Complexing Behaviour of Aromatic Thioamides (ArCSNHCOR). Transition Metal Complexes of N-Carboethoxy-4-Chlorobenzene and N-Carboethoxy-4-Bromobenzene Thioamide Ligand. *Transition Metal Chemistry*, **27**, 120-125. http://dx.doi.org/10.1023/A:1013473621826

[10]   Muresan, V., Sbirna, L.S., Sbirna, S., Lepadatu, C.I. and Muresan, N. (2001) Transition Metal Complexes with a New Thioamide of the Dibenzofuran Series. *Acta Chimica Slovenica*, **48**, 439-443.

[11]   Neagoe, C., Neagoe, S., Neagoe, O., Cercasov, C. and Lepadatu, C. (1999) Complex Compounds of Pd(II), Pt(II), Cu(I) and Hg(II) with Tertiary Thioamides. *Romanian Archives of Microbiology and Immunology*, **58**, 209-215.

[12]   Zalaru, F., Cercasov, C., Meghea, A., Zalaru, C. and Jalen, C. (1991) Copper(II) Coordination Compounds with Thioamides. *Revue Roumaine de Chemie*, **36**, 1279-1285.

[13]   Dunstan, P.O. and Dos Santos, L.C.R. (1989) Thermochemistry of Amide and Thioamide Complexes of Arsenic Trihalides. *Thermochimica Acta*, **156**, 163-177. http://dx.doi.org/10.1016/0040-6031(89)87182-5

[14]   Desseyn, H.O. (1989) Complexes of Chelates of Amides and Thioamides. *Pure and Applied Chemistry*, **61**, 867-872. http://dx.doi.org/10.1351/pac198961050867

[15]   Neagoe, S., Antonescu, L., Negoiu, M., Negoe, V., Negoiuand, D. and Cercasov, C. (1989) Transiton Metal Complexes with Thioamides. *Revue Roumaine de Chimie*, **34**, 845-854.

[16]   Carraher Jr., C.E., Chen, W. and Hess, G.G. (1988) Initial Synthesis of Palladium(II) Polyamides and Polythioamides. *Polymeric Materials: Science and Engineering*, **58**, 557-560.

[17]   Singh, B. (1987) Studies on Oxovanadium(IV) Complexes of Some Thioamides. *Indian Journal of Chemistry Section A*, **26**, 350-351.

[18]   Burns, G.R., DeRoo, C.R., Hall, D.W. and Oliver, A.R. (1987) Chromiun(III) Complexes with Thioamides. *Inorganica Chimica Acta*, **130**, 13-15. http://dx.doi.org/10.1016/S0020-1693(00)85921-6

[19]   Bret, J.M., Castan, P. and Laurent, J.P. (1983) Studies on Thioamido-Platinum Complexes: Stabilization of Platinum in Its Higher Oxidation State. *Transition Metal Chemistry*, **8**, 218-221. http://dx.doi.org/10.1007/BF00620694

[20]   Muresan, V. and Muresan, N. (1980) Thio Complex Compound Chemistry. Part VIII. Study on the Effect of Ligands and Central Metal Ion on the Nature of Chemical Bond and Structure of Thioamide Complexes. *Analete Universitatudin Craiova*, **8**, 97-101.

[21]   Kukushkin, Y.N., Vorob'ev-Desyatovsku, N.V., Sibirskaya, V.V. and Stmkov, V.V. (1980) Some Features of the Behaviour of Platinum(I) Thioamide Complexes in Solutions. *Zhurnal Obshchei Khimii*, **50**, 107-110.

[22]   Aarts, A.J., Desseyn, H.O. and Herman, M.A. (1978) Palladium(II) Complexes with Primary Thioamides. *Transition Metal Chemistry*, **3**, 144-146. http://dx.doi.org/10.1007/BF01393531

[23]   Sibirskaya, V.V. and Kukushkin, Y.N. (1978) Thioamide Complexes of Platinum Metals. *Koordinatsionnaya Khimiya*, **4**, 963-991.

[24]   Molina, M., Del'acqua, A., Melios, C.V., Azevedo, F.A. and Trabuco, E. (1977) Interaction of the Thioamides with Metals. Complexes in the Solid State and in Solution. *Eclética Química*, **2**, 87-95.

[25]   Du Preez, J.G.H. and Gibson, M.L. (1970) Phosphine and Thioamide Complexes of Oxovanadium Dichloride. *Journal of the South African Chemical Institute*, **23**, 184-190.

[26]   Niederland, J.B. and Sozzi, J.A. (1958) Microanáslisis Elemental Orgánico. Methopress, Buenos Aires.

[27]   Kolthoffand, I.M. and Sandall, E.B. (1956) Tratado de Química Analítica, Cuantitativa. 3rd Edición, Nigar, Buenos Aires.

[28]   Flaschka, H.A. (1964) Edta Titrations: An Introduction to Theory and Practice. 2nd Edition, Pergamon Press, London.

[29]   Dunstan, P.O. (1999) Thermochemistry of Adducts of Bis(2,4-Pentanedionato)Zinc with Heterocyclic Amines. *Journal of Chemical & Engineering Data*, **44**, 243-247. http://dx.doi.org/10.1021/je980113m

[30]   Herington, E.F. (1991) Recommended Reference Materials for the Realization of Physicochemical Properties (Recommendation Approved 1974). *Pure and Applied Chemistry*, **24**, 261-289.

[31]   Schofer Sr, M. and Curran, C.B. (1966) Infrared Spectra of Complexes of Metal Halides with Tetramethylurea and Tetramethylthiourea. *Inorganic Chemistry*, **5**, 265-268. http://dx.doi.org/10.1021/ic50036a023

[32]   Reedijk, J., Van Leeuwem, P.W.N.M. and Groenveld, W.L. (1968) A Semi-Empirical Energy-Level Diagram for Octahedral Nickel Complexes. *Recueil des Travaux Chimiques des Pays-Bas*, **87**, 129-141. http://dx.doi.org/10.1002/recl.19680870203

[33]   Lever, A.B.P. (1968) Electronic Spectra of Some Transition Metal Complexes. *Journal of Chemical Education*, **45**, 711-712. http://dx.doi.org/10.1021/ed045p711

[34] Bolster, M.W.G. (1972) The Coordination Chemistry of Amino-Phosphinoxide and Related Compounds. Thesis, Leiden University, Leiden.

[35] Dunstan, P.O. (2009) Thermochemistry of Morpholine Adducts of Some Bivalent Transition Metal Bromides. *Journal of Chemical & Engineering Data*, **54**, 842-846. http://dx.doi.org/10.1021/je8006315

[36] Sovast'yanova, T.N. and Suvorov, A.V. (1999) The Structure and Thermal Stability of Group III Halide Complexes with Pyridine. *Russian Journal of Coordination Chemistry*, **25**, 679-688.

[37] Dunstan, P.O. (2004) Thermochemistry of Adducts of Some Transition Metals(II) Bromides with Pyridine N-Oxide. *Thermochimica Acta*, **409**, 19-24. http://dx.doi.org/10.1016/S0040-6031(03)00333-2

[38] Wagman, D.D., Evans, W.H., Parker, V.B., Schumm, R.H., Hallow, I.S., Churney, M. and Nuttall, R.L. (1982) The NBS Table of Chemical Thermodynamic Properties. Selected Values for Inorganic and C_1 and C_2 Organic Substances in SI Units. *Journal of Physical and Chemical Reference Data*, **2**, 50-191.

[39] Kubaschewiski, O., Evans, E.L. and Alcock, C.B. (1967) Metallurgical Thermochemistry. 4th Edition, Pergamon Press, London.

[40] Inagari, S.C., Murata, S. and Sakiyama, M. (1982) Thermochemical Studies on Thioacetamide and Tetramethylthiourea. Estimation of Stabilization Energies Due to Interaction between Thiocarbonyl Group and Neighboring Nitrogen Atom. *Bulletin of the Chemical Society of Japan*, **55**, 2808-2813. http://dx.doi.org/10.1246/bcsj.55.2808

[41] Skinner, H.A. and Pilcher, G. (1963) Bond Energy-Term Values in Hydrocarbons and Related Compounds. *Quarterly Reviews, Chemical Society*, **17**, 264-288. http://dx.doi.org/10.1039/qr9631700264

20

Coherence of the Even-Odd Rule with an Effective-Valence Isoelectronicity Rule for Chemical Structural Formulas: Application to Known and Unknown Single-Covalent-Bonded Compounds

Geoffroy Auvert

CEA-Leti, Grenoble, France
Email: Geoffroy.auvert@grenoble-inp.org

Abstract

Ions or molecules are said to be isoelectronic if they are composed of different elements but have the same number of electrons, the same number of covalent bonds and the same structure. This criterion is unfortunately not sufficient to ensure that a chemical structure is a valid chemical compound. In a previous article, a procedure has been described to draw 2D valid structural formulas: the even-odd rule. This rule has been applied first to single-bonded molecules then to single-charged single-bonded ions. It covers hypovalent, hypervalent or classic Lewis' octet compounds. The funding principle of the even-odd rule is that each atom of the compound possesses an outer-shell filled only with pairs of electrons. The application of this rule guarantees validity of any single-covalent-bond chemical structure. In the present paper, this even-odd rule and its electron-pair criterion are checked for coherence with an effective-valence isoelectronic rule using numerous known compounds having single-covalent-bond connections. The test addresses Lewis' octet ions or molecules as well as hypovalent and hypervalent compounds. The article concludes that the even-odd rule and the effective-valence isoelectronicity rule are coherent for known single-covalent-bond chemical compounds.

Keywords

Isoelectronicity, Effective Valence, Molecule, Ion, Even-Odd, Rule, Structural Formula, Covalent Bond

1. Introduction

A chemical structural formula of a compound is a two-dimensional representation procedure initiated about two centuries ago. Before 1811, atoms and molecules were thought to be similar entities. Avogadro [1], using the law of multiple proportions adopted by John Dalton in 1803 [2], introduced the term "molecules" to describe basic components of gases. This conception was argued with until Albert Einstein and Jean Perrin explained the Brownian movement in a liquid by the presence of molecules. Einstein used theory [3] and Perrin experiment [4].

Following the discovery of the electrical properties of liquids, Michael Faraday coined the term "ions" to describe compounds carrying electrical charges [5].

When comparing both, the term "isoelectronicity" [6] is used for ions and molecules having nearly the same structures. With this concept, a compound is said to be isoelectronic if it has the same number of electrons [7] [8]. Unfortunately, this property does not always give a chemically valid structure.

The recently proposed "even-odd" rule [9] [10], introduced as a procedure to draw chemical structural formulas of ions and molecules, has not yet taken into account a possible isoelectronicity link between the represented compounds.

The aim of this paper is to check the coherence of the "even-odd" rule using compounds with a specific type of isoelectronicity named "effective-valence isoelectronicity". Structural formulas of compounds: cations, molecules and anions, following this rule are drawn including numbers required by the even-odd rule. These numbers, associated with each element, are calculated and discussed. The state of scientific knowledge of these compounds is taken into account as well as specific isoelectronic links and resulting even-odd structures.

2. Even-Odd Rule for Ions and Molecules

The even-odd rule is a procedure to draw chemical structural formulas of molecules and ions. The structure is composed of one or several atoms of the periodic table.

As a reminder [10], the rule is described below:

- Each atom:
  - Is an element with one or several electron shells.
  - Possesses an outer-shell filled with one to eight electrons.
  - Has a number, also called valence number, of electrons in the outer-shell as indicated in the periodic table.
  - Has a valence number of the element giving the highest number of covalent bonds that the element can form.
- A structure meets the criteria below:
  - When it is composed of only one atom, it forms no covalent bond.
  - When it is constituted of several atoms, each atom forms a single covalent bond with each of its first neighboring atoms. This covalent bond involves two electrons, one from each interconnected atom.
  - A covalent bond is represented by one line between both connected atoms.
  - An atom may have zero, one or more than one line around it.
  - In the 2D structure, each atom is represented by the letters of its element as in the periodic table.
  - Two numbers have to be evaluated and written on each side of the atom.
- The left side number and the effective valence number:
  - The left side number is the valence number as in the periodic table. It ranges from one for elements like sodium (Na) up to eight for noble gas like Argon (Ar).
  - An effective valence number has to be evaluated: For a neutral atom *i.e.* without charge, it is equal to the valence number; for a negatively charged atom *i.e.* that possesses an extra-electron, it is the valence number increased by one; for a positively charged atom, it is the valence number decreased by one.
- The right side number of an atom:
  - The right side number, the "Lewis number", is equal to the sum of the effective valence number and the number of covalent bonds of the atom. It can also be expressed as the sum of the number of electrons left in the outer-shell and twice the number of covalent bonds.
  - The Lewis number must be an even number. This is only possible when the number of bonds and the effective valence number are either both odd or both even.
  - The smallest value the Lewis number can take is zero: the atom has lost electrons from the outer-shell so it is empty and the atom has no bonds.

- The Lewis number can range up to twice the effective valence number: this is twice the maximum number of covalent bonds for this element. This number is charge dependent through the effective valence number.
- If all atoms of a compound have Lewis numbers equal to eight, the compound is compatible with Lewis' octet rule.
- Electron pairs in the outer-shell of an interconnected atom:
  - The number of electrons in the outer-shell is calculated by subtracting the effective valence-number and the number of covalent bonds. It is an even number.
  - As a consequence, the outer-shell contains electron pairs not involved in any covalent bond.
  - This electron-pair number ranges from 0 to 4 whatever the charge of the element.
  - When this electron-pair number is 0, no additional covalent bond can be formed by the element.

The even value of the right side number *i.e.* the Lewis number, and the even value of the number of electrons in the outer-shell are important keys to the validity of the even-odd rule. With these even values, molecules and ions belong to a group of electron-paired compounds [9] [10].

3. Effective-Valence Isoelectronicity Rule

An even-odd compound changes into a valid isoelectronic one when it follows the effective-valence isoelectronicity rule described below:

- The new compound configuration:
  - Has the same covalent-bond structure.
  - Has at least one different atom.
- The replacing atom has no impact on the outer-shell configuration:
  - It builds the same number of single covalent-bonds.
  - It has the same effective-valence number, as defined in the even-odd rule.
  - It has the same Lewis number *i.e.* right side number.
  - It has the same number of electrons in the outer-shell.
- But its internal configuration is different:
  - With a different name from the periodic table.
  - It meets one of the following criteria:
    - It belongs to the same column of the periodic table (same valence number).
    - It belongs to the same line of the periodic table and the valence number is corrected by a charge modification giving the same effective-valence number.
    - It is shifted in column and lines of the periodic table and with a charge modification to maintain constant the effective-valence number.

From the first criteria, one example is for BH4(−) and GaH4(−). The effective-valence isoelectronicity comes from Boron and Gallium. They are in the same column of the periodic table. Only their names are different. All other parameters of the rule are the same. Since both are in the same column of the periodic table, their isoelectronicity link is obvious. Similar cases are therefore not discussed further in this article.

From the second criteria, one example is for HF and OH(−). The effective-valence isoelectronicity comes from O(−) and F. They are on the same line in the periodic table. Both have one covalent-bond, an effective valence of seven, a right side number of 8 and six electrons in the outer-shell.

From the third criteria, one example is for CH4 and PH4(+). The effective-valence isoelectronicity comes from C and P(+). They are not on the same line of the periodic table and not in the same column. Both have 4 covalent-bonds, an effective-valence number of 4, a right-side number of 8 and no electrons in the outer-shell.

4. Isoelectronicity between Single-Charged Single-Bonded Ions and Neutral Single-Bonded Molecules

In the following, the coherence of the even-odd rule with the effective-valence isoelectronicity rule is tested for several known compounds, ions or molecules. Due to the large number of these compounds, the scope of the test is deliberately limited: firstly to compounds composed of main group elements; secondly to compounds having a star configuration. A star configuration is structured with a central atom connected to each peripheral atom via a single covalent bond. Peripheral atoms have no other bonds. To build isoelectronic compounds of such structures, it is sufficient to replace the central atom by one of its effective-valence isoelectronic atoms.

Table 1. Compounds with an effective-valence isoelectronicity. On the same line, compounds are isoelectronic. A group of lines groups compounds with central atoms having the same Lewis number *i.e.* the right side number calculated with the even-odd rule. Column 1 gives the value of the right side number of the group of isoelectronic compounds and its classical formulas. Column 2 lists cations, column 3 neutral molecules and column 4 anions. Column 5 indicates the number of electrons in the outer-shell of all three isoelectronic compounds. In column 1, when "No" stands in place of a classical formula of a compound, it means that the corresponding compounds is not referred to in any scientific literature, being in fields of physics or chemistry. The left side number (the valence number) of the element is indicated for each central atom and according to the even-odd rule. The right side number *i.e.* Lewis number, is the sum of the effective-valence number of the atom with the number of its single covalent-bonds. Note: the isoelectronicity in the same periodic column are not reported. (See text with BH4-GaH4).

-Right side number -Classical formulas -References	Positive charge	Neutral	Negative charge	Electron outer-shell number
0				
Na(+) Xe [6] [11] [12] [6] [11]-[13]	$_1\overset{+}{Na}_0$	$_0Xe_0$		0
2				
BeH(+) NaCl XeCl(−) [11] [11][13] [11][13]	$_2\overset{+}{Be}_2\!\!-\!H$	$_1Na\!-\!Cl_2$	$_0\overset{-}{Xe}_2\!\!-\!Cl$	0
No Be No No [11][13] No	$_3\overset{+}{B}_2$	$_2Be_2$	$_1\overset{-}{Na}_2$	2
4				
BF2(+) BeF2 No [11] [13] No	$F\!-\!_3\overset{+}{B}_4\!\!-\!F$	$F\!-\!_2Be_4\!\!-\!F$	$Cl\!-\!_1\overset{-}{Na}_4\!\!-\!Cl$	0
SiF(+) BH BeH(−) [11] [11] [11]	$_4\overset{+}{C}_4\!\!-\!F$	$_3B_4\!\!-\!H$	$_2\overset{-}{Be}_4\!\!-\!H$	2
No C No No [11][13] No	$_5\overset{+}{N}_4$	$_4C_4$	$_3\overset{-}{B}_4$	4
6				
CH3(+) BF3 BeF3(−) [11]-[13] [11]-[13] [11][12]	$H\!-\!_4\overset{+}{C}_6\!\!\big\langle\!\begin{smallmatrix}H\\H\end{smallmatrix}$	$H\!-\!_3B_6\!\!\big\langle\!\begin{smallmatrix}H\\H\end{smallmatrix}$	$F\!-\!_2\overset{-}{Be}_6\!\!\big\langle\!\begin{smallmatrix}F\\F\end{smallmatrix}$	0
NF2 SiH2 BF2 [11][12] [11] [11]	$F\!-\!_5\overset{+}{N}_6\!\!-\!F$	$H\!-\!_4Si_6\!\!-\!H$	$F\!-\!_3\overset{-}{B}_6\!\!-\!F$	2
OH(+) NH SiH(−) [11] [11] [11]	$_6\overset{+}{O}_6\!\!-\!H$	$_5N_6\!\!-\!H$	$_4\overset{-}{Si}_6\!\!-\!_{1}H_2$	4
No O No No [11][12] No	$_7\overset{+}{F}_6$	$_6O_6$	$_5\overset{-}{N}_6$	6

Continued

8

NH4(+) CH4 BH4(−) [11]-[13] [11]-[13] [11]-[13]	$H_2\overset{+}{N}H_2$ (5 8)	H_2CH_2 (4 8)	$H_2\overset{-}{B}H_2$ (3 8)	0
H3O(+) NH3 CH3(−) [11]-[13] [11]-[13] [11]-[13]	$H{-}\overset{+}{O}H_2$ (6 8)	$H{-}NH_2$ (5 8)	$H{-}\overset{-}{C}H_2$ (4 8)	2
F2Cl(+) H2O NF2(−) [11]-[13] [11]-[13] [11] [12]	$F{-}\overset{+}{Cl}{-}F$ (7 8)	$H{-}O{-}H$ (6 8)	$F{-}\overset{-}{N}{-}F$ (5 8)	4
XeF(+) HCl OH(−) [6] [13] [11]-[13] [11]-[13]	$\overset{+}{Xe}{-}F$ (8 8)	$Cl{-}H$ (7 8)	$\overset{-}{O}{-}H$ (6 8)	6
Xe F(−) [11] [13] [6] [11] [13]	None	Xe (8 8)	$\overset{-}{F}$ (7 8)	8

10

SF5(+) PF5 SiH5(−) [11] [12] [11]-[13] [11]	$F_3\overset{+}{S}F_2$ (6 10)	F_3PF_2 (5 10)	$H_3\overset{-}{Si}H_2$ (4 10)	0
IF4(+) SF4 PF4(−) [6] [12] [13] [11]	$F_2\overset{+}{I}F_2$ (7 10)	F_2SF_2 (6 10)	$F_2\overset{-}{P}F_2$ (5 10)	2
XeF3(+) BrF3 SeH3(−) [6] [11] [13] [12]	$F{-}\overset{+}{Xe}F_2$ (8 10)	$F{-}BrF_2$ (7 10)	$H{-}\overset{-}{Se}H_2$ (6 10)	4
XeF2 I3(−) [11]-[13] [11]-[13]	None	$F{-}Xe{-}F$ (8 10)	$I{-}\overset{-}{I}{-}I$ (7 10)	6
No	None	None	$\overset{-}{Xe}{-}F$ (8 10)	8

12

IF6(+) SeF6 AsF6(−) [6] [11]-[13] [11]-[13]	$F_3\overset{+}{I}F_3$ (7 12)	F_3SeF_3 (6 12)	$F_3\overset{-}{As}F_3$ (5 12)	0
XeF5(+) IF5 SF5(−) [6] [11]-[13] [11] [12]	$F_3\overset{+}{Xe}F_2$ (8 12)	F_3IF_2 (7 12)	$F_3\overset{-}{S}F_2$ (6 12)	2
XeF4 ICl4(−) [6] [11]-[13] [11]	None	F_2XeF_2 (8 12)	$Cl_2\overset{-}{I}Cl_2$ (7 12)	4
No	None	None	$F{-}\overset{-}{Xe}F_2$ (8 12)	6

Continued

14				
No IF7 **No** No [6] [11]-[13] No	F—Xe⟨F (8 14) +	F—I⟨F (7 14)	F—Se⟨F (6 14) −	0
XeF6 **No** [11]-[13] No	None	F—Xe⟨F (8 14)	F—I⟨F (7 14) −	2
XeF5(−) [13]	None	None	F—Xe⟨F (8 14) −	4

16				
No **No**	None	F—Xe⟨F (8 16)	F—I⟨F (7 16) −	0
XeF7(−) [13]	None	None	F—Xe⟨F (8 16) −	2

In **Table 1**, structural formulas of effective-valence isoelectronic compounds are methodically drawn. The central elements are from column 1 to 8 in the periodic table and every column of the periodic table is represented. External atoms are connected to the charged or uncharged central atom via one covalent bond. This is why surrounding atoms belong to column 1 or 7 in the periodic table.

In **Table 1**, compounds in horizontal series follow the effective-valence isoelectronicity rule. Column 1 gives the right side number of central atoms in the series and the name of each compound in the series. In one series, the left side number *i.e.* the valence number of the centers are different, according to the isoelectronicity rule. Compounds in each series appear in columns 2, 3 and 4 of **Table 1**.

Series also are arranged in groups of lines showing central atoms sharing the same Lewis number. This Lewis number is written in column 1 of **Table 1**, as well as on the right side of the center atoms in column 2 to 4. It ranges from zero for Na(+) up to 16 for XeF7(−) (on the last line).

The fifth column indicates the even number of remaining electrons in the outer-shell of the central atoms. This even number ranges from 0 for Na(+) in the first line of **Table 1**, up to 8 for F(−); with special values of 0 or 8 for Xe as a noble gas; see XeF(−).

Globally, **Table 1** has 59 known compounds and 14 unknowns. The large number of known compounds seems to validate both rules: The even-odd rule and the proposed definition of the effective-valence isoelectronicity rule. The existence of the unknown compounds is discussed in the following.

5. Discussion

5.1. Isoelectronic Compounds and Lewis' Octet Rule

In **Table 1**, only fourteen compounds, like CH4, NH4(+) and BH4(−), meet Lewis' octet rule. We notice that if one of these compounds is a Lewis compound, all its isoelectronic compounds will be as well. The number of these compounds is small in comparison with the 40 other known single-bonded even-odd compounds. We already know that the group of Lewis compounds is encompassed in the group of even-odd compounds; it seems thus possible that another rule exits, including Lewis' octet compounds, even-odd compounds and some more. This new group could include unpaired electrons or multi-bonded connections.

5.2. Similar Charge Positions between References and the Even-Odd Rule

When comparing results of the even-odd rule to those given in scientific literature, charge position is the same for the following compounds and their isoelectronic ions: BF3, SiH2, CH4, NH3, H2O, HCl and PF5. The charge is systematically located on the central atom in the structure. Twenty-one compounds are concerned. It is nearly 50% of the known compounds listed in **Table 1**. As a consequence, the existence of these 21 compounds in **Table 1**, may be interpreted as a valid confirmation of the usefulness of the even-odd rule and of the effective-valence isoelectronic rule.

5.3. Central Atoms without Covalent Bonds

Several compounds in **Table 1** are composed of single atoms, charged or neutral. About 50% of these monoatomic compounds following the even-odd and the isoelectronic rules are referenced in scientific literature.

Neutral elements in column 3, are: Be, C, O and Xe. Their structural formulas follow the even-odd rule. Information is available on each of them: their chemical and physical properties are known and they can remain monoatomic in the gas phase at sufficiently high temperature.

Monoatomic cations are Na(+), B(+), N(+) and F(+) in column 2. They are compatible with the even-odd rule. As shown on the same line, ion Na(+) is isoelectronic with Xenon, a noble gas. Both are known compounds. The three other cations, B(+), N(+) and F(+) are not known entities. Is this due to specific chemical properties that prevent them from existing in an independent form or are they inexistent? A definitive answer can only be given when these ions are synthetized and analyzed.

Among monoatomic anions in column 4, Na(−), B(−), N(−), F(−), only F(−) is known. Is it possible to produce and experiment on the three others? For example, N(−) is isoelectronic with monoatomic oxygen. As the latter is obtained at high temperature, could it be the same for N(−)?

5.4. Central Atoms with Their Highest Number of Electrons in the Outer-Shell

The highest number of electrons in the outer-shell is displayed by xenon. Compounds of this noble gas are very particular; they are not known for other noble gas. In **Table 1**, about 75% of them are referenced to in scientific literature.

Among neutral molecules listed in column 3, xenon single-bonded compounds are Xe (see group 8), XeF2, XeF4, XeF6 and XeF8. Except for XeF8, their properties are known. Is it possible to obtain information about this molecule?

Among cations in column 2, they are: XeF(+), XeF3(+), XeF5(+) and XeF7(+). Only the last one is unknown. Since its isoelectronic molecule IF7 is known, is it possible to produce it?

Among anions in column 4 at last, they are: XeF(−), XeF3(−), XeF5(−) and XeF7(−). Two of them are known: XeF5(−) and XeF7(−). An expansion of our knowledge, mainly for XeF3(−), even if it does not have an isoelectronic compounds, may be possible.

5.5. Standard Compounds

Among all other isoelectronic compounds obtained with both even-odd and isoelectronic rules, only three are not referenced in scientific literature. These are NaCl2(−), SeF7(−) and IF8(−). There are only three unknown compounds in comparison with the large number of well-known isoelectronic compounds (more than 40) shown in **Table 1**. It may be interesting to find which physical property is preventing their synthesis or their study.

5.6. Isoelectronic Group and the Even-Odd Rule

In **Table 1**, each triplet of isoelectronic compounds is drawn in agreement with the even-odd rule. In some triplets, one compound is unknown although we suppose that they can exist. An argument supporting this hypothesis is: one compound not following the even-odd rule belongs to an isoelectronic group and none of the group can be included in the even-odd rule. This argument seems acceptable even if it is in a negative form. It is therefore possible that unknown single-bonded compounds will become known compounds.

5.7. Localized Electrical Effect of Charges

On any line of **Table 1**, two isoelectronic compounds have the same covalent-bond configurations. The charge itself, located in the center, has therefore no impact on electrical properties of bonds. As a consequence, the electrical effect of the charge can only be localized in the electron inner-shell and on the nucleus. Scientific literature is unfortunately not consistent on that subject and many structures, drawn using classical methods, indicate a charge globally spread over the entire compound [14]. This apparent contradiction will probably be explained for multi-bonded compounds.

6. Conclusions

Structural formulas of several known and unknown single-bonded compounds are drawn using the even-odd rule and the effective-valence isoelectronicity rule. They are sorted by the value of the Lewis number and the number of electrons in the outer-shell. This organization forms groups of compounds and gives an independent way to test the effectiveness of the even-odd rule with the definition of the effective-valence isoelectronicity. As a consequence, isoelectronic single-bonded compounds form several groups completely included in the even-odd rule.

The validity of both rules is here confirmed for a large number of known compounds even if information is missing for about 15% of them. Each of these 15% is isoelectronic with known single-bonded compounds. These unavailable compounds like $N(+)$, XeF_4 or $XeF_3(-)$ may be under investigations to improve the number of known compounds and to confirm the global importance of both rules.

The next step to check the even-odd rule will probably be an adaptation to multiple-bonded compounds.

References

[1]   Icilio, G. (1911) Amedeo Avogadro e la sua opera scientific. Accademia delle scienze, Opere scelte di Amedeo Avogadro, Turin, i-cxl.

[2]   Greenaway, F. (1966) John Dalton and the Atom. Cornell University Press, Ithaca.

[3]   Einstein, A. (1905) Über die von der molekularkinetischen Theorie der Wärmegeforderte Bewegung von in ruhenden Flüssigkeiten suspendierten Teilchen. *Annalen der Physik*, **322**, 549-560.

[4]   Perrin, J. (1909) Mouvement brownien et réalité moléculaire. *Annales de Chimie et de Physique*, **18**, 1-114.

[5]   Faraday, M. (1859) Experimental Researches in Chemistry and Physics. Richard Taylor and William Francis, London.

[6]   Greenwood, N.N. and Earnshaw, A. (1997) Chemistry of the Elements. 2nd Edition, Butterworth-Heinemann, Oxford.

[7]   DeKock, R.L. and Gray, H.B. (1989) Chemical Structure and Bonding. University Science Books, 94.

[8]   http://en.wikipedia.org/wiki/Isoelectronicity

[9]   Auvert, G. (2014) Improvement of the Lewis-Abegg-Octet Rule Using an "Even-Odd" Rule in Chemical Structural Formulas: Application to Hypo and Hyper-Valences of Stable Uncharged Gaseous Single-Bonded Molecules with Main Group Elements. *Open Journal of Physical Chemistry*, **4**, 60-66. http://dx.doi.org/10.4236/ojpc.2014.42009

[10] Auvert, G. (2014) Chemical Structural Formulas of Single-Bonded Ions Using the "Even-Odd" Rule Encompassing Lewis's Octet Rule: Application to Position of Single-Charge and Electron-Pairs in Hypo- and Hyper-Valent Ions with Main Group Elements. *Open Journal of Physical Chemistry*, **4**, 67-72. http://dx.doi.org/10.4236/ojpc.2014.42010

[11] http://www.chemspider.com/

[12] http://www.ncbi.nlm.nih.gov/

[13] http://en.wikipedia.org/

[14] Gillespie, R.J. and Popelier, P.L.A. (2001) Chemical Bonding and Molecular Geometry. Oxford University Press, Oxford.

21

Chemical Bonds between Charged Atoms in the Even-Odd Rule and a Limitation to Eight Covalent Bonds per Atom in Centered-Cubic and Single Face-Centered-Cubic Crystals

Geoffroy Auvert[1], Marine Auvert[2]

[1]Grenoble Alps University, Grenoble, France
[2]University of Strasbourg, Strasbourg, France
Email: Geoffroy.auvert@grenoble-inp.org

Abstract

A crystal is a highly organized arrangement of atoms in a solid, wherein a unit cell is periodically repeated to form the crystal pattern. A unit cell is composed of atoms that are connected to some of their first neighbors by chemical bonds. A recent rule, entitled the even-odd rule, introduced a new way to calculate the number of covalent bonds around an atom. It states that around an uncharged atom, the number of bonds and the number of electrons have the same parity. In the case of a charged atom on the contrary, both numbers have different parity. The aim of the present paper is to challenge the even-odd rule on chemical bonds in well-known crystal structures. According to the rule, atoms are supposed to be bonded exclusively through single-covalent bonds. A distinctive criterion, only applicable to crystals, states that atoms cannot build more than 8 chemical bonds, as opposed to the classical model, where each atom in a crystal is connected to every first neighbor without limitation. Electrical charges can be assigned to specific atoms in order to compensate for extra or missing bonds. More specifically the article considers di-atomic body-centered-cubic, tetra-atomic and dodeca-atomic single-face-centered-cubic crystals. In body-centered crystals, atoms are interconnected by 8 covalent bonds. In face-centered crystal, the unit cell contains 4 or 12 atoms. For di-element crystals, the total number of bonds for both elements is found to be identical. The neutrality of the unit cell is obtained with an opposite charge on the nearest or second-nearest neighbor. To conclude, the even-odd rule is applicable to a wide number of compounds in known cubic structures and the number of chemical bonds per atom is not related to the valence of the elements in the periodic table.

Keywords

Even-Odd, Rule, Covalent Bond, Single Bond, Crystal, Solid, Centered, Face-Centered, Unit Cell

1. Introduction

In crystal structures, positions of atoms in the unit cell of a crystal are derived from experimental diffraction data [1]. Given the atoms positions, the coordination number of each atom, *i.e.* the number of its near neighbors can be evaluated [2]. For molecules, this number also indicates the number of bonds around the atom [3] [4]. Various types of bond can bind an atom and its first neighbors [5]. Unfortunately, diffraction data do not give direct information to choose the nature of bonds in a solid [6]. For the same reason, there is no direct information about atoms charges.

The recently proposed even-odd rule gives a novel procedure to evaluate and draw bonding configurations and precisely locate charges in ions [7] and molecules [8]. This procedure has been firstly successfully tested on well-known single bonded ions and molecules and secondly shown capable of turning multi-bonded ions and molecules into single-bonded compounds [9]. This procedure indicates on which atoms charges are located in charged or uncharged compounds. Applied to ions and molecules, it also gives the number of covalent bonds that each atom may form in agreement with its valence number [10].

The objective of the present paper is two-fold: to validate first an extension, of the procedure to cubic crystalline structures, during which we show that the number of bonds is not limited to the valence number; then to propose alternative bonding configurations in well-known cubic structures while limiting the number of covalent bonds per atom to a maximum of eight.

Two types of atom configurations in cubic crystalline structures are described without bonds: centered-cubic and single face-centered cubic. The even-odd rule is defined and a constraint in the number of covalent bonds for each atom is introduced: no atom should form more than eight covalent bonds. The rule and constraints are then applied to both crystalline structures taking example of a large number of known compounds. Results are discussed and the authors conclude on the validity of the even-odd rule for crystals including the maximum of 8 bonds for each element.

In the present paper, compounds and structures names are found in "crystallography open data base" [11], atoms positions in unit cells in "Wikipedia" [12]. Unit cells, atoms and bonds are represented thanks to "Avogadro V1", 3D molecular structure editor [13], completed with a drawing software to indicate charges [14].

To keep this publication consistent with the previous three [7]-[9], the writing of chemical formulas is for example: CaF_2 for calcium fluoride and HfO_2 for hafnium oxide and in a crystal, both are described as di-elements tri-atomics unit cells.

2. Centered and Face-Centered Crystals

Crystal structures are represented as a pattern of periodically repeated unit-cells. A unit cell is the base structure, classically represented as contained in a small imaginary box [15]. This cell can have different configurations [15] [16]. The present article focuses on cubic unit cells as shown in **Figure 1**, in which the cube is symbolized by red edges showing 8 corners, 6 faces and 12 edges.

Atoms can be arranged in many different ways inside cubic unit cells. In order to keep this article clear and concise, the authors resolved to focus on two of them. The first one is a centered-cubic structure, **Figure 1(a)**, and the second is a single-face-centered-cubic structure, **Figure 1(b)**. In **Figure 1(a)** and **Figure 1(b)**, each colored sphere represents an atom. Atoms can lie in the center of the cube, in the center of faces or in corners.

In **Figure 1(a)**, the central atom in the cube has a neighbor in a corner of the cube. When the unit cell is repeated, the same central atom ends up having 8 near neighbors, one in each corner of the cube. In the same way, an atom in the corner of the cube has 8 near neighbors, located in the center of each of the 8 cells it belongs to.

In **Figure 1(b)**, four atoms can be seen: one is in a corner and the other three are face centered. The corner atom seems to have three first-neighbors in the same cell but only a half of each belongs to that cell. When the unit cell is repeated, the total number of first-neighbors is 12. Atoms located in the center of a face have 12 near neighbors as well.

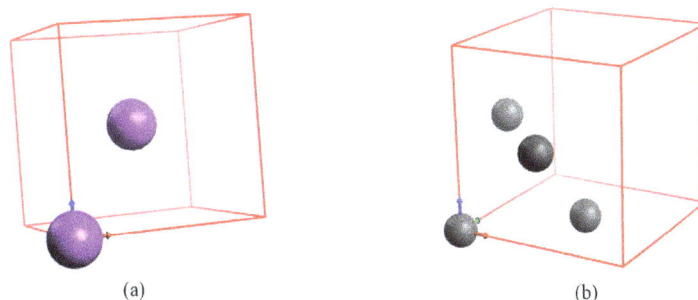

(a)   (b)

Figure 1. Cubic unit cells in crystals showing two specific arrangements of atoms. In (a) a centered-cubic unit cell with two atoms per cell and in (b), a simple-face-centered-cubic unit cell with four atoms per cell. (For a representation with all atoms participating in the unit cell, refer respectively to **Figure 2** and **Figure 4**).

Although X-ray diffraction can indicate the arrangement of atoms in a crystal, covalent bonds and potential charges cannot be derived with any known method. In the following, a rule is described to systematize how to draw bonds between atoms in cubic unit cells. This procedure uses a line to represent a single-covalent interconnection between two first-neighbor-atoms.

3. The Even-Odd Rule in a Procedure to Draw a Valid Unit Cell with Single-Covalent-Bonds between Charged or Neutral Atoms

The even-odd rule was developed for molecules and ions under liquid and gaseous phase [7]-[9]. In this chapter the even-odd rule is described when applied to solid crystals.

In the present article and in the context of the even-odd rule, the authors define the following expressions: 1) *even-atom* means an atom with an even number of electrons and 2) *odd-atom* means an atom with an odd number of electrons. For instance, reading the number of electrons in the periodic table, carbon is an *even-atom* and nitrogen is an *odd-atom*.

In scientific literature treating of crystalline structures, atom positions and their bonding arrangement in the unit cell are available for some cubic crystals [12]. Other crystal descriptions, however, totally lack information on charge or interconnection locations [17]. A six-step, clear and well-defined procedure based on the even-odd rule is proposed below, aiming to systematize how to check validity of bonding configuration in crystals:

- **Interconnections between atoms**
  - Every multiple-bonds must be replaced by a single-covalent bond as described in ref. [9].
  - When an atom has more than eight covalent bonds, remove connections in excess down to eight.
  - When an atom is charged, the charge affects the number of possible bonds: As an example: C has 4 or 2 bonds then C(+) must be with 3 or 1 bonds [7].
- **Even-odd rule**
  - Uncharged *even-atoms* must have an even number of bonds
  - Uncharged *odd-atom* must have an odd number of bonds
  - If an *even-atom* or an *odd-atom* is charged, the reverse condition must be used
  - In order to meet the above conditions, one can modify the charge or the number of bonds.
- **Unit cell on the whole neutrally charged**
  - Charged atoms must be balanced in the unit cell by opposite charges on a first or second neighbor.
  - The total charge of the crystal must be zero; when possible, this is applied to each unit cell; if not, the cell charge is neutralized by one of the nearest unit cell.
- **Conservation of the total number of electrons**
  - All bonds must be covalent, *i.e.* involve one pair of electrons.
  - The total number of electrons in the cell, once bonds are drawn, must be identical to the total number of electrons of each atom without bond as given by the periodic table.
- **Charge and bonding conditions**
  - Atoms bearing identical charges cannot be connected
  - When a unit cell contains more than one element, two atoms of the same element are not interconnected.

— The consequence on studied structures is:

  ○ In mono-element crystals with di- or tetra-atomics unit cells, each atom counts the same number of bonds and neighbors are neutral or bear opposite charges.

  ○ In di-elements crystals with di-atomics unit cells, each atom counts the same number of bonds and two neighbors are neutral or with opposite charges in first or second neighbor.

  ○ In di-elements crystals where unit cells count more than 4 atoms, the total number of bonds of the first element is equal to the total number of bonds of the second element. The neutrality of the unit cell is obtained inside one series of elements or between both series.

— When disconnecting two atoms does not provide a satisfactory solution, extend the unit cell to an even number of the base unit cell previously used.

— The number of bonds of an element in a crystal does not need to match with its valence number of the periodic table.

- **Geometry of unit cells in a crystal:**
  - — Opposing faces of a unit cell are identical.

4. The Even-Odd Rule Applied to Centered-Cubic Crystals

The even-odd rule indicates how to draw bonds in a crystal and to adapt the charge of every atom to its bonding configuration. In this chapter, the rule is applied to centered-cubic crystals with 8 covalent bonds for each atom in the unit cell.

4.1. Centered-Cubic Crystals with Mono-Element Di-Atomic Unit Cell

The objective is to build a covalent bonding configuration in a centered-cubic structure (see **Figure 1(a)** for the bondless structure). **Figure 2(a)** depicts a center-cubic structure with atoms located in the center of the unit cell and in its corners. The central atom belongs to the visible unit cell and each atom at the corner belongs to 8 unit cells, in other words only 1/8 of a corner atom belongs to the visible unit cell. **Figure 2(a)** also shows single-covalent bonds between each nearest neighbor. With 8 nearest neighbors, the central atom has eight single-covalent bonds and a corner-atom has also eight covalent bonds, one in each unit cell it belongs to. According to the even-odd rule, since atoms in the current structure have an even number of bonds:

— If the crystal is composed of *even-atoms*, (**Figure 2(a)**), atoms composing the structure are uncharged. As listed in the right-hand column of **Figure 2(a)**, main group elements falling into this category are Barium (56), Calcium (20), Radium (88), Selenium (34) and Strontium (38). Chromium, Iron and Molybdenum are few examples in the category of transition elements.

— If the crystal is composed of *odd-atoms*, (**Figure 2(b)**), they must be charged, either positively or negatively. With a charge, the number of electrons for each atom becomes evidently even and atoms can build an even number of covalent bonds. In order to neutralize the charge borne by the central atom on the scale of a unit cell, atoms located in corners of the cube must bear an opposite charge. As detailed in the right-hand column of **Figure 2(b)**, from the main group of the periodic table, six elements can form this type of structure: Antimony, Cesium, Potassium, Sodium, Rubidium and Thallium. In the group of transition elements: Europium, Manganese, Niobium, Neptunium, Vanadium and Tantalum can take this configuration.

All 26 elements listed in **Figure 2(a)** and **Figure 2(b)**, have an **Im-3m** structure [11]. They are drawn with the procedure described previously, *i.e.* their bonding and charge organizations are in agreement with the even-odd rule.

4.2. Centered Cubic-Crystals with Di-Element Di-Atomic Unit Cell

Figure 3 depicts three centered-cubic structures with a di-element di-atomic cell. The structure name is **Pm-3m** [11]. For each case, compounds from ref. [11] that can adopt this structure are listed in the right-hand column.

In **Figure 3(a)**, the binary compound is composed of neutral *even-atoms* and each atom has eight bonds, the unit cell is consequently uncharged. In **Figure 3(b)**, with a compound composed of *odd-atoms*, one atom is positively charged and the other is negatively charged; the unit cell is consequently also neutrally charged. In **Figure 3(c)**, binary compounds are composed of one *odd-atom* and one *even-atom*. Only *odd-atoms* bear a charge, since they are evenly bonded. The neighboring even atoms are also evenly bonded, so they cannot bear a charge. Compensating the charge borne by the odd-atoms is thus only possible at a greater scale, *i.e.* if the *odd-atom* of

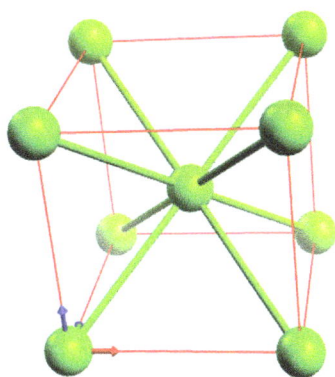

(a)

Uncharged *even-atoms*
Structure Im-3m [11]
Elements from the main group.
Ca Beta 20 c2 [11] [13]
Ba 56 [11]-[13]
Ra 88 c2 [11] [12]
Se 34 [11]
Sr 38 [11]
Transition metals
Cr 24 [11]-[13]
Fe 26 [11]-[13]
Mo 42 [11]-[13]
Pu 94 [11] [13]
Ti 22 [11] [13]
Th 90 [11]
U 92 [11]
Zr 40 [11]
W 74 [11] [12]

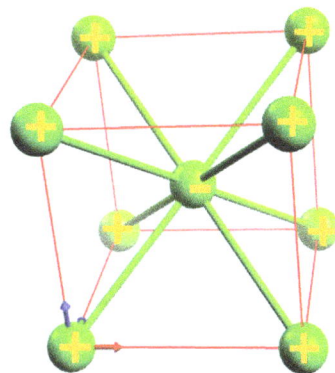

(b)

Charged *odd-atoms* **in neutral unit-cell**
Structure Im-3m [11]
Elements from the main group
Cs 55 [11]-[13]
K 19 [11]-[13]
Na 11 c1 [11]-[13]
Rb 37 [11] [12]
Sb 51 c5 [11]
Tl alpha 81 c3 [11] [12]

Transition metals
Eu 63 [11]-[13]
Mn 25 [11]-[13]
Nb 41 [11]-[13]
Np 93 [11]-[13]
V 23 [11]-[13]
Ta 73 [11] [12]

Figure 2. Unit cell of a centered-cubic structure with two atoms per cell in an **Im-3m** structure. In (a), all even-atoms are neutral and count 8 covalent bonds, in (b), all *odd-atoms* count 8 covalent bonds but bear alternatively opposite charges allowing for a neutral unit cell.

the nearest unit cell bears an opposite charge. The opposite charge is on the second nearest neighbor placed in the nearest unit cell.

In this chapter, all 75 binary compounds, found in ref. [11], adopt a **Pm-3m** structure. They have been processed with the procedure described previously and they agree with the even-odd rule. Elements studied belong to all groups of elements in the periodic table and all count 8 bonds. It seems to confirm the theoretical hypothesis according to which the number of bonds of an element in a cubic crystal is independent from the valence number of the said element.

The next structure studied is a single-face-centered cubic-structure: **Fm-3m** or **Pm-3m**.

5. Single-Face-Centered Cubic-Crystals with a Limited Number of Bonds

The even-odd rule will be applied first to be consistent with the chemical formula of the compound and second to use, when needed, a maximum of 8 covalent bonds for all atoms in the unit cell.

5.1. Single-Face-Centered Cubic-Crystals with Uncharged Mono-Element Tetra-Atomic Unit Cell

A single-face-centered cubic crystal contains 4 atoms per unit cell. (This word "single" is used to discriminate this structure from a double-face-centered crystal that involves 8 atoms in a unit cell). This single-face-centered

Even-atoms/even-atoms
Structure Pm-3m[11]
TeTh
CdCe
CeMg
HgMg
MgSr
CrZn

(a)

Odd-atoms/odd-atoms
Structure Pm-3m[11]

AuCs	AgLa	ClK
AuHo	AgPr	ClRb
AuLu	AgSc	ClTl
AuPr	AgTb	CsI
AuRb	AgTm	ITl
AuSc	AgY	NI
AuTb	BrCs	NaCl
AuTm	BrN	SbTl
AuY	BrTl	
AgGa	ClN	
AgHo	ClCs	
AgI		

(b)

Even-atoms/odd-atoms or ***odd-atoms/even-atoms***
Structure Pm-3m   [11]

AgCd	AuGd	AuZn	CsS
AgCe	AuMg	BrTi	CuPd
AgDy	AuNd	CsS	CuZn
AgEr	AuSm	CaTl	HgLi
AgGd	AuTi	CdLa	LaMg
AgMg	AuYb	CdPr	LaZn
AgNd	AlNi	CoFe	MgPr
AgYb	AuCd	CsSe	MgTl
AgZn	AuDy		NbO
AlNd	AuEr		PrZn

(c)

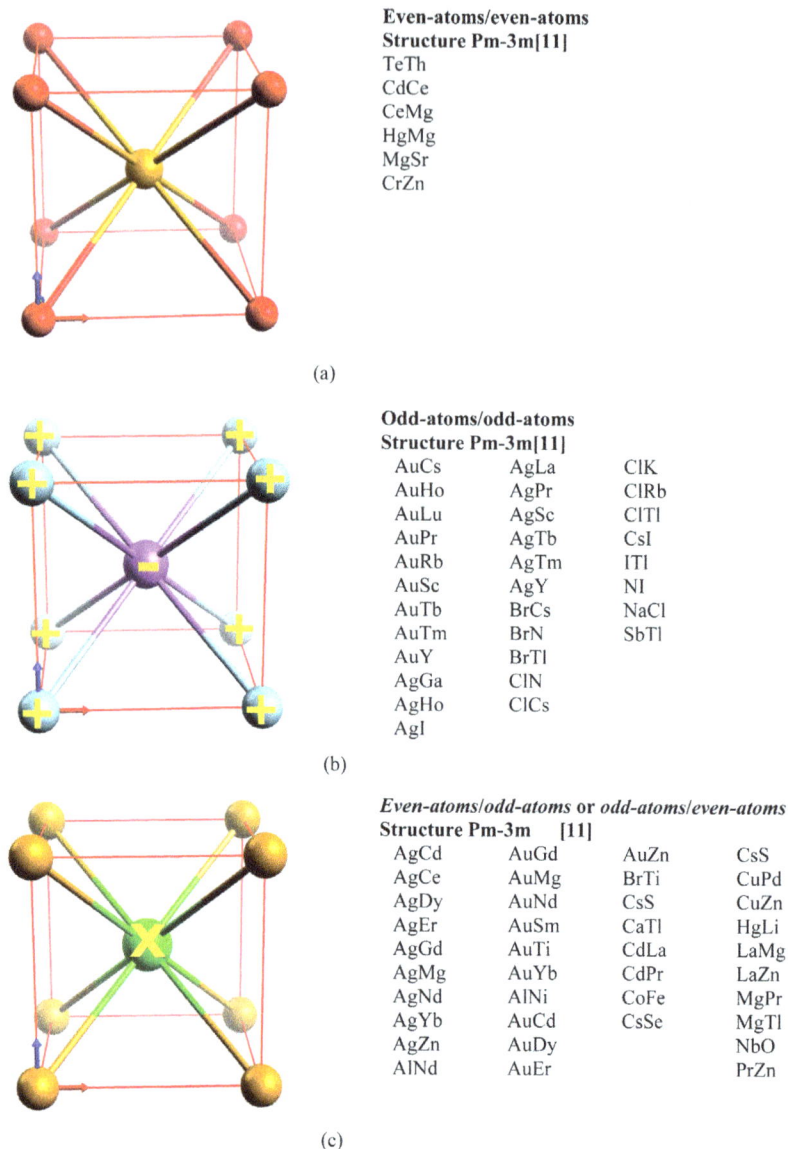

Figure 3. Centered-cubic crystals with a di-element di-atomic unit cell. In (a), both atoms are *even-atoms* and none bear charges. In (b), both atoms are *odd-atoms* and bear an opposite charge, *i.e.* + and − for a neutral unit cell. In (c), an atom in the corner of the unit cell is an uncharged *even-atom* and the center *odd-atom* bears a charge. The total charge of the cell is neutralized by the second neighbor *odd-atom* in the nearest unit cell.

cubic crystal is depicted in **Figure 4** with a complete set of identical atoms extended from the unit cell of **Figure 1(b)**.

In **Figure 4(a)**, atoms locations are displayed in the mono-element tetra-atomic unit cell, without showing covalent bonds. Each atom counts 12 neighbors. In the present article the number of covalent bonds is however proposed to be limited to a maximum of 8 covalent bonds. **Figure 4(b)** shows the bonding configuration in agreement with the even-odd rule and the limitation rule: there are no horizontal connections and each *even-atom* is uncharged.

Mono-element compounds listed in the right-hand column of **Figure 4(b)** have the same structure [11]. It includes noble gases as Argon, Krypton, Neon and Xenon, known in chemistry for not binding easily. In this chapter they build 8 covalent bonds. This shows again that, in cubic crystals studied in this article, the valence

(a)

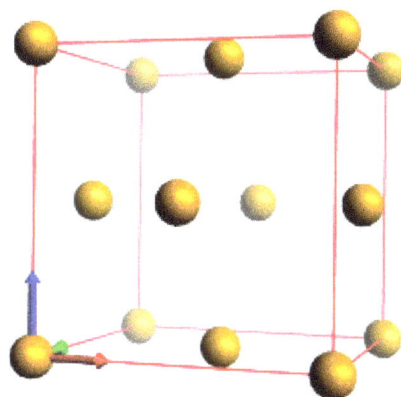

Single-face-centered cubic crystals
Structure: Fm-3m [11]
Mono-element tetra-atomic unit cell
3D view of a cubic unit cell without bonds.
Only 4 atoms belong to this unit cell.
14 atoms involved in one unit cell.
Six atoms are face-centered and eight are corners atoms.
Each atoms counts 12 first neighbors.
.

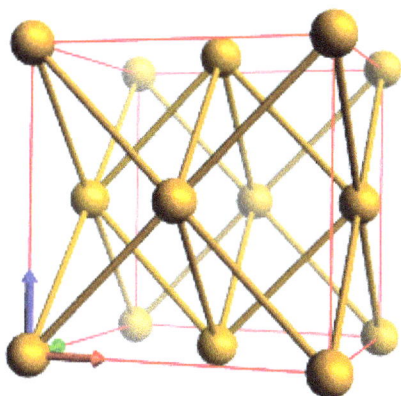

Bonding procedure
All atoms count 8 single-covalent bonds, regardless of their location in the unit cell.
Even-atoms **from the main group**
Ar 18 [11]-[13]
Ba 56 [12]
Ca alpha 20 [11]-[13]
Kr 36 [11]-[13]
Ne 10 [11]-[13]
Sr 38 [11]
Rn 86 [11]
Xe 54 [11]-[13]
Transition-Metals even-atoms
Fe gamma 26 [11]-[13]
Ni 28 [11] [12]
Pb 82 c4 [11]
Pt 78 [11]-[13]
Pd 46 [11]-[13]
Yb Beta70 [11]-[13]

(b)

Figure 4. Face-centered cubic structure in a tetra-atomic unit cell. In (a), 14 atoms are shown without any bond and in (b), each *even-atom* is interconnected to eight of its 12 neighbors in a nearly vertical direction and without charge.

number has no impact on the number of bonds between atoms.

5.2. Single-Face-Centered Cubic-Crystals with Charged Mono-Element Tetra-Atomic Unit Cell

The same single-face-centered cubic-crystal structure can be composed of *odd-atoms*. In order for the *odd-atoms* to build an even number of covalent bonds, the even-odd rule states that atoms must be charged. **Figure 5** depicts the interconnection system allowing this **Fm-3m** structure to be in agreement with the even-odd rule. Atoms belonging to a horizontal plane are not connected and bear the same charge. Each successive plane bears alternatively positive and negative charges in order to obtain neutral unit cells. **Figure 5** also shows that each covalent bond connects atoms with opposite charge, *i.e.* no two atoms bearing the same charge are bonded. The number of covalent bonds for each atom is also compliant and equal to 8. It is also noteworthy that, for all mono-element di-atomic compounds in the right-hand side column of **Figure 5**, the number of bonds is not related to the valence number of the elements involved.

5.3. Single-Face-Centered Cubic-Crystals with Di-Element Tetra-Atomic Unit Cell

Following structures with a different structure name: **Pm-3m** and with the same single-face-centered cubic organization, comes from available data [11]. Four different configurations of di-elements tetra-atomics unit cells are detailed in **Figure 6** and **Figure 7**.

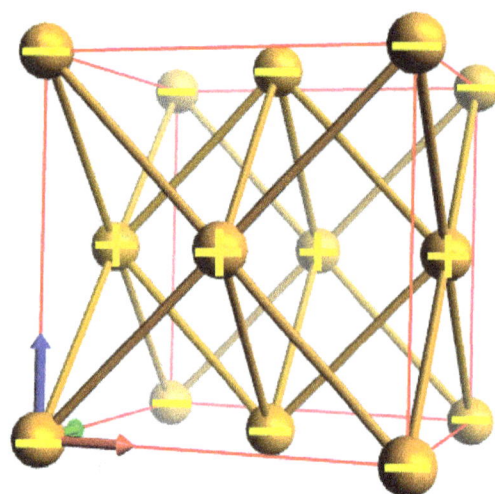

***Odd-atom*s in single-face-centered cubic crystals**
Structure: Fm-3m
Each atom counts 8 covalent bonds

Elements from the main group
Al Alpha 13 [11]-[13]
Na 11 [12]

Transition-Metals
Ac 89 [11]-[13]
Ag 47 [11]-[13]
Am Alpha 95 [11]
Au 79 [11]-[13]
Cu 29 [11]-[13]
Ir 77 [11]-[13]
Mn gamma 25 [11] [13]
Rh 45 [11]-[13]

Connections only between + and – charges.
Neutral unit cell.

Figure 5. Face-centered cubic-crystals with a mono-element tetra-atomic unit cell. 14 atoms are shown, each with 8 covalent bonds. The *odd-atoms* are charged to allow an even number of bonds. Notice that atoms with the same charge are not inter-connected.

FeNi3: Fe brown, **Ni** Green   **NaPb3: Na** brown, **Pb** Green

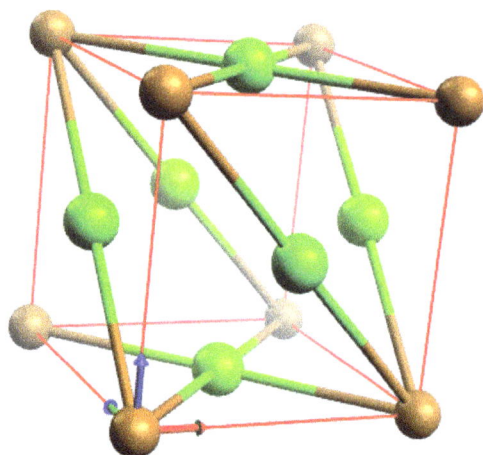

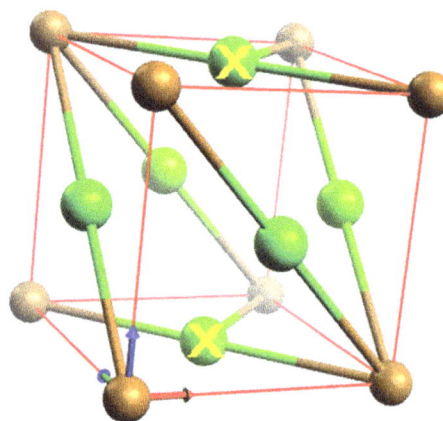

Even-even atoms
Structure: Pm-3m
Fe: 8 bonds in corners
Ni face centered
Bonds: 2*2+4 = 8
Neutral atoms

CaSn3, CaPb3, CePb3, CeSn3,
FeNi3, PdSn3, SnNi3, SnPd3
[11]

Odd-even atoms
Structure: Pm-3m
Na: 7 bonds and neutral
Pb face centered
Bonds: 1*3+2*2 = 7
One charged Pb atom

AuTi3, InPt3, LaSn3, NaPb3,
ReO3 [11]

(a)   (b)

Figure 6. Single-face-centered crystals with di-element tetra-atomic unit cell. In (a), Fe atoms (brown) have an even number of electrons and they are uncharged with an even number of bonds. Ni atoms (green) are uncharged *even-atoms*. In (b), Na atoms (brown) are *odd-atoms* with an odd number of bonds (7 for one atom). Pb atoms (green) are neutral *even-atoms* with two bonds and charged with three bonds and the same total number of bonds (7 for three green atoms).

In **Figure 6(a)**, FeNi3 is an example of the first case: the di-element tetra-atomic unit cell is composed of *even-atoms*. Inside, the iron atom (brown) counts 8 bonds. Two nitrogen atoms (green) have two bonds and one four bonds with a total of 8, which corresponds to the number of bonds of the iron atom. Eight compounds, listed below the drawing, have this configuration.

In **Figure 6(b)**, NaPb3 is an example used as a reference of an odd-even configuration. Na atoms (brown) are

PtCu3: Pt brown, Cu Green   **NNa3: N brown, Na Green**

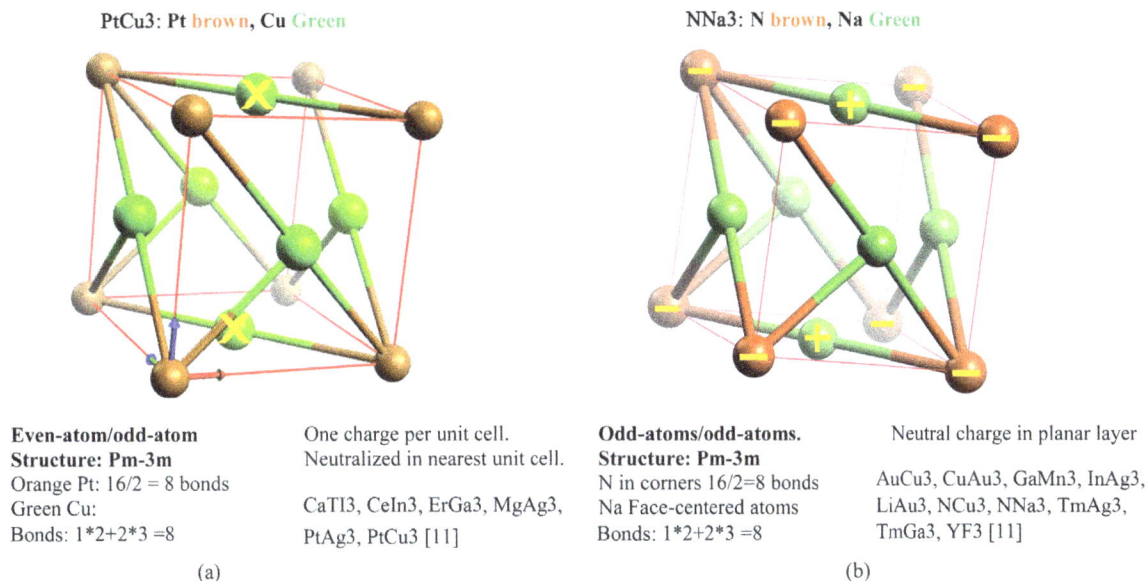

Even-atom/odd-atom	One charge per unit cell.	**Odd-atoms/odd-atoms.**	Neutral charge in planar layer
Structure: Pm-3m	Neutralized in nearest unit cell.	**Structure: Pm-3m**	
Orange Pt: 16/2 = 8 bonds		N in corners 16/2=8 bonds	AuCu3, CuAu3, GaMn3, InAg3,
Green Cu:	CaTI3, CeIn3, ErGa3, MgAg3,	Na Face-centered atoms	LiAu3, NCu3, NNa3, TmAg3,
Bonds: 1*2+2*3 =8	PtAg3, PtCu3 [11]	Bonds: 1*2+2*3 =8	TmGa3, YF3 [11]

(a)   (b)

Figure 7. Single-face-centered crystals with 4 atoms per unit cell. In (a), one *even-atom* (brown) and 3 *odd-atoms* (green) per unit cell. In (b), one *odd-atom* (brown) and 3 *odd-atoms* (green) per unit cell. The charge positions and number of bonds follow the even-odd rule.

odd-atoms and are uncharged thanks to an odd number of bonds. With 7 bonds, it is neutral in **Figure 6(b)**. Inside the unit cell, all three Pb atoms (green) should have the same total number of bonds. This is possible with 2 bonds for two Pb atoms and 3 for the third one as depicted in **Figure 6(b)**. The Pb with 3 bonds, being an *even-atom*, is charged and it cannot be neutralized in the cell. This charge, located in the center of the horizontal face, can only be compensated in the nearest unit cell by nearest second neighbor in the same horizontal plane. Five compounds, listed below the drawing, have this configuration.

After these even-even and odd-even configurations, the even-odd and odd-odd configurations are shown in **Figure 7**. In **Figure 7(a)**, PtCu3 is an example used as a reference of an even-odd configuration. Pt atoms (brown), located in corners of the cell, are *even-atoms*, have 8 bonds and are uncharged. Of the Cu *odd-atoms* (green), one is charged and bonded twice whereas both others are uncharged and bonded three times. Note that inside the unit cell, the total count of Cu-atoms bonds is 8, which corresponds to the number of bonds of the Pt atom. The atom located in the center of the lower horizontal place is charged and is compensated in the nearest unit cell. Below the PtCu3 crystal, a list of 6 compounds that can have this bonding configuration [11].

In **Figure 7(b)**, NNa3 is shown as a reference for two *odd-atoms*. Nitrogen atoms (brown), located in a corner has 8 bonds and must be charged. Of the three Nickel *odd-atoms* (green), two are uncharged with 3 bonds and one is charged with two bonds. It gives the eight expected bonds inside the unit cell. The neutrality of each horizontal planes ensures neutrality of the crystal. Ten binary compounds take this configuration and their formulas are referenced [11].

The limitation to a maximum of eight bonds per atom is very important in these four configurations. It imposes a particularly strong constraint on the second atoms present three times in the compound. In a di-elements tetra-atomics unit cell represented by AB3, the corner atoms A count 8 or 7 bonds and are interconnected with the three other elements with a total number of bonds inside the unit cell equal to 8 or 7 respectively. Each B atom can thus not build many bonds.

In the following, the total number of atoms per cell is extended to 12 atoms without any change in the single-face-centered cubic structure: **Fm-3m**.

5.4. Single-Face-Centered Cubic-Crystals with a Dodeca-Atomic Unit Cell

The present structure has the same face-centered atom positions with a tetra-atomic unit cell but a series of 8 atoms is added in the unit cell. They form a di-element dodeca-atomic unit cell with an **Fm-3m** structure as shown in **Figure 8** and **Figure 9**.

ZrO2: Zr Black 4 neutral *even-atoms*
O Red 8 neutral *even-atoms*

CaF2: Ca Green 4 neutral *even-atoms*
F Blue 8 charged *odd-atoms*

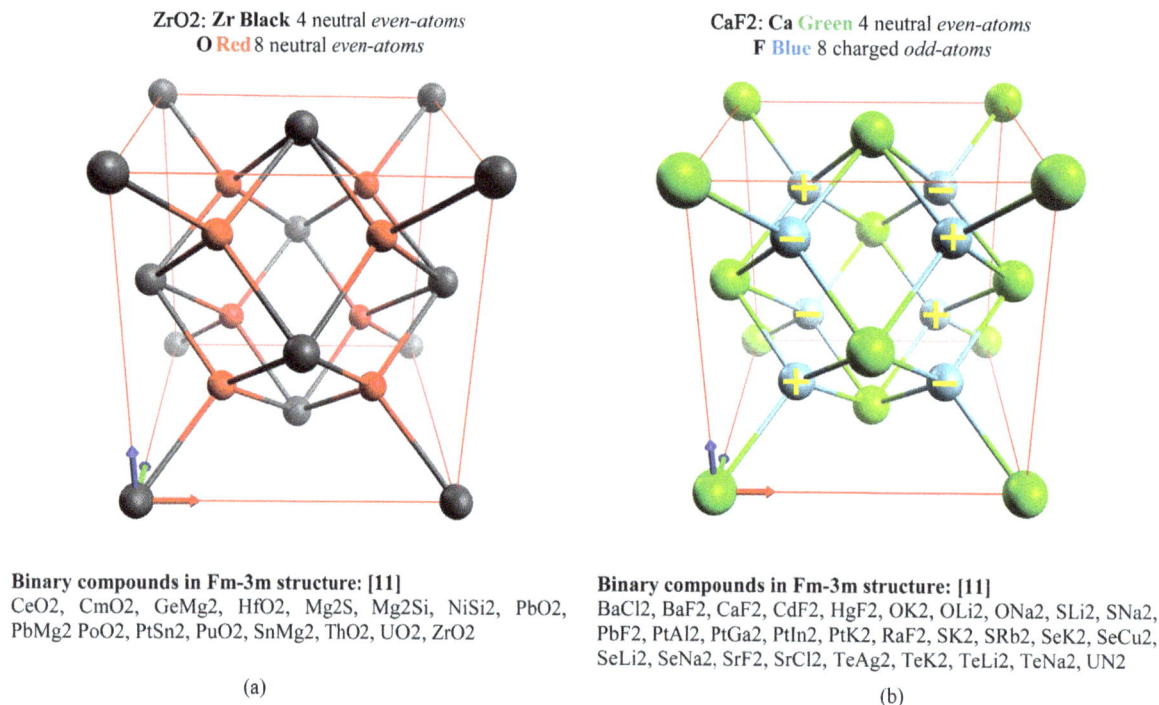

Binary compounds in Fm-3m structure: [11]
CeO2, CmO2, GeMg2, HfO2, Mg2S, Mg2Si, NiSi2, PbO2, PbMg2 PoO2, PtSn2, PuO2, SnMg2, ThO2, UO2, ZrO2

Binary compounds in Fm-3m structure: [11]
BaCl2, BaF2, CaF2, CdF2, HgF2, OK2, OLi2, ONa2, SLi2, SNa2, PbF2, PtAl2, PtGa2, PtIn2, PtK2, RaF2, SK2, SRb2, SeK2, SeCu2, SeLi2, SeNa2, SrF2, SrCl2, TeAg2, TeK2, TeLi2, TeNa2, UN2

(a)   (b)

Figure 8. Face centered unit cell with 8 added atoms in Fm-3m structure, giving a dodeca-atomic unit cell. In (a), interconnections between neutral *even-atoms*. In (b), all eight internal atoms are charged, with 4(+) and 4(−), and the total charge is neutralized.

PrO2: Pr Grey, 4 *odd-atoms* with a neutral total charge
Oxygen Blue 8 neutral *even-atoms*

EuF2: Eu Green *Odd-atoms* with a neutral total charge
Fluorine Blue *Odd-atoms*: neutral with 3 bonds

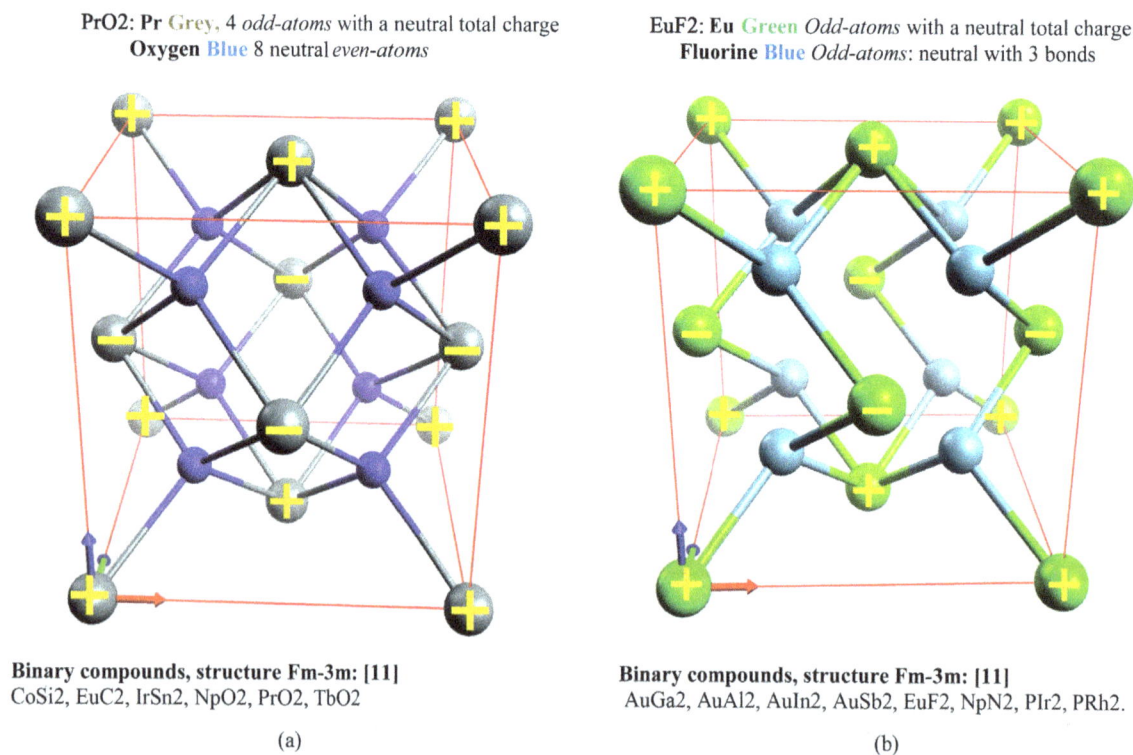

Binary compounds, structure Fm-3m: [11]
CoSi2, EuC2, IrSn2, NpO2, PrO2, TbO2

Binary compounds, structure Fm-3m: [11]
AuGa2, AuAl2, AuIn2, AuSb2, EuF2, NpN2, PIr2, PRh2.

(a)   (b)

Figure 9. Face-centered unit cell with face-centered cubic structure: Fm-3m and 8 atoms added to the unit cell with a total number of 12 atoms. (a) shows neutral internal atoms and atoms charges in the face-centered structure, (b) has nearly the same structure than (a), except with uncharged *odd-atoms* and a reduced number of covalent-bonds.

In **Figure 8(a)**, a well-known chemical formula with this face-centered structure is ZrO2. Zr *even-atoms* (black) are face-centered and oxygen *even-atoms* (red) are added inside the unit cell. Red atoms are only connected to black atoms and vice versa. Red atoms have 8 times 4 bonds and black atoms balance them with 8 times 1 bond and 6 times 4 bonds. In **Figure 8(a)**, oxygen *even-atoms* count 4 bonds each. As listed at the bottom of **Figure 8(a)**, 16 compounds follow this structure and most of them are dioxides.

The eight uncharged oxygen *even-atoms* of **Figure 8(a)** can be replaced by charged *odd-atoms*. The unit cell neutrality is only reached thanks to the even number of internal atoms. **Figure 8(b)** illustrates such a structure, of which CaF2 is an example. Ca atoms are green and F atoms are blue. The bonding condition is unchanged compared to **Figure 8(a)**, *i.e.* the structure is conserved. It can also be described like a face-centered crystal with eight internal halogen atoms. As calcium is an *even-atom* and drawn with eight single covalent bonds, it is consequently neutral and follows the even-odd rule. On the contrary, fluorine is an *odd-atom* and must be charged to allow four covalent bonds. Of all fluorine atoms, four are positively charged and four negatively. This is confirmed in **Figure 8(b)** where charges are assigned to build a neutral unit cell. There are about 30 three-atomic compounds conforming to this structure [11]. Their names appear at the bottom of **Figure 8(b)**.

The same configuration can be obtained when replacing the Zr *even-atom* of **Figure 8(a)** with an *odd-atom*. **Figure 9(a)** illustrates the result, an odd-even configuration. To conserve a neutral total charge in the unit cell of **Figure 9(a)**, horizontal planes of odd-atoms are alternatively positively and negatively charged (as in **Figure 5**). Six di-element dodeca-atomic unit cells from ref: [11] follow the proposed bonding structure.

The last possible change from **Figure 8(a)** is to replace all *even-atoms* by *odd-atoms*. A difficulty arises when attempting to keep the same bonding configuration since it would result in interconnected atoms bearing the same charge. To avoid this, each blue *odd-atom* must lose one covalent bond. As a consequence, green atoms must lose two bonds to keep the same charge and the same bonding parity, as shown in **Figure 9(b)**. An example is EuF2, in which all eight Fluorine *odd-atoms* (blue), have 3 bonds and therefore no charge. The total number of bonds from the fluorine is 24 (8 times 3). The number of bonds with Eu *odd-atoms* is 24 (8 times 1 for the corners, 2 times 4 for the horizontal faces and 4 times 2 for vertical faces centered atoms). This neutral structure is consistent with the even-odd rule. Eight di-element compounds listed below **Figure 9(b)** have this **Fm-3m** structure [11].

In Chapter 5, 45 compounds were found in scientific literature to adopt this single-face-centered-cubic structure based on a dodeca-atomic unit cell. They all are compatible with the even-odd rule and the limitation to 8 bonds per atom. With another point of view in this crystals, an element presents four times has 8 or 6 bonds interconnected directly to an element presents 8 times with 4 or 3 bonds respectively. According to the proposed procedure, the neutrality of the unit cell is obtained, when necessary via the nearest second neighbor placed in the same unit cell.

6. Discussion

More than 180 referenced configurations of atoms in body-centered cubic and simple-face-centered cubic structures have been studied in this article. This number seems great enough to form a solid base for the following discussion.

Crystals are highly organized structures in which a unit-cell is repeated over and over. Atoms are likely to form this type of structure only if there is only one possibility for them to connect. If there were more than one possible configuration, crystals would not be as regular as they are naturally found. The proposed procedure was thus built to constrain validity of bonds enough to leave only one possible bonding configuration. The limitations are: single covalent bonds only; atoms erect a number of bonds as allowed by the even-odd rule, depending on their charge and their even or odd character; no element can form more than 8 bonds; two atoms bearing the same charge, positive or negative, cannot be connected; in mono-element crystals, two atoms are connected when they are both neutral or with complementary charges; a unit cell with series of elements have connections only between different atoms; a unit cell with two series of elements have the same total number of bonds in each series; a unit cell containing charged atoms should be overall neutrally charged, or should be neutralized by the nearest neighbor or by the second nearest neighbor in the nearest unit cell.

Some of these points are now discussed below.
- **Validity of the procedure based on the even-odd rule**
The drawing procedure described above was systematically tested on single-covalent bonded structures in

centered-cubic crystals and single-face-centered cubic crystals. The list of compounds and their structures were found in ref. [11] with the following criteria: 1) di-atomic unit cell with mono-element and di-elements cubic structure (Pm-3m and Im-3m); 2) tetra-atomic unit cell with mono-element and di-elements (Fm-3m and Pm-3m), 3) dodeca-atomic unit cell with di-elements cubic structure (Fm-3m). For all listed compounds, a bonding configuration was found by applying the procedure. It seems safe to conclude that, after having been previously applied successfully to molecules [8] and ions [7], the even-odd rule is applicable to the named cubic structures.

- **Limitation to a maximum of 8 covalent bonds per atom in crystals**

The limitation to a maximum of 8 covalent bonds seems to be commonly accepted in chemistry. In solid chemistry however, this limitation is neither applied nor considered. In many crystals, atoms are surrounded by more than 8 atoms and connections are usually drawn between all nearest neighbors. In this article however, the limitation to 8 bonds per atom has gradually appeared necessary as the authors applied the drawing procedure to all typical even-odd configurations in di-atomic compounds: even-even, even-odd, odd-even and odd-odd. By limiting the number of bonds to a maximum of 8, the number of possible bonding configurations decreased, but one possible configuration always remained.

- **Particularity of alkaline earth metals.**

Elements of column 2 in the periodic table can be sorted into two groups, Ca, Sr and Ba form face-centered crystals whereas Be and Mg do not. Beryllium possesses only four electrons in total, which explain why it cannot build 8 covalent bonds as needed in centered cubic crystals. Magnesium on the other end possesses 12 electrons and no explanation for this inability to build 8 covalent bonds was satisfying at the time of writing.

- **Particularity of halogen in mono-element body-centered-crystals.**

When non-metals halogen *odd-atoms* (F, Cl, Br and I) crystallize, they never seem to adopt body-centered structures. The reason for this peculiar case could be that they cannot build more than seven bonds, as in IF7 molecules.

- **Inner shell electrons**

In chemistry, electrons located in the inner shell do not participate in the formation of bonds. A neutral atom cannot form more bonds than the number of electrons in its valence shell. Sodium has, for instance, one electron in the valence shell—it is an *odd-atom*—and ten in the inner shell. Neutrally charged, a Sodium atom only erects one bond as in NaOH. In a crystal structure like the mono-element cubic structure, however, Sodium atoms are bonded 8 times. Since the single valence electron can only participate in one covalent bond, the only possible explanation for the other bonds is a contribution of electrons from the inner shell. Because only pairs of electrons can be extracted from the inner shell, the Sodium atoms alternatively use 6 or 8 electrons from the inner shell and are charged to build an even number of bonds.

In crystals, the valence number clearly does not limit the number of bonds. The present paper proposes to extend to other elements the need of an even number of electrons coming from the inner shell to form bonds in cubic crystals.

- **Cleavage of crystals**

A crystal is cleaved along oriented planes [18]. Silicon crystals, diamond-like, have three main cleavage directions. Silicon crystals, mono-element crystals made out of *even-atoms* with four bonds, do not contain charged atoms. In other crystals containing charged atoms, charges are aligned on planes and cleavage probably occurs perpendicularly to these electrostatic planes. It would be of great interest to confirm that cleavage directions in crystals studied in the present article are the same for crystals with the same bonding configuration and differ according to the electrostatic planes.

7. Conclusions

The even-odd rule, checked previously for ions and molecules, is applied in this paper to centered-cubic and face-centered-cubic crystals. The bonding organization for mono-atomic and diatomic crystals is drawn using the even-odd rule in body-centered crystals, giving evidence of the quality of the rule. In the same way, di-element compounds in single-face-centered crystals are clearly described and, when needed, charges are assigned to atoms to build a structure of 8 covalent bonds or less.

The proposed constraint for a maximum of 8 covalent bonds per atom, is plausible for such structures, because it brings similarity to other areas of chemistry on the one hand and because it reduces the number of possible bonding configurations on the other hand. With the even-odd rule and the added constraints, every solid compounds found in literature that fit the chosen criteria were systematically studied and a bonding configuration was found. This appears to form a large support for the acceptance of the even-odd rule.

It will be interesting to extend the drawing procedure described here for single-face-centered crystals, to the very common double-face-centered structure.

References

[1]   http://en.wikipedia.org/wiki/Crystallography

[2]   http://en.wikipedia.org/wiki/Coordination_number

[3]   http://chemwiki.ucdavis.edu/Inorganic_Chemistry/Coordination_Chemistry/Coordination_Numbers

[4]   http://en.wikipedia.org/wiki/Bonding_in_solids

[5]   http://en.wikipedia.org/wiki/Chemical_bond

[6]   https://en.wikipedia.org/wiki/X-ray_crystallography

[7]   Auvert, G. (2014) Chemical Structural Formulas of Single-Bonded Ions Using the "Even-Odd" Rule Encompassing Lewis's Octet Rule: Application to Position of Single-Charge and Electron-Pairs in Hypo- and Hyper-Valent Ions with Main Group Elements. *Open Journal of Physical Chemistry*, **4**, 67-72. http://dx.doi.org/10.4236/ojpc.2014.42010

[8]   Auvert, G. (2014) Improvement of the Lewis-Abegg-Octet Rule Using an "Even-Odd" Rule in Chemical Structural Formulas: Application to Hypo and Hyper-Valences of Stable Uncharged Gaseous Single-Bonded Molecules with Main Group Elements. *Open Journal of Physical Chemistry*, **4**, 60-66. http://dx.doi.org/10.4236/ojpc.2014.42009

[9]   Auvert, G. (2014) The Even-Odd Rule on Single Covalent-Bonded Structural Formulas as a Modification of Classical Structural Formulas of Multiple-Bonded Ions and Molecules. *Open Journal of Physical Chemistry*, **4**, 173-184. http://dx.doi.org/10.4236/ojpc.2014.44020

[10] Mendelejew, D. (1869) Über die Beziehungen der Eigenschaften zu den Atomgewichten der Elemente. Zeitschrift für Chemie (in German), 405-406.

[11] http://cod.ibt.lt/

[12] https://en.wikipedia.org/

[13] Hanwell, M.D., Curtis, D.E., Lonie, D.C., Vandermeersch, T., Zurek, E. and Hutchison, G.R. (2012) Avogadro: An Advanced Semantic Chemical Editor, Visualization, and Analysis Platform. *Journal of Cheminformatics*, **4**, 17. http://dx.doi.org/10.1186/1758-2946-4-17

[14] http://www.gimp.org/downloads/

[15] https://en.wikipedia.org/wiki/Crystal_structure

[16] Haynes, W., Ed. Handbook of Chemistry and Physics. CRC 96th Edition.

[17] Hatert, F. Compléments de cristallographie www.minera.ulg.ac.be/pdf/

[18] https://en.wikipedia.org/wiki/Cleavage_(crystal)

22

Structural Study of Methylated and Non-Methylated Duplexes by IR Fourier Spectroscopy

V. G. Kunitsyn, P. A. Kuznetsov, E. N. Demchenko, O. I. Gimautdinova

Scientific Research Institute of Biochemistry, SD RAMS, Novosibirsk, Russia
Email: kunitsyn41@mail.ru, pawelkuzn@mail.ru

Abstract

Structure of the duplex consisting of 23 pairs of bases was studied before and after the methylation of two cytosine molecules from different chains of the duplex. The study was performed in a buffer solution using an IR Fourier spectrometer. The absorption bands corresponding to the duplex backbone were found to change their characteristics after the methylation. Firstly, the integrated intensity ratio of the absorption bands, S_{1044}/S_{1085}, decreased by a factor of 1.5. The absorption band at 1044 cm^{-1} corresponds to the COC bond of deoxyribose, and the band at 1085 cm^{-1} to the $PO_{2symm.vibr}$ bond. Secondly, a substantial shift of the absorption band $1085 \rightarrow 1112$ cm^{-1} ($\Delta v = 27$ cm^{-1}) was observed. In addition, pronounced changes in the absorption region of CH stretching vibrations took place. In particular, shifting of some absorption bands assigned to the stretching vibrations of CH bonds; the $2979 \rightarrow 2945.7$ shift was equal to 33.3 cm^{-1}. In addition to the indicated changes, some bands corresponding to the Z structure appeared in the methylated duplex. Thus, methylation of two cytosine molecules in the duplex leads to the order$\rightarrow$order structural transition, most likely to the B$\rightarrow$Z transition.

Keywords

Duplex, Duplm, Structural Study, FTIR Spectroscopy

1. Introduction

CpG islands in native DNA are known to exist mainly in the methylated form [1]. The methylation of DNA controls all genetic processes and serves as the mechanism of cell differentiation and gene repression [2] [3]. In some works, (CCGCC)$_n$ duplexes were studied by IR Fourier spectroscopy. Consecutive methylation of the duplex produced determinate changes in the region of 820 - 840 cm^{-1}. The observed bands were attributed to the B-structure and BI→BII structural transitions [4]. On the other hand, investigation of molecular dynamics in the methylated (CCGCC)$_n$ duplex predicted the appearance of B→Z transition [5]. Quite interesting is the IR Fourier spectroscopy study of (CCGCC)$_n$ duplex in dependence on temperature in the range of 200 - 290 K. The B→Z and ZI→ZII transitions were revealed in the indicated temperature region [6]. Methylated (CCGCC)$_n$ duplexes are certainly of great interest [2] [3]. However, we think that investigation of the minimally methylated duplexes with a single CCGCC site in the AT enriched nucleotide environment is also important. The model system was represented by oligo-deoxyribonucleotide duplex consisting of 23 pairs of bases, which had a single (GCNGC) site simulating the binding site of THC-apoA-I complex on native DNA and oligonucleotides [7]-[9]. Prior to examining the interaction of methylated duplex with THC-apoA-I complex, we studied the structure of the duplex by IR Fourier spectroscopy. The goal of the work was to investigate structural changes of methylated duplex using IR Fourier spectroscopy.

2. Materials and Methods

Deoxyribooligonucleotides:

I) 5'-GAGTTTAGCGGCTATCGATCTCT-3' and II) 5'-AGAGATCGATAGCCGCTAAACTC-3' were synthesized at the Institute of Chemical Biology and Fundamental Medicine SB RAS (Novosibirsk).

The [γ-^{32}P] ATP tracer (Isotope, St. Petersburg, 1 Ci/mol) was introduced to the 5'-end of oligonucleotides using polynucleotide kinase of T4 bacteriophage (SibEnzyme, Novosibirsk).

The OLI/OLII duplex was obtained by annealing of initial oligonucleotides taken in equimolar ratio (90°C, 3 min, gradual cooling to room temperature).

Methylation of the duplex catalyzed by M. Fsp4HI methyltransferase (SibEnzyme, Novosibirsk) was carried out according to the analytical certificate of the enzyme: 1.7×10^{-8} M of the OLI/OLII duplex, 80 μM of SAM, 1 a.u. of methyltransferase M. Fsp4HI in a methylation buffer (10 mM Tris-HCl, pH 7.6, 10 mM MgCl$_2$, 1 mM DTT). The mixture was incubated at 30°C for 1 h. Methylation of the duplex involved cytosine from each nucleotide sequence to obtain one molecule of methylated cytosineper one oligonucleotide sequence. The reaction was stopped by adding 1 μl of 0.5 m EDTA [10].

IR Fourier Spectra. Spectra of the duplexes were recorded in a Fourier spectrometer (Nicolet 6700, Thermo Scientific) using the Frustrated Total Internal Reflection (FTIR) method with a diamond cell at a resolution of 4 cm^{-1} in the frequency range of 900 - 4000 cm^{-1}, in isotonic Tris buffer with the addition of K, Na salts (pH 7.4). Volume of the tested solution was 6.0 μl. A sample was covered with a fluoroplastic cap, which was pressed to the cell. The optical chamber was sealed and purged with dry air. The time of spectra recording ranged from 30 s to 1 min. Mathematical treatment of the spectra was performed using a special OMNIC software attached to the spectrometer.

3. Results

IR Fourier spectra. Analysis of the spectra revealed the absorption bands at 1085.67 (the PO$_2$ bond) and 1044.44 (the C$_4$-O$_4$-C$_5$-O$_5$ bond) [8] [9] for the duplex (dupl_23) and 1112.56 and 1044.31 cm^{-1} bands for the methylated duplex (duplm_23) (**Figure 1(a)** and **Figure 2(a)**). It is seen that the absorption band of the PO$_2$ bond in duplm_23 is shifted to the short-wave region by 26.89 cm^{-1}. In addition, the integrated intensity ratio S$_{1044.44}$/S$_{1085.67}$ was equal to 4.38 in the duplex, while in duplm_23 the S$_{1044.31}$/S$_{1112.56}$ ratio was 2.82 (**Table 1**). Thus, the first value is greater than the second one by a factor of 1.55. A substantial shift of the absorption band of PO$_2$ bond to the short-wave region and changes in the ratio of the indicated bands testify to the conformational changes in the backbone that occur due to methylation of bases (cytosine). The methylation involved only two bases residing in different oligonucleotide chains, while the backbone changed only slightly. This suggests the effect of hyperconjugation of the methyl group with the base [11], which produces radical changes in the electron distribution and order of the bonds in bases and side radicals, and hence leads to conformational changes in the backbone and the entire molecule. The absorption bands at 1199.17 and 1218.25 cm^{-1} (dupl_23) as well as 1213.03 and 1230.93 cm^{-1} (duplm_23) can be attributed to the P=O bonds [8] [9] (**Figure 1** and **Figure 2**). In this case, a 12 - 14 cm^{-1} shift to the short-wave region is also observed. The indicated changes suggest that the pitch of a helix decreases due to an increased ordering of bases and a stronger Van der Waals interaction between them. In this case, of special interest in the analysis of CH bonds. In dupl_23, deformation vibrations of

Therrno Fisher
SCIENTIFIC

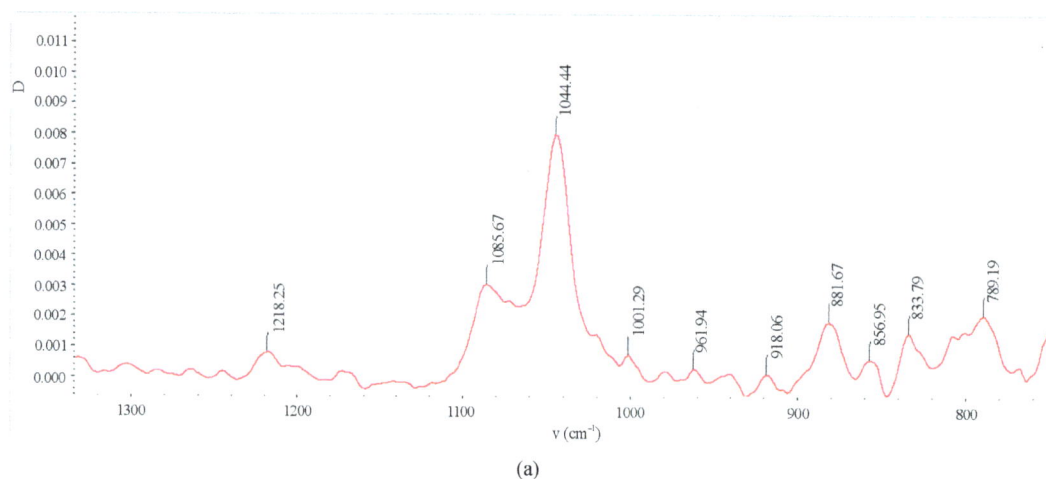

(a)

Therrno Fisher
SCIENTIFIC

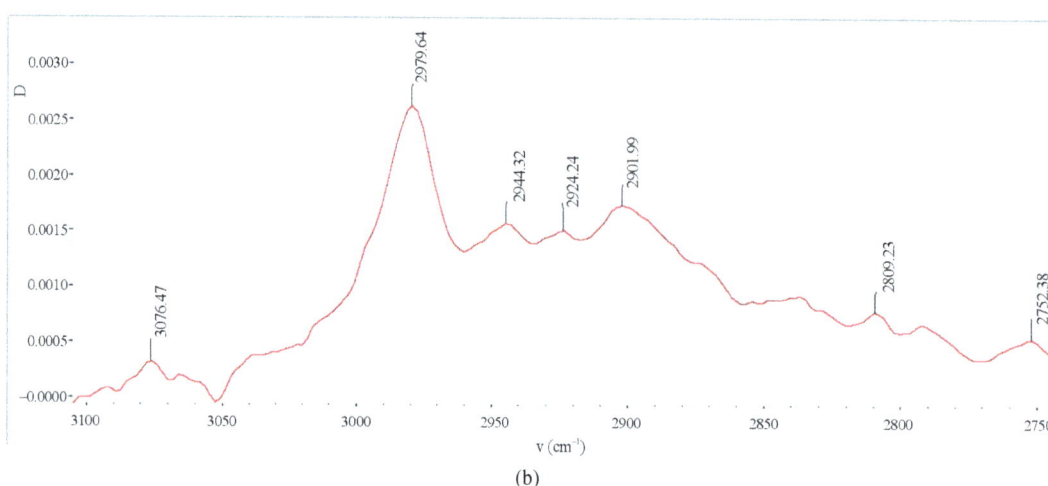

(b)

Figure 1. (a) FTIR spectra of dupl.: ($C_{tris-HClbuf}$ = 0.01 M, pH 7.4); v = 800 - 1300 cm^{-1}; (b) FTIR spectra of dupl.: ($C_{tris-HClbuf}$ = 0.01 M, pH 7.4); v = 2700 - 3100 cm^{-1}.

Table 1. Frequency characteristics of the FTIR spectra of methylated (duplm) and non-methylated (dupl) duplexes.

No.	Compound	$NH_{def.}$	$CH_{def.}$	CH_3	$CH_{stretch.}$	P=O assym.vibr.	PO_2 symm.vibr.	C-O-C deoxyrib	C-C deoxyrib
1.	Dupl	1530	1450.9 1421	1335.3	3049 2981.8 2925.3 2903.2	1271.2 1222.9	1087.3	1045.1	970.1
	Intensity	-0.0021	0.0008	0.0005		0.0005 0.0007	0.0012	0.0069	0.0003
2.	Duplm	1538.9	1459.2 1419.5*	1337.2	3063.3 2948.0 2889.9	1280.8 1228.6 1213.0	1112.5	1044.3	993.7 922.7
	Intensity	0.0002	0.0031 0.0039	0.0026	0.0013 0.0041 0.0038	0.002	0.0076	0.0169	0.0059 0.0032

*Indicates band splitting.

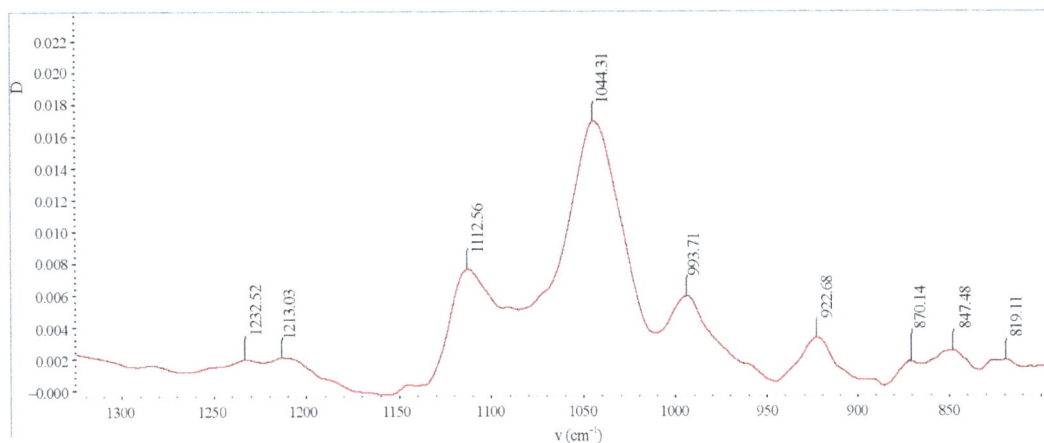

(a)

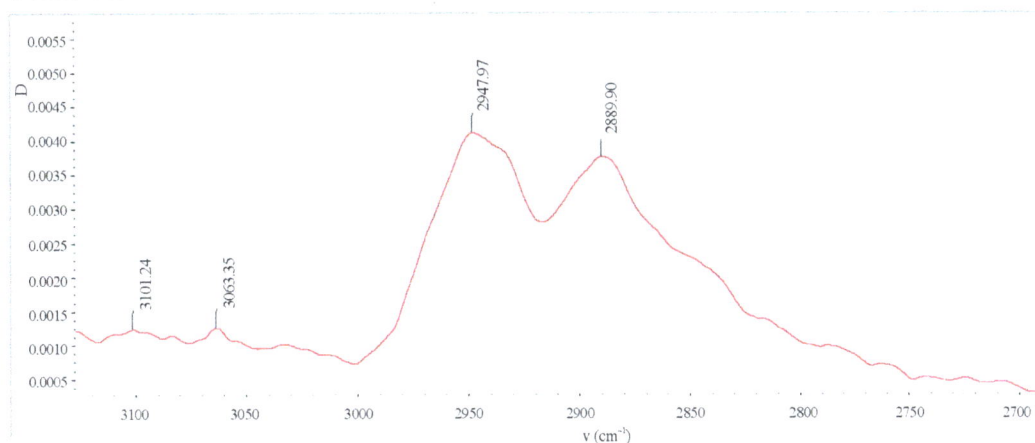

(b)

Figure 2. (a) FTIR spectra of duplm: ($C_{\text{tris-HClbuf}}$ = 0.01 M, pH 7.4); v = 800 - 1300 cm^{-1}; (b) FTIR spectra of duplm: ($C_{\text{tris-HClbuf}}$ = 0.01 M, pH 7,4); v = 2700 - 3100 cm^{-1}.

the CH bonds show up as the absorption bands at 1418.58 and 1453.88 cm^{-1}. Whereas in duplm_23 these bands (1419.50 and 1459.24 cm^{-1}) are split. In addition, quite intense absorption bands at 993.71 and 922.68 cm^{-1} (the CC skeletal vibrations) are recorded in duplm_23 (**Figure 1** and **Figure 2**), while in dupl_23 these bands are absent; note that the band at 922.68 cm^{-1} is assigned to the Z conformation.

CH stretching vibrations. Analysis of the spectra in the short-wave region revealed the presence of some absorption bands for dupl_23, in particular, 3076.47, 2979.64, 2944.32 and 2901.99 cm^{-1}. The bands at 3063.35, 2947.97, 2936.13 and 2889.90 cm^{-1} were recorded for duplm_23 (**Figure 1(b)** and **Figure 2(b)**). It is seen that all the absorption bands in duplm_23 are shifted with respect to same bands indupl_23; the band at 2979.64 is shifted by 32.33 cm^{-1} to the long-wave region. This substantial shift can be attributed to an increased ordering of bases and monosaccharides and stronger Van der Waals forces between them.

Thus, the analysis of IR Fourier spectra of dupl_23 and duplm_23 showed that their conformations differ, duplm having a more ordered structure than the duplex. This is indicated by the frequency shift in the backbone and CH bonds region as well as by changes in the ratios of the absorption bands corresponding to the backbone and changes in intensity of the bands corresponding to CH stretching vibrations. Changes in the integrated intensity ratio S_{1044}/S_{1085} for duplm testifies to the order→order structural transition that occurs due to methylation of the duplex.

Table 2. Absoption bands which correspond Z-structure [6].

No	Compound	CO$_b$ [cm^{-1}]	CH$_3$ [cm^{-1}]	PO$_{2as}$ [cm^{-1}]	PO$_{2s}$ [cm^{-1}]	C-O-C$_d$ [cm^{-1}]	C-C$_d$ [cm^{-1}]	C=N [cm^{-1}]	CH$_d$ [cm^{-1}]
1	Duplm	1712.8 1636.0	1337.2	1213.0	1112.6	1044.3	922.7	1504.6	1407
2	Dupl	–	–	–	–	+	–	–	–

4. Discussion

The study of dupl and duplm demonstrated that the integrated intensity ratio of the absorption bands S_{1044}/S_{1085} was equal to 4.4 in dupl and 2.8 in duplm. This 1.5-fold difference indicates that integrated intensity of the absorption band at 1085 cm^{-1} strongly decreases after methylation of the duplex. The band at 1085 cm^{-1} belongs to the PO$_2^-$ group (symmetric vibrations).

$$\Delta I = \partial^2 \mu / \partial q^2 \,, \qquad\qquad [12]$$

where μ is the dipole moment and q is the generalized coordinate. As follows from this expression, the dipole moment of the PO$_2^-$ group decreases, *i.e.* the duplex becomes more hydrophobic after the methylation. In addition, the duplm backbone is described by *a* quartet (the absorption bands at 1112, 1083, 1043 and 994.5), and the dupl backbone is described by a triplet (1085, 1044.4 and 1001.3 cm^{-1}). What is the reason for decreasing the dipole moment in the phosphate group? In our opinion, this is related to manifestation of the electron-donor properties of methylated bases as a result of hyperconjugation [11], which leads to partial neutralization of the positive charge of phosphorus atom and enhances the hydrophobic properties of duplm.

Essential changes occur also in the region of CH stretching vibrations. A quartet (the absorption bands at 2979, 2944, 2918 and 2901 cm^{-1}) was observed for dupl, while only two main bands (at 2945.7 and 2888.4 cm^{-1}) were seen in this region for duplm. The 2979→2945.7 cm^{-1} shift is equal to 33.3 cm^{-1}; this is a pronounced shift that can be interpreted as a structural transition from one conformation to another. The main reason is related to a considerable ordering of bases, formation of the energetically more favorable conformation of deoxyribose, and their spatial ordering with respect to the duplex axis, which produces changes in the backbone structure, in particular, the endo 2→endo 3 conformational changes of deoxyribose. Exactly these conformations radically change the structure of DNA backbone [13] [14]. We think that after methylation of the duplex, individual bases become highly cooperative and form a joint "rigid block".

It should be noted that methylation is accompanied by the order→order structural transition and, in particular, by the B→Z transition. Methylation induces conformational changes and decreases enthalpy of the transition [5]. Thus, we think that the changes observed in the region of backbone and CH stretching vibrations are caused by the B→Z transition. This is indicated also by the absorption bands revealed in methylated duplexes, which correspond to the Z structure (**Table 2**).

References

[1] Dyachenko, O.V., Shevchyk, T.V. and Buryanov, Y.I. (2010) Structural and Functional Features of the 5-Methylcytosine Distribution in the Eukaryotic Genome. *Molecular Biology*, **44**, 171-185. http://dx.doi.org/10.1134/S0026893310020019

[2] Klenov, M.S. and Gvozdev, V.A. (2005) Heterochromatin Formation: Role of Short RNAs and DNA Methylation. *Biochemistry* (*Moscow*), **70**, 1187-1198. http://dx.doi.org/10.1007/s10541-005-0247-4

[3] Vanyushin, B.F. (2007) A View of an Elemental Naturalist at the DNA World (Base Composition, Sequences, Methylation). *Biochemistry* (*Moscow*), **72**, 1289-1298. http://dx.doi.org/10.1134/S0006297907120036

[4] Banyay, M. and Graslund, A. (2002) Structural Effects of Cytosine Methylation on DNA Sugar Pucker Studied by FTIR. *Journal of Molecular Biology*, **324**, 667-676. http://dx.doi.org/10.1016/S0022-2836(02)01104-X

[5] Temiz, N.A., Donohue, D.E., Bacolla, A., Luke, B.T. and Collins, J.R. (2012) Role of Methylation in the Intrinsic Dynamics of B and Z-DNA. *PLoS One*, **7**, e35558. http://dx.doi.org/10.1371/journal.pone.0035558

[6] Rauch, C., Pichler, A., Trieb, M., Wellensohn, B., Liedt, K.R. and Mayer, E. (2005) Z-DNA's Conformer Substrates Revealed by FT-IR Difference Spectroscopy of Nonoriented Left-Handed Double Helical Poly (dG-dC). *Journal of Biomolecular Structure and Dynamics*, **22**, 595-614. http://dx.doi.org/10.1080/07391102.2005.10507029

[7]   Panin, L.E., Kunitsyn, V.G. and Tusikov, F.V. (2005) Effect of Glucocorticoids and Their Complexes with Apolipo-protein A-I on the Secondary Structure of Eukaryotic DNA. *International Journal of Quantum Chemistry*, **101**, 450-467. http://dx.doi.org/10.1002/qua.20200

[8]   Panin, L.E., Kunitsyn, V.G. and Tuzikov, F.V. (2006) Changes in the Secondary Structure of Highly Polymeric DNA and $CC(GCC)_n$-Type Oligonucleotides under the Action of Steroid Hormones and Their Complexes with Apolipo-protein A-I. *Journal of Physical Chemistry B*, **110**, 13560-13571. http://dx.doi.org/10.1021/jp068011n

[9]   Panin, L.E. and Kunitsyn, V.G. (2009) The Initiation Mechanism of Gene Expression in Ascitic Hepatoma Cells under the Action of Dehydroepiandrosterone in a Complex with Apolipoprotein A-I. *Current Chemical Biology*, **3**, 306-314.

[10]   Chmuzh, E.V., Kashirina, Yu.G., Tomilova, Yu.E., Chernukhin, V.A., Okhapkina, S.S., Gonchar, D.A., Dedkov, V.S., Abdurashitov, M.A. and Degtyarev, S.H. (2007) Gene Cloning, Comparative Analysis of the Protein Structures from Fsp4HI Restriction-Modification System and Biochemical Characterization of the Recombinant DNA Methyltransfe-rase M. Fsp4HI. *Molecular Biology*, **41**, 43-50. http://dx.doi.org/10.1134/S0026893307010062

[11]   Ingold, K. (1973) Theoretical Bases of Organic Chemistry. Nauka, Moscow.

[12]   Smith, A. (1980) Applied Spectroscopy. Mir, Moscow.

[13]   Ovchinnikov, Yu.A. (1987) Bioorganic Chemistry. Nauka, Moscow.

[14]   Rubin, B.A. (1990) Biophysics. Vyschaya Shkola.

23

How the Even-Odd Rule, by Defining Electrons Pairs and Charge Positions, Can Be Used as a Substitute to the Langmuir-Octet Rule in Understanding Interconnections between Atoms in Ions and Molecules

Geoffroy Auvert

CEA-Leti, Grenoble Alps University, Grenoble, France
Email: Geoffroy.auvert@grenoble-inp.org

Abstract

In the course of time, numerous rules were proposed to predict how atoms connect through covalent bonds. Based on the classification of elements in the periodic table, the *rule of eight* was first proposed to draw formulas of organic compounds. The later named *octet rule* exhibited shortcomings when applied to inorganic compounds. Another rule, the *rule of two*, using covalent bonds between atoms, was proposed as an attempt to unify description of organic and inorganic molecules. This rule unfortunately never managed to expand the field of application of the *octet rule* to inorganic compounds. In order to conciliate organic and inorganic compounds, the recently put forward *even-odd* and the *isoelectronicity rules* suggest the creation of one group of compounds with pairs of electrons. These rules compass the *rule of two* for covalent bonds as well as the *octet rule* for organic compounds and suggest transforming bonds of multi-bonded compounds in order to unify representations of both groups of compounds. The aim of the present paper is fourfold: to extend the *rule of two* to every atom shells; to replace the well-known *octet rule* by the *even-odd rule*; to apply the *isoelectronicity rule* to each atom and to reduce the influence range of the charge of an atom in a compound. According to both rules, the drawing of one atom with its single-covalent bonds is described with electron pairs and charge positions. To illustrate the rules, they are applied to 3D configurations of clusters.

Keywords

Even-Odd, Isoelectronicity, Rule, Effective Valence, Molecule, Chemical Formula, Covalent Bond, Ion

1. Introduction

Abegg, Lewis and Langmuir [1]-[3] have suggested a procedure to represent molecules with single or multiple connections between atoms. They also put forward a rule known as the *octet rule* to obtain valid chemical structures of organic compounds. Molecules like CH4, NH3, H2O and HCl are known compounds of this family, with eight electrons in the valence shell around the central atom. In CH4, the atom of carbon possesses 4 valence electrons: it can be connected to four atoms of hydrogen by means of eight electrons forming covalent bonds. A few incompatible molecules had however already been discovered, like BH3 or BeH2. Considering these molecules, Lewis introduced the *rule of two* [4]. This *rule of two* suggests that electrons combine in pairs, *i.e.* covalent bonds, to make each connection in organic and inorganic compounds. The *rule of two* seems to have a wider field of application [4]. Unfortunately, the *rule of two* does not expand the application field of the *octet rule* [5].

A rule named *even-odd rule* has recently been proposed [6]. This rule was validated for a large number of known ions and molecules [6] [7]. Based on these early validations, the *even-odd rule* has suggested modifying multi-bonded compounds into single-bonded compounds [8]. As a consequence, compounds classically multi-bonded and modified with the *even-odd rule*, are encompassed with all the other single-bonded molecules [8]. At the same time, an *isoelectronicity rule* has been defined and applied to single-bonded ions and molecules [9].

The aim of the present article is to submit a drawing procedure of ions and molecules, by describing how to apply the *even-odd* and the *isoelectronic rules* to single-bonded compounds. Both rules are firstly defined as local rules to determine the calculation procedure of the even number of electrons in each atom shell. In the process, the author takes example of each possible element of the main group of the periodic table, along with their bonding conditions and charge positions. 3D drawings illustrate then the *even-odd rule* applicability. Several points are finally discussed such as the importance of the number of electrons in the atom shells and the local nature of both rules when applied to charged or uncharged atoms.

2. The Even-Odd Rule for an Atom in Ions and Molecules

The even-odd rule indicates in which of the three atom shells, electrons are located: the inner shell, the inactive shell and the covalent shell. It reads as follows:
- Each element of the periodic table that is not part of a compound:
  - is electrically neutral with a number of positive charges and the same number of electrons,
  - is composed of a nucleus, where positive charges are located,
  - possesses around the nucleus, an inner shell composed of an even number of electrons,
  - possesses a valence shell with a number of electrons equal to the valence number of the atom, as defined in the periodic table,
  - has no covalent shell: it is not covalently bonded;
- In the even-odd rule, an atom, as part of an ion or a molecule:
  - is represented by the letters symbolizing its name as in the periodic table,
  - has three shells: inner, inactive and covalent, each containing an even number of electrons,
  - has as many electrons in its inner shell as indicated in the periodic table,
  - has an inactive shell surrounding the inner shell and containing 0 to 4 electron-pairs,
  - has a covalent shell around the inactive shell that:
    - contains electrons that are involved in single-covalent bonds with other atoms,
    - can be empty or contain up to eight single-covalent bonds,
    - is represented in the form of lines, one line between connected atoms symbolizing a covalent bond;
- To explain the application of the rule, the letters symbolizing the element are surrounded by four information: three numbers and one signet as follow:
  - The valence number of the periodic table:
    - is indicated near the top-left corner,
    - ranges from one for elements like sodium (Na) up to eight for noble gas like Argon (Ar);
  - The charge, assigned to one atom:
    - is represented by a symbol near the top right corner,
    - uses signs + or − for a positive or negative charge respectively,
    - in the absence of charge, no symbol is used;
  - The effective-valence number:

○ is indicated near the bottom left corner,
○ is evaluated as follows: for a neutral atom *i.e.* without charge, it is equal to the valence number; for a negatively charged atom *i.e.* that possesses an extra-electron, it is the valence number increased by one; for a positively charged atom, it is the valence number decreased by one;

■ The number of covalent bonds in the covalent shell:
○ is equal to the number of lines around the atom,
○ is less than or equal to the effective-valence number,
○ is even and starts at zero, if the effective-valence number is even,
○ is odd and starts at one, if the effective-valence number is odd;

■ The number of electrons in the inactive shell of an atom:
○ is indicated near the bottom right corner,
○ is calculated by subtracting the effective valence-number with the number of covalent bonds,
○ is even; this is only possible when the number of bonds and the effective valence number are both odd or both even;

■ As a consequence for the inactive shell:
○ It contains electron pairs that are not involved in any covalent bond,
○ The number of electron pairs ranges from 0 to 4 whatever the charge of the element,
○ When the number of electron pairs is 0, no additional covalent bond can be formed by the atom.

An important key to the validity of the even-odd rule is that the number of electrons in every shell is an even number. Specifically for the inactive shell, the even number of electrons imposes that molecules and ions belong to a group of electron-paired compounds [6] [7] *i.e.* molecules at standard energy scale [5].

In this rule, the charge translates directly into the number of electron pairs located either in the inactive shell or in the covalent shell of the atom. In the covalent shell, the charge only link to the immediate surroundings of the atom is through the covalent bonds. Its influence does not even extend to the first neighbors inactive shell. In other words, it means that a positive or negative charge is always localized on one specific atom in an ion or a molecule. It does not interfere with the compound on the whole.

In addition to the even-odd rule, an isoelectronicity rule is defined.

3. Isoelectronicity Rule

Isoelectronicity is better illustrated using two atoms like the Cl(−) ion and the Argon rare gas [9]. The number of electrons in both inactive shells and the number of covalent bonds are the same: eight and zero respectively. They are isoelectronic. They additionally both follow the even-odd rule [7].

More generally, an atom following the even-odd rule in a compound, can be replaced by another atom by using the isoelectronicity rule detailed below:

● The atom to be transformed has:
■ An inner shell containing an even number of electrons,
■ An inactive shell containing an even number of electrons,
■ A covalent shell containing covalent bonds,
■ A charge localized on the atom: negative (one added electron), positive (one removed electron) or neutral (with no local modification);

● The new atom bears similarities to the original atom:
■ the same number of covalent-bonds (number of lines surrounding the atom),
■ the same effective-valence number (bottom left corner of the atom),
■ the same number of electrons in the inactive shell (bottom right corner of the atom);

● But its internal configuration is different:
■ A different name from the periodic table,
■ A different valence number,
■ It meets one of the following criteria:
○ It belongs to the same column of the periodic table with a different inner shell.
○ It is the next one in the same row of the periodic table; the one-column shift is compensated by a charge to keep the effective-valence number constant (bottom left corner of the atom) and to have the same number of electrons in the inactive shell.
○ It is shifted in row and by column of the periodic table in agreement with both above conditions.

○ If the atom is charged, the sign of its charge can be reversed; in which case the replacement atom is 2 columns further in the periodic table.

Example: BH4(−) and GaH4(−) are isoelectronic compounds under the first criterion. The isoelectronicity comes from Boron and Gallium, which belong to the same column of the periodic table. As stated in the isoelectronicity rule, both atoms have the same number of covalent bonds, the same effective valence number and the same number of electrons in the inactive shell. They bear different names and don't have the same number of electrons in the inner shell.

HF and OH(−) are isoelectronic compounds under the second criterion. The isoelectronicity comes from O(−) and F, which belong to the same row and are one column apart in the periodic table. Both have one covalent-bond, an effective-valence of seven and six electrons in the inactive shell.

CH4 and PH4(+) are isoelectronic compounds under the third criterion. The isoelectronicity comes from C and P(+), which are neither in the same row of the periodic table nor in the same column. Both have 4 covalent bonds, an effective-valence number of 4 and no electron in the inactive shell.

BH4(−) and NH4(+) are isoelectronic compounds under the last criterion. B(−) and N(+) are in the same row of the periodic table and are two columns apart. Both atoms have 4 covalent bonds, an effective-valence number of 4 and no electron in the inactive shell.

4. Compounds with Elements from the Main Group of the Periodic Table

The even-odd and the isoelectronicity rules are here applied to elements of the periodic table.

Table 1 presents several atoms embedded in compounds, ions or molecules. They belong to column 1 to 8 in the main group of the periodic table: Na, Be, B, C or Si, N or P, O or S, Cl or I, Xe respectively.

Table 1. Atom of ions or molecules, in gaseous or liquid phase, compatible with the even-odd rule. Each structure represents an atom of a compound and is classified in groups having the same bottom-right number *i.e.* the same number of electrons in the inactive shell. In each group, they are ordered by increasing valence number like in the periodic table. Column one presents formulas of compounds that are composed of the atoms listed in the three other columns. Column 2 presents the positively charged element, column 3 its neutral form and column 4 its negative form. The signification of the four numbers around the letters of the element is detailed in the even-odd rule presented in chapter 2.

			Number of electrons in the inactive shell = 0		
Valence number			Positive charge	neutral	Negative charge
0	---	Xe [10]-[12]	XeCl(−)	0 Xe 0 0	0 Xe — 1 0
1	Na(+) [10] [11]	NaF [12]	NaF2(−) ..	1 Na⁺ 0 0	1 Na — 1 0	1 — Na — 2 0
2	BeH(+) [10]	BeF2 [12]	BeF3(−) [10] [11]	2 Be⁺ — 1 0	2 — Be — 2 0	2 — Be ═ 3 0
3	BF2(+) [10]	BH3 [10]	BH4(−) [11]	3 — B⁺ — 2 0	3 — B ═ 3 0	3 ═ B⁻ ═ 4 0
4	CF3(+) [10]	CH4 [10]	SiH5(−) [10]	4 — C⁺ ═ 3 0	4 ═ C ═ 4 0	4 ═ Si⁻ ═ 5 0
5	PH4(+) [10]	PCl5 [10]	PCl6(−) [10]	5 ═ P⁺ ═ 4 0	5 ═ P ≡ 5 0	5 ═ P⁻ ≡ 6 0

Continued

6					
	SF5(+)	SF6	**SF7(−)**		
	[10]	[10]	..		

Positive: $^6_5 S^+_0$ (2 left, 3 right) neutral: $^6_6 S_0$ negative: $^6_7 S^-_0$

7					
	ClF6(+)	IF7	**IF8(−)**		
	[10]	[10]	..		

Positive: $^7_6 Cl^+_0$ neutral: $^7_7 I_0$ negative: $^7_8 I^-_0$

8				
	XeF7(+)	Xe		
	[12]	[12]		

Positive: $^8_7 Xe^+_0$ neutral: $^8_8 Xe_0$

Number of electrons in the inactive shell = 2			
Valence number	Positive charge	neutral	Negative charge
1			
Na(−)			$^1_2 Na^-_2$
2			
Be [10] BeH(−) [10]		$^2_2 Be_2$	$^2_3 Be^-_2$
3			
B(+) BH [10] BF2(−) [10]	$^3_2 B^+_2$	$^3_3 B_2$	$^3_4 B^-_2$
4			
SiF(+) [10] CH2 [12] CF3(−) [10]	$^4_3 C^+_2$	$^4_4 C_2$	$^4_5 Si^-_2$
5			
PH2(+) [11] PCl3 [10] PH4(−) [10]	$^5_4 P^+_2$	$^5_5 P_2$	$^5_6 P^-_2$
6			
H3O(+) [10]-[12] SeF4 [10]-[12] SF5(−) [10]	$^6_5 S^+_2$	$^6_6 S_2$	$^6_7 S^-_2$
7			
ClF4(+) [10] IF5 [10] IF6(−) [13]	$^7_6 Cl^+_2$	$^7_7 I_2$	$^7_8 Cl^-_2$
8			
XeF5(+) [12] XeF6 [10] [12] XeF7(−) [12]	$^8_7 Xe^+_2$	$^8_8 Xe_2$	

Number of electrons in the inactive shell = 4			
Valence number	Positive charge	neutral	Negative charge
3	**B(−)**		$\overset{3}{\underset{4\quad 4}{B}}^{-}$
4	C [10] SiH(−) [10]	$\overset{4}{\underset{4\quad 4}{C}}$	$\overset{4}{\underset{5\quad 4}{C}}^{-}{=}$
5	**N(+)** N2 [8] PH2(−) [11]	$\overset{5}{\underset{4\quad 4}{N}}^{+}$	$\overset{5}{\underset{5\quad 4}{P}}{-}$	${-}\overset{5}{\underset{6\quad 4}{P}}^{-}{-}$
6	NO(+) [12] H2O [10]-[12] SeH3(−) [11]	$\overset{6}{\underset{5\quad 4}{O}}^{+}{-}$	${-}\overset{6}{\underset{6\quad 4}{S}}{-}$	${-}\overset{6}{\underset{7\quad 4}{S}}^{-}{=}$
7	ClF2(+) [10] ClF3 [10] ClF4(−) [10]	${-}\overset{7}{\underset{6\quad 4}{Cl}}^{+}{-}$	${-}\overset{7}{\underset{7\quad 4}{Cl}}{=}$	${=}\overset{7}{\underset{8\quad 4}{Cl}}^{-}{=}$
8	XeF3(+) [13] XeF4 [10] [12] XeF5(−) [12]	${-}\overset{8}{\underset{7\quad 4}{Xe}}^{+}{=}$	${=}\overset{8}{\underset{8\quad 4}{Xe}}{=}$

Number of electrons in the inactive shell = 6			
Valence number	Positive charge	neutral	Negative charge
5	**P(−)**		$\overset{5}{\underset{6\quad 6}{P}}^{-}$
6	O [10]-[12] OH(−) [10]-[12]	$\overset{6}{\underset{6\quad 6}{S}}$	$\overset{6}{\underset{7\quad 6}{S}}^{-}{-}$
7	**Cl(+)** HCl [10]-[12] ClF2(−) [10]	$\overset{7}{\underset{6\quad 6}{Cl}}^{+}$	$\overset{7}{\underset{7\quad 6}{Cl}}{-}$	${-}\overset{7}{\underset{8\quad 6}{Cl}}^{-}{-}$
8	XeF(+) [12] XeF2 [10]-[12] **XeF3(−)**	$\overset{8}{\underset{7\quad 6}{Xe}}^{+}{-}$	${-}\overset{8}{\underset{8\quad 6}{Xe}}{-}$

Number of electrons in the inactive shell = 8			
Valence number	Positive charge	neutral	Negative charge
7	Cl(−) [10]-[12]		$\overset{7}{\underset{8\quad 8}{Cl}}^{-}$
8	Xe [10]	$\overset{8}{\underset{8\quad 8}{Xe}}$

Atoms are represented with single-covalent bonds, as if they would connect to other atoms in the compound. As defined in the even-odd rule, each atom is surrounded by one symbol, indicating the charge, and three numbers to calculate the number of electrons in the inactive shell.

In **Table 1**, atoms are grouped according to the number of electrons in the inactive shell. This even number ranges from zero to 8. Inside each group, each row corresponding to an element, is ordered by growing valence number from zero up to eight as in the periodic table. In the left column, the said valence number is indicated along with three possible compounds formulas. In the next three columns, the focus is on the element, once positively charged, once neutral and then negatively charged.

Still in **Table 1**, the number of bonds of a charged atom is deduced using the even-odd rule and the isoelectronicity rule. For example, the neutral atom in the top row is Xenon (Xe) with no bond. Adding a charge to it gives Xe(−)-right column. According to the even-odd rule, the charge is compensated with one covalent bond. In a second step, the isoelectronicity rule is applied: Xe(−) can be replace by a neutral sodium atom (Na) that is located in the middle column just below Xe. The number of bonds is maintain during the isoelectronic modification and the neutral Na has one bond. The even-odd rule is then again used to build Na(+) without bond and Na(−) with two bonds. To build the row underneath, the isoelectronicity rule is applied to give the beryllium atom (Be). The same method alternating the even-odd and isoelectronic procedures, is used to build the other rows of **Table 1**.

In **Table 1**, the bottom right number gives the number of electrons in the inactive shell. The number is always even, implying that electrons go in pair. There cannot be more than 4 pairs in the inactive shell to ensure the coherence of the even-odd rule with the periodic table.

In order to show the potential of the even-odd rule, it is in the following applied to the atoms of 3D molecules.

5. The Even-Odd Rule in 3D Molecules

In previous papers, the even-odd drawing procedure has been applied to every atom of many ions or molecules [6]-[8]. In this chapter, the rule is applied to atoms in 3D molecules: two examples are displayed in **Table 2** and **Table 3**. **Table 2** relates to the P2O3 family and **Table 3** to the Si3N4 family. In the gas phase, these molecules are typically coupled [13]. The drawing software used is referenced in [14].

In the P2O3 family (**Table 2**), four molecules with numerous references in scientific literature are listed in the right column: P2O3, P2(NMe)3, (PO)2O3 and (SiR)2(PH)3 [13]. On the left side of **Table 2**, the 3D drawing of P2O3 is shown as a reference. The oxygen is symbolized in red with two bonds and the phosphorus in grey with three bonds.

Table 2. Molecules, in gaseous phase, compatible with the even-odd rule. P2O3 family.

Triangle-based pyramid in 3D			
(P2O3)2 Neutral	With 3 bonds	With two bonds	Molecule
Red Oxygen **Grey** Phosphorus			P2O3 Neutral [13]
			P2(NMe)3 Neutral [13]
			(P(+)O(−))2O3 Charged [12] [13]
As2O3, P2S3, Sb2O3, N2(CH2)3, P4S10, P4O10 [12] Mn2O3, Me2H3, N2(CH2)3, As2S3, (PS)2O3, (PS)2S3 [13]			(RSi)2(PH)3 Neutral [13]

In the middle columns, atoms composing the molecules are represented using the even-odd rule: P2O3, P2(NMe)3, (PO)2O3 and (SiR)2(PH)3. In P2(NMe)3, phosphorus P remains and the oxygen atom is replaced by NMe without charge. In (PO)2O3, the oxygen is unchanged and the phosphorus atom is replaced by PO. In this PO, the oxygen is negatively charged and the new P is positively charged to obtain a neutral molecule. In (SiR)2(PH)3, P is replaced without charge by SiR and O by the uncharged PH.

Below the 3D picture of **Table 2** is a list of 12 compounds of the P2O3 family. All of them are referenced in literature with 3D constructions [12] [13].

Table 2 illustrates clearly how the even-odd rule is an essential intermediate step between the formula and the 3D construction.

In **Table 3**, the silicon nitride family is processed in the same way. The family formula: Si3N4, is well known and serves as a reference [13]. Unfortunately, a 3D drawing of this compound is not available in literature. The author proposes a 3D representation when this molecule is coupled in a cluster: (Si3N4)2. This cluster has naturally the shape of a cube and is shown in the left column of **Table 3**: 8 nitrogen atoms in the corners and 6 silicon atoms in the face-centers. By applying the even-odd rule to the atoms of this molecule, we find that: nitrogen has 3 bonds and silicon has 4 bonds, thus giving the 3D structure.

The bonding configurations of each atom in the cluster are presented in the middle columns of **Table 3**.

Other molecules that follow the same pattern also appear on the right side of **Table 3** [12]. Al4C3 has neutral atoms like in Si3N4. As3O4 has charged atoms and the molecules underneath are uncharged. When comparing As3O4 and As4O3, it can be seen that an oxygen atom can have 3 or 4 bonds. Both forms agree with the even-odd rule. For As3O4, the atoms are charged: half of the As atoms are positively charged and half negatively, with the same for the oxygen. By alternating the sign of charges, the overall neutrality of the molecule is guaranteed.

6. Isoelectronicity between Atoms and the Even-Odd Rule

In the present paper, the isoelectronicity rule is only applicable to single bonded compounds [8]. It also allows for an atom to be replaced by another only if no modifications occur in its interconnections. In **Table 1**, groups of three isoelectronic elements can be found by taking a diagonal going upward and to the right.

Examples of isoelectronic atoms in the second group (2 electrons in the inactive shell) are B(+), Be and Na(−). They have no bond and the same effective valence number, equal to 2. The only difference lies in the charge type. Another example with neutral carbon: N(+), C and B(−) all with 4 bonds and an empty inactive shell.

Table 3. Molecules, in gas phase, compatible with the even-odd rule. Si3N4 family.

3D cubic cluster with 24 bonds = 4*6			
(Si3N4)2 cluster	With 4 bonds	With 3 bonds	Molecule

	$\equiv^4 Si_{4\ 0}$	$\equiv^5 N_{5\ 2}$	(Si3N4)2 [13]
	$\equiv^4 C_{4\ 0}$	$\equiv^3 Al_{3\ 0}$	(Al4C3)2 [13]
	$\equiv^5 As_{4/6\ 0/2}^{+/-}$	$\equiv^4 O_{3/5\ 0/2}^{+/-}$	(As3O4)2 Charged [12]
	$\equiv^6 O_{6\ 2}$	$\equiv^5 As_{5\ 2}$	(As4O3)2 [12] (Sb4Sn3)2 [15] Neutral

Grey 4 bonds for: silicon, carbon, oxigen or charged arsenic
Orange 3 bonds for: nitrogen, aluminum, arsenic or charged oxigen

In the extreme positions in **Table 1**, some atoms do not belong to triplets of isoelectronic atoms. At the top are Na(+) and Xe without bond and an empty inactive shell. At the bottom, Xe and Cl(−) have no bonds and a full inactive shell. Just above, another couple Xe, Cl(−) have 3 electrons pairs in the inactive shell and two bonds. The isoelectronic candidates for these atoms are limited, exactly as the valence number in the periodic table when they are at the end of a row.

As both rules are applicable individually to each atom of a compound, the charge located on an atom does not influence the other atoms of the compound. For this reason, knowing the number of bonds an atom forms is sufficient to know the atom configuration as shown in **Table 1**.

7. Discussion

The main objective in the discussion is to compare the even-odd and the isoelectronicity rules to the classical octet and multi-bonded rules. This discussion is more general than in previous paper, which was focused on the validity of the rules [6]-[8].

7.1. The Number of Electron Pairs in the Inactive Shell and Classical Multi-Bonded Compounds

The *rule of two* for covalent bonds is classically used in molecules with single and multi-bonded atoms [2]. As a difference, the *even-odd rule* states that atoms are only bonded through single bonds. This rule also imposes that electrons go in pairs in the inactive shell. This shell has 0 electron pair, like in Na(+), up to 4 electron pairs, like in Cl(−). The difference may be interpreted as a transfer of the extra bonds in the multi bond from the covalent shell into a given number of electron pairs in the inactive shell. In the classical model, the type of bond can only be derived from the knowledge of both connected atoms whereas in the even-odd model, only one atom is needed. This different seems fundamental. Therefore, the *multi-bonds rule* and the *even-odd rule* do not have any correspondence.

7.2. The Langmuir-Octet Rule and the Empty Inactive Shell Rule

The CH_4 molecule is known as being the main representative of the octet rule [12]. Other molecules are well known, such as PH_3, H_2S and HF. In these molecules, central atoms all have 8 electrons around their inner shell (*i.e.* in the classical valence shell). This forms the fundament of the *octet rule*: atoms tend to surround themselves in the valence shell. We here would like to explore the reversed statement: what if one of the stable states of an atom in a compound is when its shell is empty? Atoms meeting this criterion are in the first group of **Table 1**. In this group, NaF, BeF_2, BH_3, CH_4, PCl_5 and SF_6 all have an empty inactive shell. Since these atoms cover every columns of the main group in the periodic table, this would seem to be an acceptable hypothesis for the *even-odd rule*. As a consequence, replacing the *octet rule* by the empty inactive shell rule seems possible. Unfortunately, this rule has a similar limitation to the *octet rule*: it is not applicable to the other groups in **Table 1**. This hypothesis to reduce the rule to about one third of possible chemical structures seems possible but too restrictive.

A better hypothesis would be to replace the *octet rule* by the *even-odd rule* including the (0 to 4) electron pairs in the inactive shell of the atoms. With this definition, this rule covers a much larger number of compounds [6]-[8], including the compounds used in **Tables 1-3**.

For example with this proposed rule, halogen can have one, three, five or seven covalent bonds without any restrictions.

7.3. Known or Unknown Compounds

Sixty-nine compounds are presented in **Table 1**. Among them, many are well documented in scientific literature: CH_4, OH(−), Na(+), H_2O, Xe and all neutral compounds. However, in the ions columns of **Table 1**, about 10% of the compounds remain unreferenced. The even-odd and isoelectronic rules help predict their existence, which has never been suggested before. They could be experimentally observed in a near future. It might also be possible that these compounds are so reactive that they rapidly form other compounds. CH_2 for instance is very unstable and rapidly reacts to form C_2H_4.

7.4. Dissociation of Diatomic Gas at High Temperature

Halogen can form diatomic molecules like for instance Cl2. By applying the even-odd rule and at high temperature, Cl2 is a gas and atoms are bonded with one bond, as mentioned in **Table 1**. Experimentally, these halogen molecules start to dissociate between 500°C or 1000°C-depending on the halogen [13]. Their dissociation formulas are unknown but for this dissociation, the even-odd rule proposes the existence of ions like Cl(−) and Cl(+).

These ions are important by-products of the even-odd rule and it would be very interesting to run an analysis to prove or investigate their existence.

7.5. Charge Delocalization and Charge Position

A well-known molecule like Benzene, C6H6 is classically represented with alternating single and double bonds between carbon atoms. A circle is often used to replace the double bonds, indicating that some charges are delocalized [12]. With the even-odd rule, double bonds-or the circle-must be replaced by single bonds between charged atoms, alternatively positively and negatively charged [8]. Resulting effective valence numbers of carbon atoms are either 3 or 5, depending on the atom charge, but all atoms have 3 bonds. If negatively charged, the carbon atoms have an electron pair in the inactive shell. If positively charged, the inactive shell is empty and no extra bond can be built.

As mention above with the even-odd rule, the effective valence is used to calculate the number of electron pairs near the element. The effective valence takes into account the charge of the element. A negative charge corresponds to a pair of electrons that can be located in two different shells. Either the pair is in the covalent shell, meaning an extra bond to a distinct atom or it is in the inactive shell. A positive charge corresponds to an absence of a pair also either in the covalent shell or in the inactive shell. As both shells are located near an atom, the charge is localized on a specific atom. There is no need to use the concept of delocalized electron.

This local property of the electron pairs ensures the coherence of the even-odd rule and allows us to apply the isoelectronicity rule [9].

7.6. Cubic Cluster

Page 1022 of reference [13] cites a (Me3X4)2 cluster-Me Metal and X halide-that takes a cubic form similar to that in **Table 3**. In the (Me3X4)2 cluster, atoms are in the same location as in **Table 3**. Each atom in the compound has however 8 covalent bonds unlike in **Table 3** where the cluster has 3 or 4 bonds per atom. These bonding configurations are not compatible and the isoelectronicity rule is not applicable on such a bonding configuration. A better understanding of this difference is under investigation.

7.7. Isoelectronicity Rule and Classical Multi-Bonds

Two molecules, single bonded or multi-bonded, are isoelectronic under the usual definition when they have the same total number of valence electrons [16]. In the proposed even-odd rule, compounds are single bonded and the isoelectronicity rule can be independently applied to each atom.

A classical isoelectronicity between binary compounds is mentioned by Greenwood on page 412 [13]. It lists N2, NO(+), CN(−), CO as being isoelectronic. They all have 10 valence electrons and 3 bonds. Unfortunately, the charge position is not given. If we apply the even-odd and the isoelectronicity rules described in the present paper, binary compounds must have single bonds and charges are consequently assigned to specific atoms: N2, NO(+), C(−)N and C(−)O(+). (In these four compounds, atoms are bonded through single bonds, not triple bonds like in the classical approach). All atoms have respectively an effective valence of 5 and 4 electrons in the inactive shell. From this example, we can conclude that both rules are similar except for the number of bonds. In the same way, for single-bonded compounds, classical and present isoelectronic compounds are similar: B(−)H4, CH4, N(+)H4 [9].

This similarity cannot be found for double-bonded compounds. As an example, O2, classically with a double bond has a valence number of 12. They share 4 electrons-double bond-and each oxygen atom is surrounded with 8 electrons. Classical isoelectronicity states that the diatomic compound NCl also with a double bond and 12 electrons, is an isoelectronic of O2. In the even-odd rule though, the NCl compound has one bond and both atoms are neutral as shown in **Table 1**. Again from **Table 1**, N and Cl are isoelectronic with O(+) and O(−) respectively. As a consequence, with double bonds, compounds do not have the charges on the same atoms.

According to this last example, both isoelectronicity rules are not completely similar.

8. Conclusion

In the present article, the even-odd and the isoelectronicity rules used on chemical formulas of compounds were described. More specifically, the rules give a way to compute three numbers that are written as a convention around the name of the element. If the element is charged, the charge appears as a sign, like in classical representation. These three numbers indicate how many electron pairs are located in the three different shells of the atom. A specificity of the rules is the inactive shell, containing electrons that are available for bonding. The inactive shell only contains pairs of electrons, between 0 and 4 pairs. Another peculiarity is that both rules only deal with single-covalent bonds, indicating that multiple bonds are not needed to explain compounds configuration. More precisely, the present paper aimed to put forward both rules as a substitute to the too restrictive Langmuir-octet-rule. In order to support this claim, several atom structures from the main group in the periodic table were processed using the rules and the number of electrons pairs in the inactive shell was calculated. As presented, the structures could easily be grouped in triplets following the isoelectronic rule. As a final conclusion, the article illustrated that the even-odd and isoelectronicity rules, applied to each atom in a compound, seem to give a plausible description of single-bonded ionic or molecular structures.

References

[1]   Abegg, R. (1904) Die Valenz und das periodische System. Versuch einer Theorie der Molekularverbindungen. (The Valency and the Periodical System—Attempt on a Theory of Molecular Compound.) *Zeitschrift für anorganische Chemie*, **39**, 330-380. http://dx.doi.org/10.1002/zaac.19040390125

[2]   Lewis, G.N. (1916) The Atom and the Molecule. *Journal of the American Chemical Society*, **38**, 762-785. http://dx.doi.org/10.1021/ja02261a002

[3]   Langmuir, I. (1919) The Arrangement of Electrons in Atoms and Molecules. *Journal of the American Chemical Society*, **41**, 868-934. http://dx.doi.org/10.1021/ja02227a002

[4]   Lewis, G.N. (1923) Valence and the Structure of atoms and Molecules. *American Chemical Monograph Series*, The Chemical Catalog Co., Inc., New York.

[5]   Gillespie, R.J. and Robinson, E.A. (2007) Gilbert N. Lewis and the Chemical Bond: The Electron Pair and the Octet Rule from 1916 to the Present Day. *Journal of Computational Chemistry*, **28**, 87-97. http://dx.doi.org/10.1002/jcc.20545

[6]   Auvert, G. (2014) Improvement of the Lewis-Abegg-Octet Rule Using an "Even-Odd" Rule in Chemical Structural Formulas: Application to Hypo and Hyper-Valences of Stable Uncharged Gaseous Single-Bonded Molecules with Main Group Elements. *Open Journal of Physical Chemistry*, **4**, 60-66. http://dx.doi.org/10.4236/ojpc.2014.42009

[7]   Auvert, G. (2014) Chemical Structural Formulas of Single-Bonded Ions Using the "Even-Odd" Rule Encompassing Lewis's Octet Rule: Application to Position of Single-Charge and Electron-Pairs in Hypo- and Hyper-Valent Ions with Main Group Elements. *Open Journal of Physical Chemistry*, **4**, 67-72. http://dx.doi.org/10.4236/ojpc.2014.42010

[8]   Auvert, G. (2014) The Even-Odd Rule on Single Covalent-Bonded Structural Formulas as a Modification of Classical Structural Formulas of Multiple-Bonded Ions and Molecules. *Open Journal of Physical Chemistry*, **4**, 173-184. http://dx.doi.org/10.4236/ojpc.2014.44020

[9]   Auvert, G. (2014) Coherence of the Even-Odd Rule with an Effective-Valence Isoelectronicity Rule for Chemical Structural Formulas: Application to Known and Unknown Single-Covalent-Bonded Compounds. *Open Journal of Physical Chemistry*, **4**, 126-133. http://dx.doi.org/10.4236/ojpc.2014.43015

[10]   http://www.chemspider.com/

[11]   http://www.ncbi.nlm.nih.gov/pccompound/

[12]   http://en.wikipedia.org/wiki/Main_Page

[13]   Greenwood, N.N. and Earnshaw, A. (1998) Chemistry of the Elements. 2nd Edition, Butterworth-Heinemann, Oxford.

[14]   Hanwell, M.D., Curtis, D.E., Lonie, D.C., Vandermeersch, T., Zurek, E. and Hutchison, G.R. (2012) Avogadro: An Advanced Semantic Chemical Editor, Visualization, and Analysis Platform. *Journal of Cheminformatics*, **4**, 17. http://dx.doi.org/10.1186/1758-2946-4-17

[15]   http://pubchem.ncbi.nlm.nih.gov/compound/86205385?

[16]   http://en.wikipedia.org/wiki/Isoelectronicity

Thermal and Photochemical Effects on the Fluorescence Properties of Type I Calf Skin Collagen Solutions at Physiological pH

Julian M. Menter, Latoya Freeman, Otega Edukuye

Department of Microbiology, Biochemistry and Immunology, Morehouse School of Medicine, Atlanta, GA, USA
Email: jmenter@msm.edu

Abstract

Mammalian collagens exhibit weak intrinsic UV fluorescence that depends on the age and previous history of the sample. Post-translational modifications result in additional fluorescent products (e.g. DOPA, dityrosine, and advanced glycation end products (AGE)). UV radiation can cause longer wavelength fluorescent oxidative bands. These alterations can assess the extent of photolysis. We describe the ground- and excited-state oxidative transformations of newly-purchased type I calf skin collagens (samples #092014 and #072012) and a 7-year-old sample (#072005). We compare the effects of UV radiation (mainly 254 nm) and age on the photochemical reaction kinetics and fluorescence spectral distribution of type I calf skin collagen at pH 7.4. The fluorescence spectra of samples #072012 and #092014 were similar but not identical to pure tyrosine, whereas #072005 indicated significant "dark" oxidation at the expense of tyrosine. Fading of oxidized product(s) at 270/360 nm is second-order. Build-up of 325/400 nm (dityrosine) fluorescence is linear with time. Rate parameters r_2 and r_1 were respectively proportional to second order disappearance of ground state oxidation products and the quasi–first-order photochemical formation of dityrosine. There is a reciprocal relationship between the rates of decrease in the 270/360 nm fluorescence and concomitant increase in 325/400 nm fluorescence. Their relative rates depend on the age of the collagen sample. There is a reciprocal relationship between r_1 and r_2. This relationship results because both ground state autoxidation and excited state photo-dimerization proceed via a common tyrosyl radical intermediate. Water of hydration appears to play a role in generating tyrosyl radical.

Keywords

Collagen, Fluorescence, Oxidation, Photodimerization, Reaction Kinetics

1. Introduction

It is well known that mammalian collagens exhibit weak intrinsic UV fluorescence. Observed fluorescence spectra depend on the age and previous history of the sample. In nascent collagen, the only fluorescent chromophore is tyrosine (peak absorption/emission maxima at 275/305 nm). However, post-translational ground- and/or excited-state modifications can result in additional fluorescent products derived from tyrosine oxidation that absorb and emit at longer wavelengths (e.g. DOPA (285/315 nm) [1], dityrosine (325/400 nm) [2], DOPA-oxidation products (emitting from 330 - 470 nm) [3] and advanced glycation end products, involving a reducible sugar and lysine (370/450 nm) (AGE) [4]-[8]). We do not expect to find AGEs in these experiments, because there are no sugars in our reaction mixtures. We have found that oxidative changes can be observed even under storage in the solid state at 4°C in the dark.

These compounds are photolabile, the predominant photo-product being photo-dimerization to dityrosine [9] [10]. Photo-induced dityrosine formation and fluorescence disappearance ("fading") of DOPA oxidation products ("oxidation products") attending exposure to UV radiation have been used as convenient markers of the extent of photolysis [11] [12]. This photolability suggests the possibility of collagen or its oxidation products acting as *in situ* photosensitizers and/or phototoxic agent.

These ground- and excited-state processes, and the relationships between them are not well described, and a better understanding of them is necessary. We are therefore systematically investigating the time- and UV-dependent fluorescence properties of type I calf skin collagen solutions buffered at pH = 7.4. Our initial results are reported in this communication.

2. Methods

2.1. Collagens

Three commercial purified collagen samples from the same manufacturer were used; newly-purchased calf-skin collagens (#092014), a "new" sample purchased in 2012 (#072012) and an older sample purchased in 2005 (#072005) that had been stored in the solid state in the dark at 4°C for ~7 years before analysis in 2012. Samples #072012 and sample #072014 were analyzed in June-July 2012 and September 2014 respectively as soon as they were received, and their spectral and kinetic properties were virtually identical. Sample #072005 was analyzed at the same time as #072012 (June-July 2012).

2.2. Fluorescence Spectroscopy

Fluorescence emission spectra of collagen samples were recorded on a Perkin-Elmer 650 - 40 fluorescence spectrometer, equipped with a thermostatted sample compartment, in conjunction with a circulating bath (Lauda, K-2R; Brinkmann Instruments, Westbury, NY, USA). Horizontal entrance and exit slits in the sample compartment allow measurement of as little as 0.3 ml, and optimize measurement of turbid solutions. Optical quartz 1.0 cm cells (Hellma Cells, Inc., Plainview, NY, USA) were used to collect fluorescence emitted at right angles to the excitation beam. Because of the weak fluorescence, the band widths of both excitation and emission monochromators were set at 5 nm. Fluorescence spectra were recorded manually at excitation wavelength 270 nm. Possible photochemical activity was limited by filling the cuvettes to capacity (3.0 ml), and minimizing exposure to excitation light. The time constant was 2.0 s. All spectra were corrected for instrumental distortions.

Photochemical changes occurring in these solutions were studied as follows: one (1.0) ml of 0.05% collagen solutions were irradiated in a thermostatted cuvette (Hellma Cells, Plainview, NY, 11803, USA). Readings were taken at excitation/emission wavelengths of 270/300 (tyrosine), 270/330 (DOPA), 270/360 ("interacting" tyrosine or DOPA oxidation product(s)), 270/400 (link between excitation at 270 nm and 325 nm), 325/400 (dityrosine), and 370/450 nm (AGE's) (not relevant here); DOPA oxidation product(s) [12] [13]. The choice of excitation/emission wavelength pairs was chosen to limit interference from overlapping bands. Irradiations were carried out with a 4 W UVG-11 hand lamp (emitting primarily 254 nm superimposed on a UV continuum; (output 6.6 W/m^2)). The geometry was such that the irradiation impinged on the entire sample.

The resulting fluorescence intensity, $I_f(t)$ vs. irradiation time t plots were analyzed kinetically. The 270/360 nm fluorescence fading was analyzed as a second-order plot of $1/I(t)$ vs. t to afford a second order rate parameter [11] r_2, whose slope is proportional to a molecular rate constant for disappearance of the (unknown) fluorophore at 270/360 nm, is:

$$r_2 \equiv \Delta \left[1/I_f(t) \right] \big/ \Delta t \tag{1}$$

The build-up of 325/400 nm fluorescence attending photolysis (dityrosine formation) was plotted as $I(t)$ vs. t to give a quasi-linear plot [9] whose slope affords a first order rate parameter, proportional to a molecular rate constant for build-up of the dityrosine fluorophore at 325/400 nm, is:

$$r_1 \equiv \Delta \left[\mathbf{I_f}(\mathbf{t}) \right] \big/ \Delta t \tag{2}$$

Both r_1 and r_2 are proportional to fluorophore concentration [9] [11]. Changes in the other fluorescence pairs were kinetically complicated alterations and were not amenable to simple analysis (see [12] for results with Skh-1 hairless mouse collagen).

Although the dimensions of r_1 and r_2 are different from each other, the ratio of slopes r_2/r_1 serves as a convenient empirical tool for comparing the individual collagen samples.

2.3. Statistics

Data were analyzed by the Student's t test as *Mean* $\pm$ *S.D.* of the slopes of *fluorescence fading/build-up* curves from 3 separate experiments.

3. Results

The only fluorescent chromophore in nascent collagen is tyrosine. However, post-translational modifications in amino acid structure result in additional fluorescent oxidation products, so that collagen fluorescence spectra depend on the age and previous history of the sample. This result is illustrated in **Figure 1**. The fluorescence spectra of new samples #072012 and #092014 (not shown) are very similar to each other, showing tyrosine fluorescence at 300 - 305 nm as the predominant peak. However comparison of these spectra with that of pure tyrosine [14] shows a slight broadening of the overall fluorescence envelope, due to small amounts of DOPA and its oxidation products (**Figure 2**). The older sample, #072005, shows significant diminution of the tyrosine peak at 305 nm ($I_{(072005}/I_{(072012)} = 0.57$) as well as the more obvious appearance of additional longer wavelength bands that can be ascribed to a mixture of DOPA, DOPA oxidation products and dityrosine [11]-[13].

UVC (254 nm) irradiation of collagen causes a rapid *decrease* of the 270/360 nm fluorescence ("photolabileoxidation product") and concomitant *increase* in the 325/400 nm band (dityrosine formation). Other fluorescence

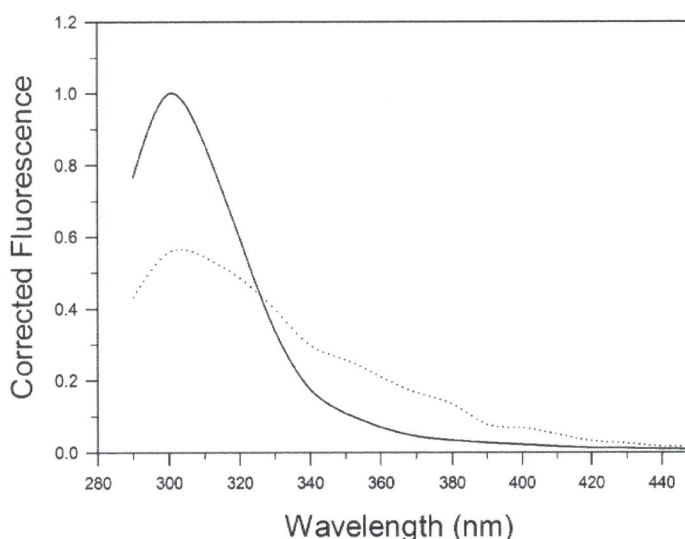

Figure 1. Corrected fluorescence emission spectra of calf skin type I collagen samples 0.05% in 0.1 M phosphate buffer, pH = 7.4, T = 24°C - 26°C, Excitation at 270 nm. Solid line: fresh sample #072012, Dotted line: old sample #072005 that had been stored in the dark for ~7 years prior to analysis. Spectra are normalized to the 305 nm peak of #072012.

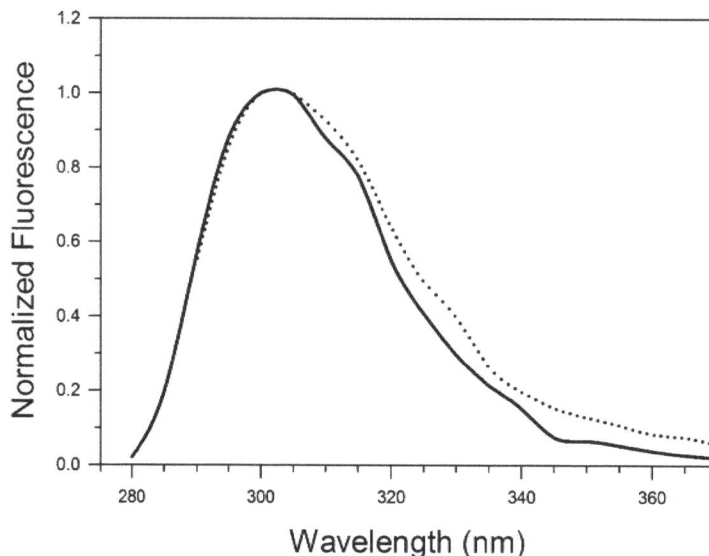

Figure 2. Corrected normalized fluorescence spectra of purified tyrosine [14] (solid line) and of sample #072012 shown in **Figure 1** above (dotted line). Oxidation products in #072012 are revealed by spectra broadening at long-wavelength edge.

pairs are kinetically complicated alterations and not amenable to simple analysis. Similar results were previously obtained with Skh-1 hairless mouse collagen [12] [13].

The 270/360 nm band shows a linear relationship between reciprocal fluorescence intensity, $1/I_f(t)$ vs. time, t. **Figure 3(a)** compares fading results of (270)360 nm fluorescence from lot #072005 and #072012. The value of $r_2/r_1 = 2.31 \pm 0.40$. Covalent dityrosine (325/400 nm fluorescence) is formed from a metastable steady-state tyrosyl radical, where the tyrosyl steady state concentration is proportional to the amount of tyrosine [9]. In such cases, plots of $I_f(t)$ vs. t are quasi-linear, [9]; (**Figure 3(b)**), with $r_2/r_1 = 0.246 \pm 0.097$ (see **Table 1** for results).

Table 1 clearly shows the reciprocal relationship between the rates of *decrease* in the 270/360 nm fluorescence (oxidation products) and concomitant increase in 325/400 nm fluorescence (dityrosine) where the relative rates depend on the age of the collagen sample. Thus, $r_2 > r_1$ in sample #072005, whereas $r_2 < r_1$ in #072012. It therefore appears that the oxidation products are formed at the expense of tyrosine.

4. Discussion

We have investigated the thermal and photochemical stability of type I calf skin collagen at physiological pH. Preparations lot #072012 and #092014 had fluorescence spectra that were virtually identical, and were similar to the spectrum of pure tyrosine. Although both collagen samples were analyzed on arrival, there was nonetheless a small trace of DOPA oxidation products in them that were lacking in pure tyrosine (**Figure 2**). Since these collagen samples were purified from calf skin, it is quite possible that some oxidation had occurred *in vivo*. Lot #072005 had been stored at 4°C in the dark and there was evidently a significant amount of oxidation at the expense of tyrosine. Since these additional oxidation products are photolabile, their overall effect is to decrease the stability of collagen with time.

Dityrosine formation by UV can be analyzed as a pseudo-first-order steady-state reaction [9], where [Tyr]$_{SS}$ is directly proportional to the initial concentration of tyrosine in the collagen sample.

The second order fading of the (270)360 nm species suggests a "double molecule", which we have previously attributed to an "excimer-like" interaction between two tyrosine molecules in close proximity [12]. Alternatively, it may be a stable, as yet uncharacterized, DOPA oxidation product. One might expect such an excimer-like interaction to favor dityrosine formation. In fact, the observed reciprocal relationship between the 325/400 nm and the 270/360 nm species suggests that the latter is formed at the *expense* of dityrosine. This finding argues against an interaction involving two proximal tyrosine molecules, but is consistent with the formation of a stable oxidized fluorescent product.

(a)

(b)

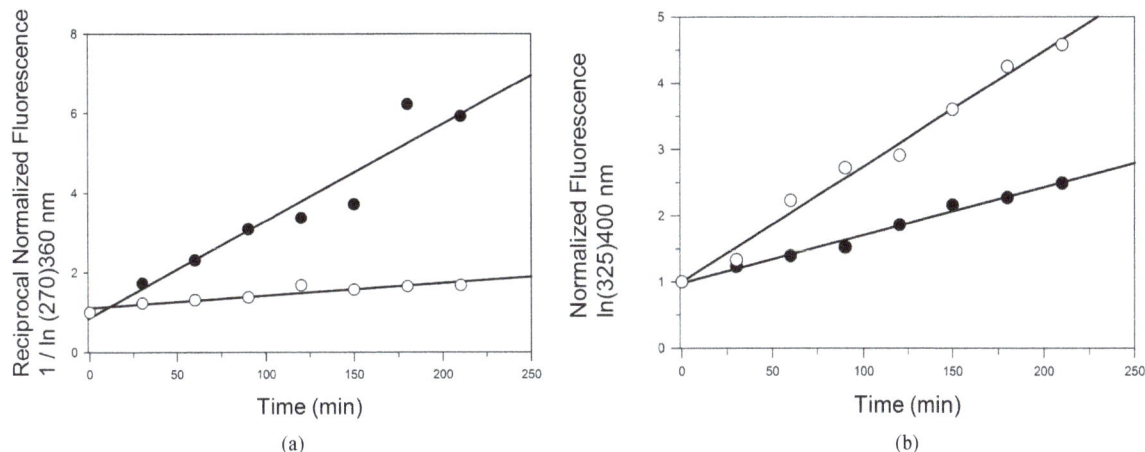

Figure 3. (a) Second-order fading of 270/360 nm fluorescence (oxidized product(s)) as reciprocal plot vs. time for sample #072005 (black dots) and #072012 (white dots). This species fades significantly faster in the older sample #072005; (b) First-order build-up of 325/400 nm fluorescence (dityrosine) as fluorescence vs. time for sample #072005 (black dots) and #072012 (white dots). This species accumulates significantly slower in the older sample #072005.

Table 1. Normalized fluorescence fading of the (270)360 nm band as $ln/I_f(t)$ vs. t and normalized fluorescence build-up of (325)/400 nm as $I_f(t)$ vs. t in calf-skin collagen sample #072012 (fresh sample) and #072005 (~7 years old) in 0.1 M PO_4 buffer, pH = 7.4 (see text). Parameters r_1 and r_2 are defined in the text. The ratio r_2/r_1 is an empirical measurement proportional to the relative amounts of oxidation product(s)/dityrosine.

SAMPLE	$r_1 \times 10^{3*}$	$r_2 \times 10^{3*}$	r_2/r_1*	REMARKS
#072012	13.6 ± 1.88	3.37 ± 1.33	0.246 ± 0.097	Fresh sample promptly analyzed
#072005	7.68 ± 0.71	17.6 ± 2.20	2.31 ± 0.40	Sample stored in dark at 4°C for ~7 years

*Mean ± S.D. of three experiments.

Both photochemical photo-dimerization [9] [10] and ground state reactions [3] of tyrosine involve formation of a tyrosyl radical intermediate, which decays either by forming covalent dityrosine, or by reaction with O_2 and consequent formation of DOPA, and higher oxidation products [3]. In the absence of UV radiation, the relative abundance of oxygen favors the latter route, but a small amount of dityrosine was also formed in the dark in sample #072005.

Tyrosyl radical formation usually requires a highly oxidative agent because of the high redox potential of the tyrosyl/ tyrosine couple. The photochemical formation of dityrosine uses the absorbed photonic energy to accomplish tyrosyl radical formation. Ground state oxidation of tyrosine residues in collagen is most likely to involve water of hydration [15]. Our collagen samples, no doubt, were hydrated even in the solid state. The electrical properties of hydrated collagen have been studied by several workers (for quick review, see [15] and references therein). Tomaselli *et al.* [15] found that the adsorption of water onto solid bovine Achille's tendon (BAT) caused a dramatic increase on the values of dc conductivity σ that is both temperature- and hydration parameter (h)-dependent (as much as 8 orders of magnitude at $h = 0.24$). Increasing hydration decreases the activation energy, and moderate hydration levels, confer semiconductor-like properties on the collagen, with water possibly acting as an impurity [15]. Although the exact mechanism of tyrosyl formation is unclear, we can speculate that moderate hydration lowers the activation energy sufficiently to allow tyrosine oxidation via electron transfer (ET), proton transfer (PT), or proton-coupled electron transfer (PCET) (for review of PCET, see [16]). In buffered solution, tyrosyl formation could also have been facilitated by the presence of phosphate buffer [17].

5. Conclusion

Collagen fluorescence spectra and reaction kinetics depend on the age and previous history of the sample. Autoxidation can take place even in the solid state at 4°C in the dark. There is a reciprocal relationship between the rates of decrease in the 270/360 nm fluorescence and concomitant increase in 325/400 nm fluorescence, resulting in a "Ying-Yang" relationship between r_1 and r_2. This relationship results because both ground state autoxidation and excited state photo-dimerization proceed via a common tyrosyl radical intermediate. Water of hydration appears to play a role in generating tyrosyl radical, particularly in dark autoxidation reactions.

Acknowledgements

This work was supported in part by DOD Grant #911 NF-10-1, MBRS Grant #GM08248, and RCMI Grant #. RR 3034. There are no conflicts of interest.

References

[1] Smith, G.J. (1999) The Fluorescence of Dihydroxyphenylalanine; The Effects of Protonation-Deprotonation. *Journal of Studies in Dynamics and Change*, **115**, 346-349. http://dx.doi.org/10.1111/j.1478-4408.1999.tb00351.x

[2] Gross, A.J. and Sizer, I.W. (1959) The Oxidation of Tyramine, Tyrosine, and Related Compounds by Peroxidase. *Journal of Biological Chemistry*, **234**, 1611-1614.

[3] Guilivi, C., Traaseth, N.J. and Davies, K.J.A. (2003) Tyrosine Oxidation Products: Analysis and Biological Relevance. *Amino Acids*, **25**, 227-232. http://dx.doi.org/10.1007/s00726-003-0013-0

[4] Nagaraj, R.H. and Monnier, V.M. (1992) Isolation and Characterization of a Blue Fluorophore from Human Lens Eye Crystallins: *In Vitro* Formation from Maillard Reaction with Ascorbate and Ribose. *Biochimica et Biophysica Acta*, **1116**, 34-42. http://dx.doi.org/10.1016/0304-4165(92)90125-E

[5] Sell, D.R. and Monnier, V.M. (1989) Structure Elucidation of a Senescence Cross-Link from Human Extracellular Matrix. *Journal of Biological Chemistry*, **264**, 21597-21602.

[6] Sell, D.R., Ramanakoppa, H., Grandhee, S.K., Odetti, P., Lapolla, A., Fogarty, J. and Monnier, V.M. (1991) Pentosidine: A Molecular Marker for the Cumulative Damage to Proteins in Diabetes, Aging and Uremia. *Diabetes/Metabolism Reviews*, 7, 239-251. http://dx.doi.org/10.1002/dmr.5610070404

[7] Verzijl, N., DeGroot, J., Oldehinkel, E., Bank, R.A., Thorpe, S.R., Baynes, J.W., Bayliss, M.T., Bijlsma, J.W.J., Lafeber, F.P.G. and TeKoppele, J.M. (2000) Age-Related Accumulation of Maillard Reaction Products in Human Articular Cartilage Collagen. *Biochemical Journal*, **350**, 381-387. http://dx.doi.org/10.1042/0264-6021:3500381

[8] Lamore, S.D., Azimian, S., Horn, D., Anglin, B.L., Uchida, K., Cabello, C.M. and Wondrak, G.T. (2010) The Malondialdehyde-Derived Fluorophore DHP-Lysine Is a Potent Photosensitizer of UVA-Induced Photosensitized Oxidative Stress in Human Skin Cells. *Journal of Photochemistry and Photobiology*, **101**, 261-264. http://dx.doi.org/10.1016/j.jphotobiol.2010.07.010

[9] Shimizu, O. (1973) Excited States in Photodimerization of Aqueous Tyrosine at Room Temperature. *Photochemistry and Photobiology*, **18**, 125-133. http://dx.doi.org/10.1111/j.1751-1097.1973.tb06402.x

[10] Lehrer, S.S. and Fasman, G.D. (1967) Ultraviolet Irradiation Effects in Poly-L-Tyrosine and Model Compounds. Identification of Bityrosine as a Photoproduct. *Biochemistry*, **6**, 757-767. http://dx.doi.org/10.1021/bi00855a017

[11] Menter, J.M., Williams, G.D., Carlyle, K., Moore, C.L. and Willis, I. (1995) Photochemistry of Type I Acid-Soluble Calf Skin Collagen: Dependence on Excitation Wavelength. *Photochemistry and Photobiology*, **62**, 402-408. http://dx.doi.org/10.1111/j.1751-1097.1995.tb02360.x

[12] Menter, J.M., Chu, E.G. and Martin, N.V. (2009) Temperature-Dependence of Photochemical Fluorescence Fading in Skh-1 Hairless Mouse Collagen. *Photodermatology, Photoimmunology & Photomedicine*, **25**, 128-131. http://dx.doi.org/10.1111/j.1600-0781.2009.00421.x

[13] Menter, J.M., Abukhalaf, I.K., Patta, A.M., Silvestrov, N.A. and Willis, I. (2007) Fluorescence of Putative Chromophores in Skh-1 and Citrate-Soluble Calf Skin Collagens. *Photodermatology, Photoimmunology & Photomedicine*, **23**, 222-228. http://dx.doi.org/10.1111/j.1600-0781.2007.00312.x

[14] Permyakov, E.A. (1993) Luminescent Spectroscopy of Proteins. CRC Press, Inc., Boca Raton, 38.

[15] Tomaselli, V.P. and Shamos, M.H. (1974) Electrical Properties of Hydrated Collagen II. *Biopolymers*, **13**, 2423-2434. http://dx.doi.org/10.1002/bip.1974.360131203

[16] Barry, B.A., Chen, J., Keough, J., Jenson, D., Offenbacher, A. and Paqba, C. (2012) Proton-Coupled Electron Transfer and Redox-Active Tyrosines: Structure and Function of the Tyrosyl Radicals in Ribonucleotide Reductase and Photosystem II. *The Journal of Physical Chemistry Letters*, **3**, 543-554. http://dx.doi.org/10.1021/jz2014117

[17] Chen, Z., Concepcion, J.A., Hu, X., Yang, W., Hoertz, P.G. and Meyer, T.J. (2010) Proton Transfer in the Excited State-Concerted O Atom-Proton Transfer in the O-O Bond Forming Step in Water Oxidation. *Proceedings of the National Academy of Sciences*, **107**, 7225-7229. http://dx.doi.org/10.1073/pnas.1001132107

Abbreviations and Acronyms

Fluorescence excitation and emission wavelengths (nm): e.g. excitation at 270, emission at 360 nm = (270)/360 nm; DOPA: 3,4-dihydroxyphenylalanine; AGE: advanced glycation end products; $r_2 \equiv$ rate of 1/(270) 360 nm fluorescence disappearance; $r_1 \equiv$ rate of (325)/400 nm fluorescence build-up; BAT: Bovine Achille's Tendon; ET: Electron Transfer; PT: Proton Transfer; PCET: Proton-Coupled Electron Transfer.

Permissions

All chapters in this book were first published in OJPC, by Scientific Research Publishing; hereby published with permission under the Creative Commons Attribution License or equivalent. Every chapter published in this book has been scrutinized by our experts. Their significance has been extensively debated. The topics covered herein carry significant findings which will fuel the growth of the discipline. They may even be implemented as practical applications or may be referred to as a beginning point for another development.

The contributors of this book come from diverse backgrounds, making this book a truly international effort. This book will bring forth new frontiers with its revolutionizing research information and detailed analysis of the nascent developments around the world.

We would like to thank all the contributing authors for lending their expertise to make the book truly unique. They have played a crucial role in the development of this book. Without their invaluable contributions this book wouldn't have been possible. They have made vital efforts to compile up to date information on the varied aspects of this subject to make this book a valuable addition to the collection of many professionals and students.

This book was conceptualized with the vision of imparting up-to-date information and advanced data in this field. To ensure the same, a matchless editorial board was set up. Every individual on the board went through rigorous rounds of assessment to prove their worth. After which they invested a large part of their time researching and compiling the most relevant data for our readers.

The editorial board has been involved in producing this book since its inception. They have spent rigorous hours researching and exploring the diverse topics which have resulted in the successful publishing of this book. They have passed on their knowledge of decades through this book. To expedite this challenging task, the publisher supported the team at every step. A small team of assistant editors was also appointed to further simplify the editing procedure and attain best results for the readers.

Apart from the editorial board, the designing team has also invested a significant amount of their time in understanding the subject and creating the most relevant covers. They scrutinized every image to scout for the most suitable representation of the subject and create an appropriate cover for the book.

The publishing team has been an ardent support to the editorial, designing and production team. Their endless efforts to recruit the best for this project, has resulted in the accomplishment of this book. They are a veteran in the field of academics and their pool of knowledge is as vast as their experience in printing. Their expertise and guidance has proved useful at every step. Their uncompromising quality standards have made this book an exceptional effort. Their encouragement from time to time has been an inspiration for everyone.

The publisher and the editorial board hope that this book will prove to be a valuable piece of knowledge for researchers, students, practitioners and scholars across the globe.

List of Contributors

Li Chen, Mitsugi Hamasaki, Hirotaka Manaka and Kozo Obara
Country Graduate School of Science and Engineering, Kagoshima University, Kagoshima, Japan

Mohamed El Miz, Samira Salhi, Ali El Bachiri and Abdesselam Tahani
LACPRENE, Faculté des Sciences, Université Mohamed 1er, Oujda, Morocco

Ikrame Chraibi
Département de Géologie, Faculté des Sciences, Université Mohamed 1er, Oujda, Morocco

Marie-Laure Fauconnier
Unité de Chimie Générale et Organique, Gembloux Agro-Bio Tech, Université de Liège, Gembloux, Belgique

Vitalyi P. Malyshev and Astra M. Makasheva
Chemical and Metallurgical Institute, Karaganda, Kazakhstan

Kinshuk Raj Srivastava
Department of Physics and Astronomy, Michigan State University, East Lansing, USA
Department of Chemistry, Indian Institute of Technology Bombay, Mumbai, India

Susheel Durani
Department of Chemistry, Indian Institute of Technology Bombay, Mumbai, India

Nimibofa Ayawei and Seimokumo Samuel Angaye
Department of Chemical Sciences, Niger Delta University, Wilberforce Island, Nigeria

Donbebe Wankasi and Ezekiel Dixon Dikio
Department of Chemical Sciences, Niger Delta University, Wilberforce Island, Nigeria
Applied Chemistry and Nanoscience Laboratory, Department of Chemistry, Vaal University of Technology, Vanderbijlpark, South Africa

Hisham A. Abo-Eldahab
Chemistry Department, Faculty of Science, Alexandria, Egypt
Umm Al-Qura University, University College, Makkah, KSA

Henk M. Buck
Kasteel Twikkelerf 94, Tilburg, The Netherlands

Geoffroy Auvert
CEA-Leti, Grenoble, France

Bing He
Molecular Design Institute, Chengdu Normal University, Chengdu, China
State Key Laboratory of Biotherapy and Cancer Center, West China Hospital, Sichuan University, and Collaborative Innovation Center for Biotherapy, Chengdu, China

Hongwei Zhou
Molecular Design Institute, Chengdu Normal University, Chengdu, China
IPNL, UMR IN2P3-CNRS-UCBL 5822, Villeurbanne, France

Fan Yang
Molecular Design Institute, Chengdu Normal University, Chengdu, China

Wai-Kee Li
Department of Chemistry, The Chinese University of Hong Kong, Hong Kong, China

John H. Summerfield
Department of Chemistry and Physical Sciences, Missouri Southern State University, Joplin, USA

Hussam El Desouky
Chemistry Department, University of Umm Al-Qura, Makka, KSA

Hisham A. Aboeldahab
Chemistry Department, University of Umm Al-Qura, Makka, KSA
Chemistry Department, Faculty of Science, University of Alexandria, Alexandria, Egypt

S. Abdel Aal
Department of Chemistry, Faculty of Science, Benha University, Benha, Egypt

Kholmirzo T. Kholmurodov
Frank Laboratory of Neutron Physics, JINR, Dubna, Russia
Dubna International University, Dubna, Russia

Sagille A. Ibragimova, Pavel P. Gladishev and Tatyana Yu. Zelenyak
Dubna International University, Dubna, Russia

Anatoly V. Vannikov and Alexey R. Tameev
A.N. Frumkin Institute of Physical Chemistry and Electrochemistry, Russian Academy of Sciences, Moscow, Russia

Jaime González Velasco
Universidad Autónoma de Madrid, Facultad de Ciencias, Cantoblanco, Madrid, Spain

Victor F. Waingeh
School of Natural Sciences, Indiana University Southeast, New Albany, IN, USA

Felix N. Ngassa
Department of Chemistry, Grand Valley State University, Allendale, MI, USA

Jie Song
Department of Chemistry and Biochemistry, University of Michigan-Flint, Flint, MI, USA

Sanaa T. Arab and Huda M. Alghamdi
Department of Physical Chemistry, KAU, Jeddah, KSA

Khadijah M. Emran
Department of Chemistry, Taibah University, Al-Madinah Al-Monawarah, KSA

Mohammed I. Abdulsalam
Department of Chemical and Materials Engineering, KAU, Jeddah, KSA

Nobuaki Tanaka
Department of Environmental Science and Technology, Faculty of Engineering, Shinshu University, Nagano, Japan

M. B. Ibrahim
Department of Pure and Industrial Chemistry, Bayero University, Kano, Nigeria

S. Sani
Department of Applied Chemistry, Federal University, Dutsin-Ma, Nigeria

Pedro Oliver Dunstan
Instituto de Química, Universidade Estadual de Campinas, Campinas, Brazil

Geoffroy Auvert
CEA-Leti, Grenoble, France

Geoffroy Auvert
Grenoble Alps University, Grenoble, France

Marine Auvert
University of Strasbourg, Strasbourg, France

V. G. Kunitsyn, P. A. Kuznetsov, E. N. Demchenko and O. I. Gimautdinova
Scientific Research Institute of Biochemistry, SD RAMS, Novosibirsk, Russia

Geoffroy Auvert
CEA-Leti, Grenoble Alps University, Grenoble, France

Julian M. Menter, Latoya Freeman and Otega Edukuye
Department of Microbiology, Biochemistry and Immunology, Morehouse School of Medicine, Atlanta, GA, USA